AF251964

# MULTIGROUP EQUATIONS FOR THE DESCRIPTION OF THE PARTICLE TRANSPORT IN SEMICONDUCTORS

Series on Advances in Mathematics for Applied Sciences – Vol. 70

# MULTIGROUP EQUATIONS FOR THE DESCRIPTION OF THE PARTICLE TRANSPORT IN SEMICONDUCTORS

## Martin Galler

Graz University of Technology, Austria

**World Scientific**

NEW JERSEY • LONDON • SINGAPORE • BEIJING • SHANGHAI • HONG KONG • TAIPEI • CHENNAI

*Published by*

World Scientific Publishing Co. Pte. Ltd.

5 Toh Tuck Link, Singapore 596224

*USA office:* 27 Warren Street, Suite 401-402, Hackensack, NJ 07601

*UK office:* 57 Shelton Street, Covent Garden, London WC2H 9HE


**Library of Congress Cataloging-in-Publication Data**
Galler, Martin, 1977–
   Multigroup equations for the description of the particle transport in
semiconductors / Martin Galler.
     p. cm. -- (Series on advances in mathematics for applied sciences ; v. 70)
ISBN 981-256-355-5 (alk. paper)
  1. Transport theory--Mathematics. 2. Semiconductors--Mathematics. I.
Title. II. Series.

QC793.3.T7 G35 2005
530.13'8--dc22

                                           2005049431


**British Library Cataloguing-in-Publication Data**
A catalogue record for this book is available from the British Library.




Printed in Singapore by World Scientific Printers (S) Pte Ltd

Für den wichtigsten Menschen in meinem Leben.

# Preface

Accurate semiconductor device simulation is mainly based on Monte Carlo methods. However, there are essential advantages gained by directly solving the Bloch-Boltzmann-Peierls equations, which govern the dynamics of carriers and phonons in semiconductors. In this book, an attempt is made to introduce such deterministic solution techniques, called multigroup model equations, especially for describing the particle transport in III-V compound semiconductors.

First, we present a multigroup model to the Boltzmann equations governing the transient transport regime in polar semiconductors. Special effort is invested in an accurate description of the coupled hot-electron hot-phonon system. The related conservation laws for the electron density and the total energy density of the multigroup model equations are deduced. This physically motivated, discrete model is used for studying the transport properties of indium phosphide and gallium arsenide in response to a time-depending external electric field. The results are compared to experimental and theoretical data.

Second, a multigroup transport model for describing degenerated carrier gases is deduced. These model equations are based on a general carrier dispersion law and contain the full quantum statistics of both, the carriers and the phonons. We prove the boundedness of the solution according to the Pauli principle and study the conservational properties of the multigroup equations. Moreover, the existence of a Lyapounov functional to the proposed model equations is proved and expressions for the equilibrium solution are given.

Furthermore, the two-dimensional electron transport at an AlGaN/GaN heterojunction in the presence of strain polarization fields is simulated with the help of a multigroup model. The envelope wave functions for the confined electrons are calculated using a self-consistent Poisson-Schrödinger solver. The electron gas degeneracy and hot phonons are included in these transport equations.

Finally, a multigroup-WENO solver for the non-stationary Boltzmann-Poisson system for semiconductor device simulation is constructed. The proposed numerical technique is applied for investigating the carrier transport in bulk silicon, in a silicon $n^+ - n - n^+$ diode, in a silicon MESFET and in a silicon MOSFET as well as in bulk GaAs, in a GaAs $n^+ - n - n^+$ diode and in a GaAs MESFET. Additionally, the obtained results are compared to those of a full WENO solver and Monte Carlo simulations.

This book is based in the doctoral thesis, which I wrote at the Institute of Theoretical and Computational Physics of the Graz University of Technology. First of all I would like to thank my supervisors, Prof. Dr. Ferdinand Schürrer and Prof. Dr. Armando Majorana. It was Prof. Schürrer who encouraged me to write both the diploma thesis and the doctoral thesis in the field of transport theory. His special way of motivating persons, his critical reading of my work and the various discussions, which often exceeded his personal frame of time, proves to be invaluable for me. Thank you.

During my doctoral studies I got the possibility to participate in the IHP-project "Hyperbolic and Kinetic Equations: Asymptotics, Numerics, Analysis (HYKE)" of the European Community. In the course of this project, I had the pleasure to enjoy the Italian hospitality of Prof. Majorana and his colleagues at the Dipartimento di Matematica e Informatica dell'Università di Catania. My very exciting and interesting three months stay in Catania submontane the Etna greatly enriched both my mathematical experience and the knowledge of the Italian culture. Thank you very much.

Moreover, I would like to thank my family and my friends who supported me in a way only they can do.

Finally, I acknowledge the financial support of my doctoral thesis by the Fond zur Förderung der wissenschaftlichen Forschung, Vienna, contract numbers P14669-TPH and P17438-N08, and by the European community program IHP, under the contract number HPRN-CT-2002-00282 on behalf of the CNR.

M. Galler

# Contents


*Preface*   vii

1. Introduction   1

2. The Bloch-Boltzmann-Peierls Equations   5

   2.1 Introduction   5
   2.2 Electrons in Semiconductors   5
   2.3 Phonons in Semiconductors   9
   2.4 Scattering Mechanisms   11
      2.4.1 General Theory of Scattering   12
      2.4.2 Phonon Scattering   14
         2.4.2.1 Non-polar Phonon Scattering   15
         2.4.2.2 Polar Phonon Scattering   19
      2.4.3 Ionized Impurity Scattering   22
   2.5 Semiclassical Dynamics of Electrons   24
   2.6 The Bloch-Boltzmann-Peierls Equations   26
   2.7 Mathematical Properties of the BBP Equations   32

3. Multigroup Model Equations for Polar Semiconductors   37

   3.1 Introduction   37
   3.2 Multigroup Equations to the Bloch-Boltzmann-Peierls Equations   38
      3.2.1 The Electron Boltzmann Equation   40
      3.2.2 The LO Phonon Boltzmann Equation   45
      3.2.3 The Coupling POP Interaction Term   47
      3.2.4 The Evaluation of the Collision Coefficients   51

3.3 Conservation Laws . . . . . . . . . . . . . . . . . . . . . . 53

4. Particle Transport in Indium Phosphide       61

   4.1 Introduction . . . . . . . . . . . . . . . . . . . . . . . . . . 61
   4.2 Two-valley Model . . . . . . . . . . . . . . . . . . . . . . . 61
      4.2.1 Validation of the Method . . . . . . . . . . . . . . 64
      4.2.2 Electron Distribution Function . . . . . . . . . . . 66
      4.2.3 Phonon Distribution Function . . . . . . . . . . . . 68
      4.2.4 Transport Parameters . . . . . . . . . . . . . . . . 69
   4.3 Three-valley Model . . . . . . . . . . . . . . . . . . . . . . 71

5. Particle Transport in Gallium Arsenide       77

   5.1 Introduction . . . . . . . . . . . . . . . . . . . . . . . . . . 77
   5.2 Transport in a Time-dependent Electric Field . . . . . . . . 78
   5.3 The Stationary-state Electron Distribution . . . . . . . . . 82

6. Multigroup Equations for Degenerated Carrier Gases       87

   6.1 Introduction . . . . . . . . . . . . . . . . . . . . . . . . . . 87
   6.2 The Bloch-Boltzmann-Peierls Equations . . . . . . . . . . . 88
   6.3 The Multigroup Model Equations . . . . . . . . . . . . . . . 89
   6.4 Mathematical Aspects of the Multigroup Model Equations . 93
      6.4.1 Boundedness of the Solution . . . . . . . . . . . . . 93
      6.4.2 Conservation Laws . . . . . . . . . . . . . . . . . . 95
      6.4.3 H-theorem . . . . . . . . . . . . . . . . . . . . . . . 96
      6.4.4 Equilibrium Solution . . . . . . . . . . . . . . . . . 98
   6.5 Numerical Results . . . . . . . . . . . . . . . . . . . . . . . 100

7. The Two-dimensional Electron Gas       107

   7.1 Introduction . . . . . . . . . . . . . . . . . . . . . . . . . . 107
   7.2 General Theory of Transport in Confined Systems . . . . . 107
      7.2.1 Dispersion Laws . . . . . . . . . . . . . . . . . . . 107
      7.2.2 Scattering Mechanisms . . . . . . . . . . . . . . . . 111
         7.2.2.1 Acoustic Deformation Potential Scattering . 112
         7.2.2.2 Piezoelectric Scattering . . . . . . . . . . . 114
         7.2.2.3 Polar Optical Phonon Scattering . . . . . . 115
         7.2.2.4 Screening Effects . . . . . . . . . . . . . . 118
      7.2.3 BBP Equations for 2D Systems . . . . . . . . . . . 120
   7.3 Multigroup Equations to the 2D-BBP Equations . . . . . . 124

7.4   Transport in $Al_xGa_{1-x}N/GaN$ .............................. 132

    7.4.1   Self-consistent Solution for Confining Potential ... 135

    7.4.2   Transport Properties ........................ 139

    7.4.3   Distribution Functions ....................... 145

8.  The Multigroup-WENO Solver for Semiconductor Device Simulation     147

    8.1   Introduction .............................. 147

    8.2   The Boltzmann-Poisson System ................ 148

    8.3   The Multigroup-WENO Scheme ............... 150

9.  Simulation of Silicon Devices     155

    9.1   Introduction .............................. 155

    9.2   Transport in Bulk Silicon ..................... 157

    9.3   The Silicon $n^+ - n - n^+$ Diode ................. 158

    9.4   The Si-MESFET ........................... 162

    9.5   The Si-MOSFET .......................... 172

10.  Simulation of Gallium Arsenide Devices     191

    10.1  Introduction .............................. 191

    10.2  Bulk GaAs ............................... 193

    10.3  The GaAs $n^+ - n_i - n^+$ Diode ................. 195

    10.4  The GaAs-MESFET ........................ 200

11.  Conclusion     213

*Bibliography*     217

*Related Publications of the Author*     223

*Index*     225

# Chapter 1

# Introduction

Very large scale integration is the forthcoming design in semiconductor technology. This implies that in modern integrated electron devices the scale length of individual components becomes comparable with the distance between successive carrier interactions with the crystal, and the well-established drift-diffusion models describing the carrier transport lose their accuracy [Markowich *et al.* (1990)]. Consequently, to cope with high-field and sub-micron phenomena, Boltzmann transport equations (BTEs) must be applied [Ferry (1991)]. In femtosecond laser experiments non-equilibrium longitudinal-optical (LO) phonons have been found to affect strongly the electron distribution function. Thus, for a unified treatment, one has also to include kinetic equations for the evolution of phonons in a realistic description [Vaissiere *et al.* (1992); Vaissiere *et al.* (1996)].

Deterministic as well as stochastic procedures can be considered as solution approaches to these extremely sophisticated equations, the so-called Bloch-Boltzmann-Peierls (BBP) equations. So far, mainly stochastic solution methods have been applied to solve the Boltzmann transport equations [Jacoboni and Lugli (1989); Jacoboni and Reggiani (1983); Fiscetti (1991); Jungemann and Meinzerhagen (2003)]. Although Monte Carlo (MC) methods combined with drift-diffusion or hydrodynamic models can be considered as an approved method for device simulation, purely deterministic procedures are characterized by some essential advantages. Standard Monte Carlo methods are unable to resolve almost empty regions of two-dimensional devices, e.g., areas close to the gate of a MESFET, while deterministic approaches can do. Hence, deterministic results should be used as benchmarks for Monte Carlo, hydrodynamic or drift-diffusion results, even though they are not competitive with Monte Carlo schemes with respect to the computation time in two dimensions. In addition, direct solution techniques to the BTEs exhibit high efficiency in computing transient processes. They are featured by the knowledge of the particle distribution functions and not only of their moments and they give results without numerical noise even close to regions between different boundary conditions.

Because of the very complicated mathematical structure of the BBP equations and the imperative of high computer power, deterministic solution methods which allow an efficient and physically relevant description of the dynamics of the coupled electron-phonon system were rare in the past. However, with increasing power and memory of modern computers, the development of such methods is becoming an interesting task.

Early alternative approaches to Monte Carlo techniques coping with strongly anisotropic distribution functions are, e.g., the iterative technique [Budd (1967); Nougier and Rolland (1973)], the scattering matrix approach [Alam *et al.* (1993)] or a direct matrix method [Aubert *et al.* (1984)] based on a complete discretization of the Boltzmann transport equations in the reciprocal space. In this context, the path integral solution to the Boltzmann transport equation should by mentioned [Reggiani (1985)]. This method takes advantage of the electron motion along the trajectories in the momentum space in response to the effective electric field. Based on the increasing power of modern computers, a very efficient finite difference approach to the Boltzmann transport equation was proposed by Fatemi and Odeh [Fatemi and Odeh (1993)]. They developed an upwind finite difference approximation for the Boltzmann-Poisson system. Another useful technique for solving the Boltzmann transport equation is based on the expansion of the distribution function in terms of orthogonal polynomials [Ventura *et al.* (1995)]. Hennacy *et al.* [Hennacy and Goldsman (1993); Hennacy *et al.* (1995)] and Gnudi *et al.* [Gnudi *et al.* (1993)] discussed numerical solutions by Legendre polynomials and spherical harmonics expansions. Other deterministic approaches, based on series expansion methods for the Boltzmann transport equation, were derived by Ringhofer [Ringhofer (1997)]. The used Galerkin method leads to a hyperbolic system solved by finite difference methods in space-time variables [Ringhofer (2000); Ringhofer *et al.* (2001)].

Majorana and Pidatella [Majorana and Pidatella (2001)] solved the Boltzmann-Poisson system by the help of a box method in the energy and angle variables and combined this approach with a classical discretization technique for advection equations based on upwinding in the spatial variable. Recently, Carrillo *et al.* [Carrillo *et al.* (2002);

Carrillo *et al.* (2003)a] succeeded in introducing a deterministic high-order finite difference WENO solver for the solution of the one-dimensional Boltzmann-Poisson system for semiconductor devices. Moreover, they extended their numerical technique to cope with spatially two-dimensional geometries [Carrillo *et al.* (2003)b]. Finally, the works of Niclot *et al.* [Niclot *et al.* (1988)] and Cáceres *et al.* should be mentioned.

This book is intended to take a further step in developing deterministic solution methods to the Boltzmann transport equation. Special attention is paid to the design of numerical schemes for handling the nonlinear Bloch-Boltzmann-Peierls equations. To this purpose, multigroup model equations (MMEs) to this set of evolution equations are formulated. Several variations of multigroup approaches have been published for an approximative description of the dynamics of rarefied gases. Hence, the multigroup method can be regarded as well-established in the classical kinetic theory [Caraffini *et al.* (1995); Galler *et al.* (2003); Galler *et al.* (2004)]. However, an attempt to apply the multigroup formalism to the Boltzmann equations of semiconductors has not been made up to now.

The multigroup approach of discretizing the Boltzmann transport equations is motivated on physical grounds. The wave vector spaces of electrons and phonons are divided into tiny cells and the full Boltzmann transport equations are transformed into a system of coupled transport equations balancing the particle transfer among these cells. From a mathematical point of view, this approach is based on the method of weighted residuals [Lapidus and Pinder (1982)], which is basically a finite element technique.

This book is organised as follows. Chapter 2 deals with the solid-state physics relevant for the description of the particle transport in semiconductors. In chapter 3, the multigroup formalism is adapted to the special requirements related to the Boltzmann transport equations in polar semiconductors. In such materials, the polar-optical interaction between electrons and longitudinal optical phonons is the main relaxation mechanism at room temperature. As a consequence, the deviation of the phonon distribution function from thermal equilibrium, which cannot be neglected for sufficiently high doping concentrations, affects the electron distribution function significantly. Therefore, the calculations are performed with the coupled hot-electron hot-phonon Boltzmann transport equations. This procedure allows us to study the modifications of the main transport properties in III-V semiconductor compounds due to non-equilibrium phonons by investigating the transient transport regime in bulk indium phosphide in chapter 4 and gallium arsenide in chapter 5.

In chapter 6, deterministic multigroup model equations to the Bloch-Boltzmann-Peierls equations are introduced, which are based on a general carrier dispersion law. They contain the full quantum statistics of both, the carriers and the phonons. A deterministic transport model is the more reliable, the more mathematical properties the model equations and the original Bloch-Boltzmann-Peierls equations possess in common grounds. Hence, it must be a main goal to investigate the most important features of the multigroup transport model and to compare them to those of the original Bloch-Boltzmann-Peierls equations. Therefore, we show the boundedness of the distribution coefficients, which reflects the Pauli principle. The conservational properties of the multigroup model are discussed and a Boltzmann H-theorem for the obtained evolution equations is proved. The equilibrium distribution of the multigroup equations is given by a set of discretized Fermi-Dirac and Bose-Einstein distributions for non-drifting particles, which corresponds to the features of the continuous equations.

Another application, where a coupled electron-phonon system plays a major role, is the transport of a two-dimensional electron gas along a heterojunction formed by polar semiconductors. Here, the main difficulty lies in the combination of the only numerically given quantities describing the heterojunction and the multigroup model equations, which are based on analytical expressions. This complicated problem is solved in chapter 7 by constructing transport equations for the coupled system of degenerated two-dimensional electrons and longitudinal optical phonons formed at a $Al_x Ga_{1-x} N/GaN$ heterojunction.

In modern highly integrated devices, a consistent description of the dynamics of carriers is essential for a deeper understanding of the observed transport properties. For the simulation of such devices on a mesoscopic level, a deterministic multigroup-WENO solver to the coupled Boltzmann-Poisson system is proposed in chapter 8. This numerical scheme is based on the combination of the multigroup method for treating the dependence of the electron distribution function on the three-dimensional wave vector and a fifth-order WENO solver [Carrillo *et al.* (2003)a; Carrillo *et al.* (2003)b] for dealing with the two-dimensional physical space. The resulting transport equations are used for simulating the charge transport in bulk silicon, in a silicon $n^+ - n - n^+$ diode, in a silicon MESFET and in a silicon MOSFET in chapter 9 as well as in bulk GaAs, in a GaAs $n^+ - n - n^+$ diode and in a GaAs MESFET in chapter 10.

# Chapter 2

# The Bloch-Boltzmann-Peierls Equations

## 2.1 Introduction

The transport of carriers in semiconductors can be understood as the propagation of charged particles in an almost periodic lattice potential. The description of such transport phenomena from a mesoscopic point of view must be based on solid-state physics.

It is the aim of this chapter to give an overview of the quantum mechanical foundations of the particle transport in semiconductors. We present the Bloch-Boltzmann-Peierls equations, which constitute the governing set of evolution equations for the carrier and phonon distribution functions in such materials, and study their main properties. More detailed information on these topics are found, for example, in the books [Markowich *et al.* (1990); Lundstrom (2000); Ziman (2001); Tomizawa (1993); Fetter and Walecka (1971); Weißmantel and Hamann (1995); Ashcroft and Mermin (1976)].

## 2.2 Electrons in Semiconductors

Electrons in a semiconductor crystal move in a periodic crystal potential, which is formed by the potential of the atomic nuclei and that due to the other electrons. When studying the transport of electrons in such a crystal, one must consider an extremely complicated many-body problem [Fetter and Walecka (1971); Czycholl (2000)]. However, if attention is only paid to the motion of an electron in the crystal by assuming that the effects of the atomic nuclei and the remaining electrons on the selected electron can be approximated by a prescribed potential $V(\mathbf{r})$ depending on the position $\mathbf{r}$, the many-body problem reduces to the problem of a single electron. The potential $V(\mathbf{r})$ must be periodic with the same periodicity as that of the lattice. This property is mathematically expressed by

$$V(\mathbf{r} + l\mathbf{a} + m\mathbf{b} + n\mathbf{c}) = V(\mathbf{r}), \tag{2.1}$$

where $\mathbf{a}$, $\mathbf{b}$ and $\mathbf{c}$ are the primitive basis vectors of the considered crystal lattice, and $l, m, n \in \mathbb{N}$.

To determine the electronic states for a periodic potential $V$, we must solve the Schrödinger equation

$$\left[ -\frac{\hbar^2}{2m_0}\Delta + V(\mathbf{r}) \right] \psi(\mathbf{r}) = E\,\psi(\mathbf{r}) \tag{2.2}$$

with the eigenfunctions $\psi(\mathbf{r})$ to be determined, the eigenvalues of the energy $E$, the electron mass in free space $m_0$ and the reduced Planck constant $\hbar$. According to the Bloch theorem [Fetter and Walecka (1971)], these solutions are of the form

$$\psi(\mathbf{k}, \mathbf{r}) = u_j(\mathbf{k}, \mathbf{r})\, e^{i\mathbf{k}\cdot\mathbf{r}} \tag{2.3}$$

for a perfectly periodic potential, where $\mathbf{k}$ and $j$ label the wave vector of the electron and the index of the band, respectively. The Bloch functions $u_j(\mathbf{k}, \mathbf{r})$ are periodic in $\mathbf{r}$ with the same periodicity as $V(\mathbf{r})$, i. e.,

$$u_j(\mathbf{k}, \mathbf{r} + l\mathbf{a} + m\mathbf{b} + n\mathbf{c}) = u_j(\mathbf{k}, \mathbf{r}). \tag{2.4}$$

The energy eigenvalues $E_j(\mathbf{k})$ are periodic with the periodicity of the reciprocal lattice; thus,

$$E_j(\mathbf{k} + \mathbf{G}) = E_j(\mathbf{k}), \quad \mathbf{G} = g\mathbf{a}^* + h\mathbf{b}^* + k\mathbf{c}^*. \tag{2.5}$$

Here, $\mathbf{G}$ is a vector of the reciprocal lattice with $g, h, k \in \mathbb{N}$ and the basis vectors $\mathbf{a}^*$, $\mathbf{b}^*$ and $\mathbf{c}^*$ of the reciprocal lattice defined by

$$\mathbf{a}^* = 2\pi \frac{\mathbf{b} \times \mathbf{c}}{\mathbf{a} \cdot (\mathbf{b} \times \mathbf{c})}, \quad \mathbf{b}^* = 2\pi \frac{\mathbf{c} \times \mathbf{a}}{\mathbf{a} \cdot (\mathbf{b} \times \mathbf{c})}, \quad \mathbf{c}^* = 2\pi \frac{\mathbf{a} \times \mathbf{b}}{\mathbf{a} \cdot (\mathbf{b} \times \mathbf{c})}. \tag{2.6}$$

All information on the $E_j(\mathbf{k})$ relation, the so-called energy band structure or dispersion law, can be expressed in one period of the reciprocal lattice because of the periodicity of $E_j(\mathbf{k})$. Usually, the first Brillouin zone $\mathcal{B}$, which is a period centred around the origin of the $\mathbf{k}$-space, is employed to show the energy band structures along some important crystallographic orientations $(\Lambda, \Delta, \Sigma, \ldots)$. The determination of the energy band structure as a function of $\mathbf{k}$ and $j$ is an important problem and has been studied theoretically using a variety of numerical methods. For more details, we refer to [Fetter and Walecka (1971); Yu and Cardona (1991)].

For common semiconductors, the band structures are well known from various experiments and from numerical solutions to the wave equation. The dispersion laws of semiconductors are characterized by an energy region where electronic states are not found. This is a forbidden energy interval, which is called the energy gap and typical for semiconductors and isolators. Electronic states are completely empty above (conduction bands) and completely occupied below (valence bands) of this energy gap in the limit of temperature $T \to 0$. The energy separation between the lowest conduction band minimum and the highest valence band maximum is called the band gap energy $\Delta E_{\mathrm{G}}$, which is one of the important parameters in semiconductor physics.

For studying the carrier transport, information on the band structure near the conduction band minima and the valence band maxima is of special interest, since carriers are usually located near the band edges. These regions of the energy bands can be described with the help of analytical approximations for the full dispersion laws. A schematic diagram of such an approximative band model for GaAs is shown in Fig. 2.1. Gallium arsenide is called a direct semiconductor, since both, the lowest conduction band minimum and the highest valence band maximum are situated at the $\Gamma$ point. However, GaAs also has minima at the $L$ points and near the $X$ points on the $\Delta$ lines. They are usually called $L$ valleys and $X$ valleys, respectively. The $L$ valleys and the $X$ valleys are about a few tenth of an electron-volt higher than the band at $\Gamma$ ($\Gamma$ valley) and play an important role when high-field transport takes place in GaAs. In addition, Fig. 2.1 displays the two highest valence bands, which feature degeneracy at the $\Gamma$ point.

Model band structures are determined by various band parameters, which represent, for instance, the energy levels of the band extremes and the relations between the electron energy $E_j(\mathbf{k})$ and the electron wave vector $\mathbf{k}$. The conduction band near the minimum is often approximated by a quadratic function of $\mathbf{k}$ according to a truncated Taylor series expansion of the real $E_j(\mathbf{k})$ relation. When the band minimum lies at $\mathbf{k} = 0$, $E_j(\mathbf{k})$ can be written in the form (parabolic band approximation)

$$E_j(\mathbf{k}) = \frac{\hbar^2 k^2}{2m_j^*} \tag{2.7}$$

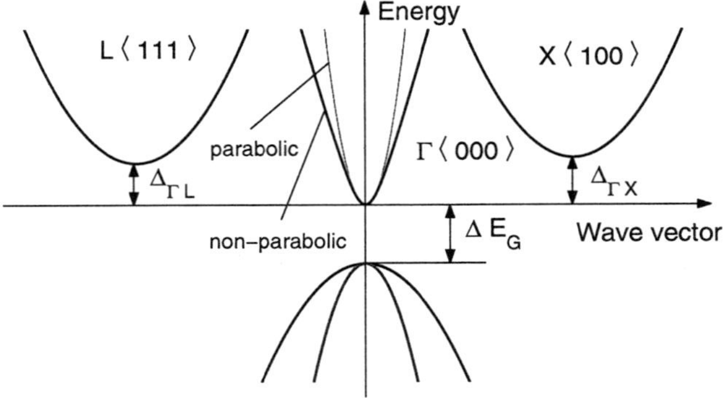


Fig. 2.1    Schematic illustration of a three valley model of GaAs.

with $k = |\mathbf{k}|$ and the effective mass $m_j^*$ for the valley $j$, which is obtained from the real dispersion law $E_j^{\text{real}}(\mathbf{k})$ via

$$\frac{1}{m_j^*} = \frac{1}{\hbar^2} \frac{\partial^2 E_j^{\text{real}}(\mathbf{k})}{\partial \mathbf{k}^2}\bigg|_{\mathbf{k}=0} \tag{2.8}$$

The dispersion law (2.7) shows that $\hbar\mathbf{k}$ plays the role of momentum, and that the electrons just behave like electrons in free space, except for the change in the electron mass. The quantity $\hbar\mathbf{k}$ is termed the crystal momentum and $E_j(\mathbf{k})$ represents the kinetic energy of the electron measured from the bottom of the conduction band. The simple parabolic band approximation is widely used to simplify the calculation of carrier transport. However, for high applied electric fields, the energy of electrons can be far from a band edge, and the approximation (2.7) loses its validity. Hence, more sophisticated band approximations must be applied. The $\mathbf{k} \cdot \mathbf{p}$ perturbation theory based on a two-band model [Lundstrom (2000); Ridley (1982); Nag (1980); Datta (1989)] provides a relation for the non-parabolicity in the so-called non-parabolic band approximation (Kane model in the case of silicon),

$$E_j(\mathbf{k})[1 + \alpha_j E_j(\mathbf{k})] = \frac{\hbar^2 k^2}{2m_j^*} \tag{2.9}$$

by approximating the non-parabolicity factor $\alpha_j$ by

$$\alpha_j = \frac{1}{\Delta E_G}\left(1 - \frac{m_j^*}{m_0}\right)^2, \qquad (2.10)$$

which is a constant approximately equal to the inverse of the energy gap.

## 2.3 Phonons in Semiconductors

Lattice vibrations contribute substantially to the momentum and energy relaxation of carriers in semiconductors at room temperature. Therefore, the scattering of electrons by phonons is one of the most important interaction mechanism between carriers and real structure of the crystal. In this section, we summarize some features of quantized lattice vibrations (phonons) [Ziman (2001); Czycholl (2000); Datta (1989)].

Lattice vibrations are collective oscillations of ions, which are tightly connected with each other in a crystal. These collective oscillations can be described as normal mode oscillations. Therefore, the displacement $\mathbf{u}$ of an ion at the lattice site $\mathbf{r}$ can be expressed by the superposition of these normal modes as

$$\mathbf{u}(\mathbf{r}, t) = \sum_{\mathbf{q}}\left(\frac{\hbar}{2\rho V \omega(\mathbf{q})}\right)^{\frac{1}{2}} \mathbf{e}_{\mathbf{q}}[a_{\mathbf{q}}e^{i\mathbf{q}\cdot\mathbf{r}} + a_{\mathbf{q}}^{\dagger}e^{-i\mathbf{q}\cdot\mathbf{r}}], \qquad (2.11)$$

where $\mathbf{q}$ is the phonon wave vector, $\omega(\mathbf{q})$ is the angular frequency, $\rho$ is the mass density of the crystal with volume $V$, $a_{\mathbf{q}}$ and $a_{\mathbf{q}}^{\dagger}$ are the phonon dynamical variables and $\mathbf{e}_{\mathbf{q}}$ is the phonon unit vector of polarization [Kittel (1963)]. In isotropic crystals, there are one longitudinal mode ($\mathbf{e}_{\mathbf{q}}\|\mathbf{q}$) and two transverse modes ($\mathbf{e}_{\mathbf{q}}\perp\mathbf{q}$) for a given $\mathbf{q}$. For $N$ atoms in the unit cell, as it is the case for the common semiconductor materials, there are $3N - 3$ optical modes besides the three acoustic modes, which are characterized by the fact that all $N$ atoms oscillate in phase in the limit $\mathbf{q} \to 0$. The relation $\omega(\mathbf{q})$ versus $\mathbf{q}$ is called the phonon dispersion law. Numerical methods for calculating these dispersion laws are presented in [Baroni *et al.* (2001)].

By rearranging the summation over $\mathbf{q}$, (2.11) can be written as

$$\mathbf{u}(\mathbf{r}, t) = \sum_{\mathbf{q}}\left(\frac{\hbar}{2\rho V \omega(\mathbf{q})}\right)^{\frac{1}{2}} \mathbf{e}_{\mathbf{q}}(a_{\mathbf{q}} + a_{-\mathbf{q}}^{\dagger})e^{i\mathbf{q}\cdot\mathbf{r}}. \qquad (2.12)$$

Moving into the realm of quantum mechanics, we can interpret $a_\mathbf{q}$ and $a_\mathbf{q}^\dagger$ as annihilator and creator operators satisfying the commutator relation

$$[a_\mathbf{q}, a_{\mathbf{q}'}^\dagger] = a_\mathbf{q} a_{\mathbf{q}'}^\dagger - a_{\mathbf{q}'}^\dagger a_\mathbf{q} = \delta_{\mathbf{q},\mathbf{q}'}. \tag{2.13}$$

This relation implies that the energy operator can be expressed by

$$H = \sum_\mathbf{q} \hbar\omega(\mathbf{q}) \left( a_\mathbf{q}^\dagger a_\mathbf{q} + \frac{1}{2} \right). \tag{2.14}$$

If $|n_\mathbf{q}\rangle$ denote the eigenfunctions of this Hamilton operator and if we label the eigenvalues of $a_\mathbf{q}^\dagger a_\mathbf{q}$ by $g(\mathbf{q})$, the equation

$$a_\mathbf{q}^\dagger a_\mathbf{q} |n_\mathbf{q}\rangle = g(\mathbf{q})|n_\mathbf{q}\rangle, \tag{2.15}$$

holds, and we obtain the following energy eigenvalues:

$$\varepsilon_\mathbf{q} = \hbar\omega(\mathbf{q}) \left( g(\mathbf{q}) + \frac{1}{2} \right). \tag{2.16}$$

Since the energy of the quantum state $|n_\mathbf{q}\rangle$ is given by $g(\mathbf{q})$ multiples of $\hbar\omega(\mathbf{q})$, we conclude that there are $g(\mathbf{q})$ phonons in this state.

The non-vanishing matrix elements of $a_\mathbf{q}$ and $a_\mathbf{q}^\dagger$ are given by

$$\langle n_\mathbf{q} - 1|a_\mathbf{q}|n_\mathbf{q}\rangle = \sqrt{g(\mathbf{q})}, \tag{2.17a}$$

$$\langle n_\mathbf{q} + 1|a_\mathbf{q}^\dagger|n_\mathbf{q}\rangle = \sqrt{g(\mathbf{q}) + 1}, \tag{2.17b}$$

which are important relations when calculating transition rates for carrier-phonon interactions. Since phonons are bosons, the number of these particles with the wave vector $\mathbf{q}$ at the lattice temperature $T_\mathrm{L}$ in equilibrium is given by the Bose-Einstein distribution

$$g_\mathrm{BE}(\mathbf{q}) = \left[ \exp \frac{\hbar\omega(\mathbf{q})}{k_\mathrm{B} T_\mathrm{L}} - 1 \right]^{-1}. \tag{2.18}$$

If the phonon energy is so small or the temperature so high that $\hbar\omega(\mathbf{q}) \ll k_\mathrm{B} T_\mathrm{L}$, the case of equipartition is applicable, which means that the equilibrium phonon number can be approximated by

$$g_\mathrm{EP}(\mathbf{q}) = \frac{k_\mathrm{B} T_\mathrm{L}}{\hbar\omega(\mathbf{q})}. \tag{2.19}$$

The actual structures of the phonon dispersion laws are very complicated. However, when considering electron-phonon scattering, either

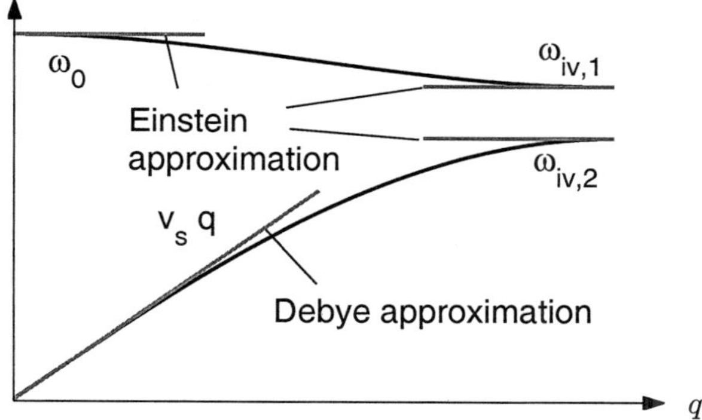


Fig. 2.2  Approximations to the dispersion laws of acoustic and optical phonons.

phonons with small $\mathbf{q}$ or such with $\mathbf{q}$ near the boundary of the Brillouin zone can be involved in a scattering event. In these cases, the following approximations for the $\omega(\mathbf{q})$ relation are valid [Lundstrom (2000); Ziman (2001)]. For small $\mathbf{q}$, the dispersion relation for acoustic modes can be described by the Debye approximation, i. e.,

$$\omega(\mathbf{q}) = v_{\mathrm{s}}\,q, \qquad (2.20)$$

where $v_{\mathrm{s}}$ is the sound velocity. On the other hand, the dispersion laws of optical phonons with small $\mathbf{q}$ and both the acoustic and the optical phonons with $\mathbf{q}$ at the boundary of the Brillouin zone can be approximated as suggested by Einstein:

$$\omega(\mathbf{q}) = \omega_0. \qquad (2.21)$$

Here, $\omega_0$ is a constant with its value depending on the type of the considered phonons. A schematic illustration of the relation between the Debye and the Einstein approximation and the full phonon dispersion law is given in Fig. 2.2.

## 2.4  Scattering Mechanisms

In this section, we summarize the most important scattering mechanisms that carriers undergo during their motion in the host crystal. We assume that the dynamics of the electronic interaction is independent of the applied electric field. All scattering calculations presented are carried out with

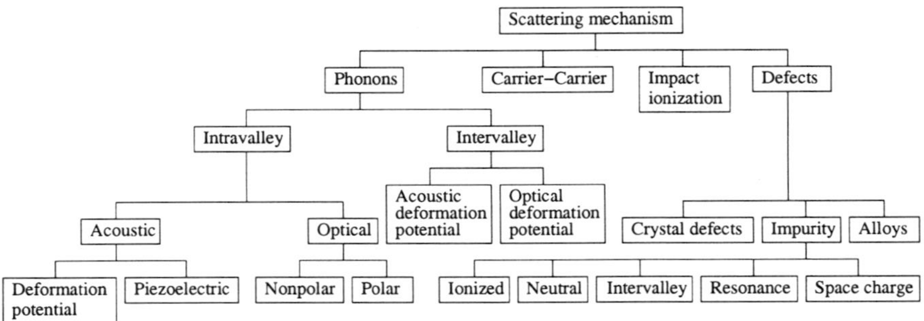

Fig. 2.3    Scattering mechanisms in cubic semiconductors.

first-order and time-dependent perturbation theory (Fermi's golden rule); consequently, only two-body interactions can be analyzed.

### 2.4.1  *General Theory of Scattering*

The electronic transitions of interest in charge transport in semiconductors can be classified as intravalley transitions, when initial and final states of the electrons lie in the same valley, or intervalley transitions, when initial and final states belong to different valleys. The most important scattering sources, which determine these transitions, are

- phonons;
- defects;
- impact-ionization;
- carrier-carrier.

A more detailed analysis of the possible scattering mechanisms in cubic semiconductors is found in Fig. 2.3. In this book, only the scattering of carriers by phonons and ionized impurities is considered.

The scattering theory is based on Fermi's golden rule, which is derived from the time-dependent perturbation theory of first order [Datta (1989)]. This rule gives the transition rate $S(\mathbf{k}, \mathbf{k}')$, with is the transition probability per unit time between the two electron eigenstates $\mathbf{k}$ and $\mathbf{k}'$ of the unperturbed Hamiltonian $H_0$ caused by the perturbation potential $H'$:

$$S(\mathbf{k}, \mathbf{k}') = \frac{2\pi}{\hbar} |\langle \mathbf{k}' | H' | \mathbf{k} \rangle|^2 \delta[E(\mathbf{k}') - E(\mathbf{k}) \mp \hbar\omega]. \tag{2.22}$$

The matrix elements of the perturbation potential are defined as

$$\langle \mathbf{k}'|H'|\mathbf{k}\rangle = \int_V d^3r\, \psi^*(\mathbf{k}',\mathbf{r})\, H'\, \psi(\mathbf{k},\mathbf{r}) \qquad (2.23)$$

with the eigenfunctions $\psi(\mathbf{k},\mathbf{r})$ of $H_0$. The frequency $\omega$ is related to the demanded harmonical time-dependence of $H'$:

$$H'(t) = H'e^{\mp i\omega t}. \qquad (2.24)$$

Fermi's golden rule (2.22) is the basic result of the scattering theory, which we will apply to scattering calculations of carriers in semiconductors. We note that for the validity of (2.22), the interaction must be very weak. This ensures that the free time between two successive collisions is large enough to be consistent with the preconditions for the derivation of Fermi's rule as a first order approximation.

For applying the golden rule, the perturbation potential $H'$ must be identified so that the matrix elements (2.23) can be evaluated. For simplifying this procedure, we proceed as follows. We assume that the perturbation potential $H'$ may be expanded in a Fourier series,

$$H' = \sum_{\mathbf{q}} U_{\mathbf{q}} e^{i\mathbf{q}\cdot\mathbf{r}}, \qquad (2.25)$$

where the summation traverses from minus to plus infinity. The Fourier coefficients $U_{\mathbf{q}}$ to the perturbation potential $U(\mathbf{r},t)$ are given by

$$U_{\mathbf{q}} = \int_V d^3r\, U(\mathbf{r},t)e^{-i\mathbf{q}\cdot\mathbf{r}}. \qquad (2.26)$$

We note that for real $U(\mathbf{r},t)$, we have the relation $U_{\mathbf{q}}^* = U_{-\mathbf{q}}$. The combination of (2.23) and (2.25) and the evaluation of the matrix elements for Bloch waves leads to

$$\langle \mathbf{k}'|H'|\mathbf{k}\rangle = \int_V d^3r\, u^*(\mathbf{k}',\mathbf{r})e^{-i\mathbf{k}'\cdot\mathbf{r}} \sum_{\mathbf{q}} U_{\mathbf{q}} e^{i\mathbf{q}\cdot\mathbf{r}} u(\mathbf{k},\mathbf{r})e^{i\mathbf{k}\cdot\mathbf{r}} \qquad (2.27)$$

$$= \sum_{\mathbf{q}} U_{\mathbf{q}} \int_V d^3r\, u^*(\mathbf{k}',\mathbf{r})u(\mathbf{k},\mathbf{r})e^{i(\mathbf{q}-\mathbf{k}'+\mathbf{k})\cdot\mathbf{r}}.$$

Because of the periodicity of the exponential function, the integral vanishes except for $\mathbf{q} = \mathbf{k}' - \mathbf{k} + \mathbf{G}$ with the reciprocal lattice vector $\mathbf{G}$ (cf. (2.5)). Setting $\mathbf{G}=0$, i.e., neglecting umklapp processes, the matrix element reduces to

$$\langle \mathbf{k}'|H'|\mathbf{k}\rangle = \sum_{\mathbf{q}} U_{\mathbf{q}}\delta_{\mathbf{q},\mathbf{k}'-\mathbf{k}} I(\mathbf{k},\mathbf{k}') = U_{\mathbf{k}'-\mathbf{k}} I(\mathbf{k},\mathbf{k}') \qquad (2.28)$$

with the overlap integral

$$I(\mathbf{k},\mathbf{k}') = \int_V d^3r\, u^*(\mathbf{k}',\mathbf{r}) u(\mathbf{k},\mathbf{r}). \qquad (2.29)$$

Since $I(\mathbf{k},\mathbf{k}') \approx 1$ in good approximation for nearly parabolic bands, we find

$$\langle \mathbf{k}'|H'|\mathbf{k}\rangle = \int_V d^3r\, e^{-i\mathbf{k}'\cdot\mathbf{r}} U(\mathbf{r},t) e^{i\mathbf{k}\cdot\mathbf{r}}. \qquad (2.30)$$

This result means that we can use plane waves as the undisturbed states for evaluating matrix elements.

Having found the matrix elements $\langle \mathbf{k}'|H'|\mathbf{k}\rangle$ and the transition rates $S(\mathbf{k},\mathbf{k}')$, one can easily evaluate the scattering rate $1/\tau(k)$, which is an important measure for quantifying the strength of an interaction mechanism and when dealing with relaxation time models and Monte Carlo simulations to the Boltzmann transport equations. This quantity is defined as the integral of $S(\mathbf{k},\mathbf{k}')$ given by (2.22) with respect to the final state $\mathbf{k}'$:

$$\begin{aligned}
\frac{1}{\tau(k)} &= \frac{V}{8\pi^3} \int d^3k'\, S(\mathbf{k},\mathbf{k}') \qquad (2.31)\\
&= \frac{V}{4\pi^2\hbar} \int d^3k'\, |\langle \mathbf{k}'|H'|\mathbf{k}\rangle|^2 \delta[E(\mathbf{k}') - E(\mathbf{k}) \mp \hbar\omega].
\end{aligned}$$

In the following sections, we use the scattering rates for illustrating the dependence of the scattering probability on the electron energy for several scattering mechanisms.

### 2.4.2 *Phonon Scattering*

Bloch electrons are the eigenstates of the perfect crystal. Hence, electrons are not scattered by the purely periodic potential associated with the periodic array of ions constituting the crystal. However, electrons are scattered by lattice vibrations, causing small deviations of the crystal potential from its perfect structure. Because of the difficulty in knowing the crystal potential itself, this deviation from the periodicity is treated in an almost phenomenological way by introducing deformation potentials. Since the wave nature of lattice vibrations is quantized as phonons, the influence of these oscillations on the motions of electrons must be expressed by quantum processes, which are called electron-phonon interactions.

The scattering of electrons by phonons can be caused by the deformation of the otherwise perfect crystal structure produced by the lattice vibrations, or by the electrostatic forces associated with the polarization waves which accompany the phonons. The first kind of mechanism, which is typical for covalent semiconductors, is called deformation potential interaction for both, the acoustic and the optical phonons. The electrostatic interaction, typical for polar materials, is named piezoelectric interaction in the case of acoustic phonons, and polar interaction for optical phonons.

### 2.4.2.1 *Non-polar Phonon Scattering*

Vibrations of ions about their equilibrium position, described by their displacements $\mathbf{u}$ (cf. 2.12), produce instantaneous changes in the energy band, and, thus, cause the scattering of electrons. For a small change of the lattice constant, we expect a small change of the energy band structure. For acoustic phonons, this change can be considered proportional to the change of the lattice spacing, which can be expressed by the induced strain $\nabla \cdot \mathbf{u}$. Hence, we write the interaction potential for acoustic phonons as

$$H'_{\mathrm{ADP}} = D_{\mathrm{A}} \, \nabla \cdot \mathbf{u} \tag{2.32}$$

with the deformation potential $D_{\mathrm{A}}$ [Bardeen and Shockley (1950)]. Here, we note that $\nabla \cdot \mathbf{u}$ vanishes for transverse acoustic modes, which implies that acoustic deformation potential scattering is only caused by longitudinal acoustic phonons.

Substituting (2.12) into (2.32) yields

$$H'_{\mathrm{ADP}} = \sum_{\mathbf{q}} iq D_{\mathrm{A}} \left( \frac{\hbar}{2\rho v_{\mathrm{s}} q V} \right)^{\frac{1}{2}} (a_{\mathbf{q}} + a^{\dagger}_{-\mathbf{q}}) e^{i\mathbf{q}\cdot\mathbf{r}}, \tag{2.33}$$

with $q = \mathbf{q} \cdot \mathbf{e}_{\mathbf{q}}$ and the use of the Debye approximation for the acoustic phonon dispersion law (2.20). The non-vanishing matrix elements to this perturbation potential are

$$\langle \mathbf{k}', n_{\mathbf{q}}-1 | H'_{\mathrm{ADP}} | \mathbf{k}, n_{\mathbf{q}} \rangle = iq D_{\mathrm{A}} \left( \frac{\hbar}{2\rho v_{\mathrm{s}} q V} \right)^{\frac{1}{2}} \sqrt{g_{\mathrm{ac}}(\mathbf{q})} \delta_{\mathbf{k}'-\mathbf{k},\mathbf{q}}, \tag{2.34a}$$

$$\langle \mathbf{k}', n_{\mathbf{q}}+1 | H'_{\mathrm{ADP}} | \mathbf{k}, n_{\mathbf{q}} \rangle = iq D_{\mathrm{A}} \left( \frac{\hbar}{2\rho v_{\mathrm{s}} q V} \right)^{\frac{1}{2}} \sqrt{g_{\mathrm{ac}}(\mathbf{q})+1} \delta_{\mathbf{k}'-\mathbf{k},-\mathbf{q}} \tag{2.34b}$$

according to (2.17). Inserting (2.34) into Fermi's golden rule (2.22) and combining both, the absorption (upper sign) and the emission (lower sign) processes, results in the transition rate for acoustic deformation potential (ADP) scattering

$$S_{\mathrm{ADP}}(\mathbf{k}, \mathbf{k}') = \sum_{\mathbf{q}} \frac{\pi D_{\mathrm{A}}^2 q}{\rho v_{\mathrm{s}} V} \left[ g_{\mathrm{ac}}(\mathbf{q}) + \frac{1}{2} \mp \frac{1}{2} \right] \delta_{\mathbf{k} \pm \mathbf{q}, \mathbf{k}'} \qquad (2.35)$$

$$\times \delta[E(\mathbf{k}') - E(\mathbf{k}) \mp \hbar v_{\mathrm{s}} q].$$

We observe that the momentum $\mathbf{k}' = \mathbf{k} \pm \mathbf{q}$ as well as the energy $E(\mathbf{k}') = E(\mathbf{k}) \pm \hbar v_{\mathrm{s}} q$ are conserved for acoustic deformation potential scattering.

For simplifying the calculations for the room temperature case, we take into account that the acoustic phonon energy $\hbar v_{\mathrm{s}} q$ is much smaller than the electron energy on the order of $k_{\mathrm{B}} T_{\mathrm{L}}$. Hence, acoustic deformation potential scattering can be considered elastic. Moreover, when we assume an equipartitioned equilibrium acoustic phonon distribution, $g_{\mathrm{ac}}(\mathbf{q})$ can be approximated by $g_{\mathrm{ac}}(\mathbf{q}) \approx g_{\mathrm{ac}}(\mathbf{q}) + 1 \approx k_{\mathrm{B}} T_{\mathrm{L}} / \hbar v_{\mathrm{s}} q$ according to (2.19). These approximations allow us to rewrite (2.35) as

$$S_{\mathrm{ADP}}(\mathbf{k}, \mathbf{k}') = \frac{2\pi D_{\mathrm{A}}^2 k_{\mathrm{B}} T_{\mathrm{L}}}{\rho \hbar v_{\mathrm{s}}^2 V} \delta[E(\mathbf{k}') - E(\mathbf{k})]. \qquad (2.36)$$

The associated scattering rate $1/\tau_{\mathrm{ADP}}$ is evaluated with the help of (2.31):

$$\frac{1}{\tau_{\mathrm{ADP}}(k)} = \frac{2\pi D_{\mathrm{A}}^2 k_{\mathrm{B}} T_{\mathrm{L}}}{\hbar \rho v_{\mathrm{s}}^2} \mathcal{Z}[E(\mathbf{k})] \qquad (2.37)$$

with the density of states $\mathcal{Z}$ without the spin factor depending on the used energy band model. For instance, $\mathcal{Z}$ is given by

$$\mathcal{Z}(E) = \frac{m^*}{2\pi^2 \hbar^3} \sqrt{2m^* E} \qquad (2.38)$$

in the parabolic band approximation. In Fig. 2.4, we display the acoustic deformation potential scattering rate for GaAs using the parabolic band structure. The material parameters used are taken from Tab. 5.1.

Carrier scattering due to non-polar optical phonons can be treated similarly to the acoustic deformation potential scattering. Here, we introduce

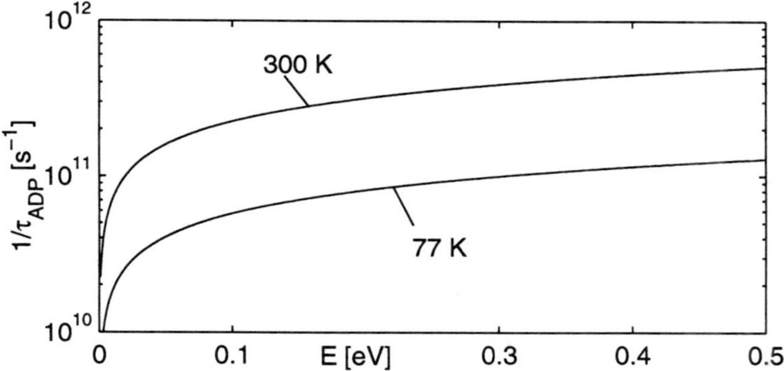

Fig. 2.4  Acoustic deformation potential scattering rate for electrons in the $\Gamma$ valley of GaAs at 77 K and 300 K.

the interaction potential of the form [Pötz and Vogl (1981)]

$$H'_{\mathrm{ODP}} = \mathbf{D_O} \cdot \mathbf{u} \tag{2.39}$$

$$= \sum_{\mathbf{q}} D_O \left( \frac{\hbar}{2\rho\omega_O V} \right)^{\frac{1}{2}} (a_{\mathbf{q}} + a^{\dagger}_{-\mathbf{q}}) e^{i\mathbf{q}\cdot\mathbf{r}} \tag{2.40}$$

with (2.12) and the Einstein approximation for the optical phonon energy $\hbar\omega_O$ (cf. Fig. 2.2). A procedure as for the acoustic deformation potential scattering leads to the transition rate for optical deformation potential (ODP) scattering

$$S_{\mathrm{ODP}}(\mathbf{k}, \mathbf{k}') = \frac{\pi D_O^2}{\rho\omega_O V} \left\{ g_O[\pm(\mathbf{k}' - \mathbf{k})] + \frac{1}{2} \mp \frac{1}{2} \right\} \tag{2.41}$$
$$\times \delta[E(\mathbf{k}') - E(\mathbf{k}) \mp \hbar\omega_O]$$

with the optical phonon distribution function $g_O$, and the corresponding scattering rate

$$\frac{1}{\tau_{\mathrm{ODP}}(k)} = \frac{\pi D_O^2}{\rho\omega_O} \left\{ g_O[\pm(\mathbf{k}' - \mathbf{k})] + \frac{1}{2} \mp \frac{1}{2} \right\} \mathcal{Z}[E(\mathbf{k}) \pm \hbar\omega_O]. \tag{2.42}$$

Figure 2.5 displays schematically the scattering of an electron by optical phonon displacements.

Non-polar phonons can also contribute to the intervalley transfer of carriers. Intervalley scattering can, for example, take place between equivalent valleys lying along the $\langle 100 \rangle$ direction in silicon, between $\Gamma$ and $L$ valleys

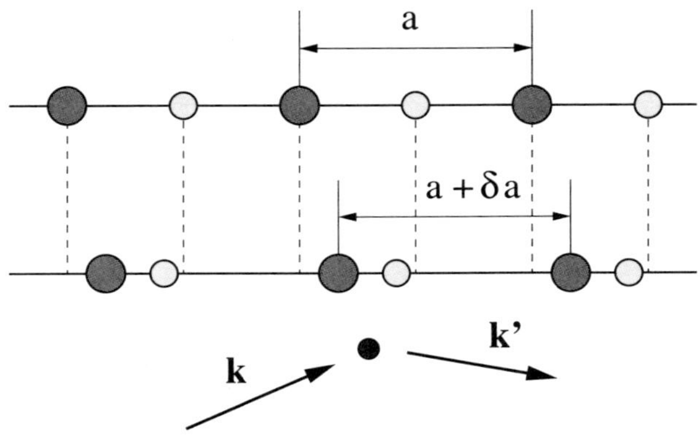


Fig. 2.5   Schematic illustration of optical deformation potential scattering.

in gallium arsenide and so on. Since large momentum transfer is needed for intervalley scattering, phonons with wave vectors near the boundary of the Brillouin zone only contribute to this type of carrier transfer. For large $\mathbf{q}$, both, the acoustic and the optical phonons can be described in the Einstein approximation (2.21), as it is illustrated in Fig. 2.2. In the following, the frequency of phonons transferring electrons from valley $i$ to valley $j$ is labeled by $\omega_{ij}$.

For intervalley scattering, we assume that the interaction potential is of the form

$$H'_{\mathrm{IV}} = \mathbf{D}_{ij} \cdot \mathbf{u} \tag{2.43}$$

with the deformation potential $D_{ij}$ for scattering between valley $i$ and $j$ [Pötz and Vogl (1981)]. By analogy to (2.41), we find for the intervalley (IV) transition rate

$$S_{\mathrm{IV}}^{i \to j}(\mathbf{k}, \mathbf{k}') = \frac{\pi D_{ij}^2 Z_j}{\rho \omega_{ij} V} \left\{ g_{\mathrm{IV}}[\pm(\mathbf{k}' - \mathbf{k})] + \frac{1}{2} \mp \frac{1}{2} \right\} \tag{2.44}$$
$$\times \delta[E(\mathbf{k}') - E(\mathbf{k}) \mp \hbar\omega_{ij} + \Delta E_{ji}],$$

where $Z_j$ is the number of equivalent final valleys, $g_{\mathrm{IV}}$ labels the intervalley phonon distribution function and $\Delta E_{ji}$ is the bottom energy of the $j$ valley measured from the bottom of the $i$ valley. The intervalley scattering rate

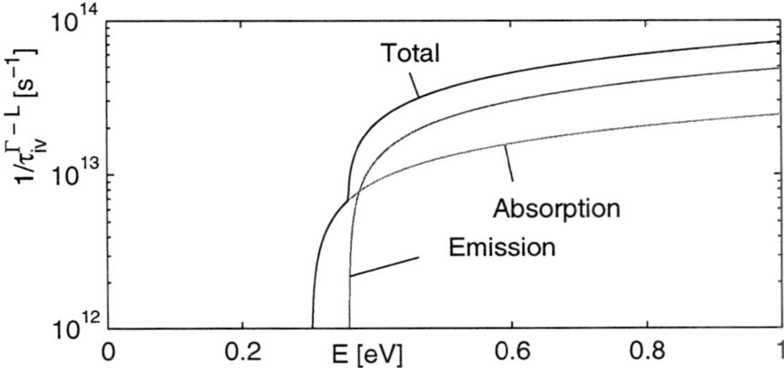


Fig. 2.6   Intervalley scattering rate for electrons in GaAs from $\Gamma$ valley to $L$ valleys.

reads

$$\frac{1}{\tau_{\mathrm{IV}}^{i\rightarrow j}(k)} = \frac{\pi D_{ij}^2 Z_j}{\rho\omega_{ij}} \left\{ g_{\mathrm{IV}}[\pm(\mathbf{k}' - \mathbf{k})] + \frac{1}{2} \mp \frac{1}{2} \right\} \tag{2.45}$$
$$\times \mathcal{Z}[E(\mathbf{k}) \pm \hbar\omega_{ij} - \Delta E_{ji}].$$

Figure 2.6 shows the intervalley scattering rate for electrons in GaAs for scattering from $\Gamma$ to $L$ valley by equilibrium phonons at 300 K with material parameters as given in Tab. 5.1.

### 2.4.2.2   *Polar Phonon Scattering*

Longitudinal lattice vibrations cause polarization waves in compound semiconductors such as GaAs or InP. These polarization waves strongly interact with carriers, resulting in the polar scattering of electrons. Polar acoustic scattering, termed piezoelectric scattering, is important at low temperatures in very pure semiconductors, while it is marginal at room temperature. On the other hand, polar optical phonon scattering is very strong and represents the dominant scattering process in compound semiconductors near room temperature.

The strength of polar interaction is proportional to the change of the dipole moment $\mathbf{p}$ (cf. Fig. 2.8), which in turn is related to the effective charge $e^*$. Fröhlich succeeded in deriving a formula for the effective charge in terms of static and optical dielectric constants $\kappa_0$ and $\kappa_\infty$ of the crystal

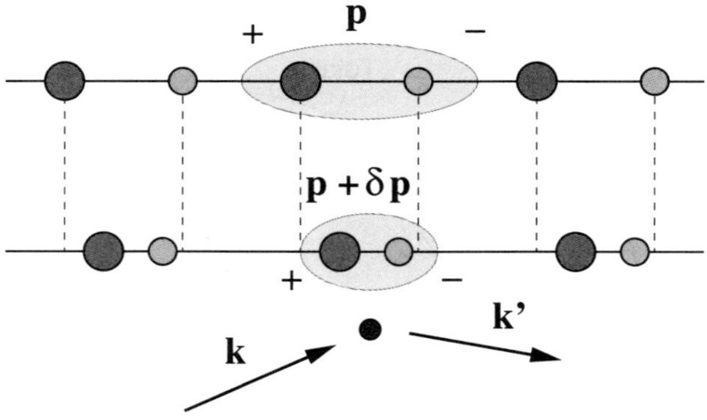


Fig. 2.7    Schematic illustration of polar optical scattering.

[Fröhlich and Mott (1939)]:

$$e^* = \left(\frac{VM}{\varepsilon_0 N}\right)^{\frac{1}{2}} \omega_0 \kappa_0 \varepsilon_0 \left(\frac{1}{\kappa_\infty} - \frac{1}{\kappa_0}\right)^{\frac{1}{2}}. \tag{2.46}$$

Here, $\omega_0$ is the frequency of the optical phonons and $N$ is the number of the pairs of ions with the masses $M_1$ and $M_2$, combined to the reduced mass $M$ via $1/M = 1/M_1 + 1/M_2$. For longitudinal optical phonons, the relative displacement $\Delta \mathbf{u} = \mathbf{u}_1 - \mathbf{u}_2$ with the displacements of the ions $\mathbf{u}_1$ and $\mathbf{u}_2$ perturbs the dipole moment directly. The quantity $\Delta \mathbf{u}$ can be expressed by

$$\Delta \mathbf{u}(\mathbf{r}, t) = \sum_{\mathbf{q}} \left(\frac{\hbar}{2NM\omega_0}\right)^{\frac{1}{2}} \mathbf{e}_{\mathbf{q}}(a_{\mathbf{q}} + a^\dagger_{-\mathbf{q}}) e^{i\mathbf{q}\cdot\mathbf{r}}. \tag{2.47}$$

The polarization $\mathbf{P}$ accompanied by the relative displacement of ions contributes to the dielectric displacement $\mathbf{D}$ via

$$\mathbf{D} = \varepsilon_0 \mathbf{E} + \mathbf{P}_{\text{ion}} + \mathbf{P}, \tag{2.48}$$

where $\mathbf{E}$ is the electric field and $\mathbf{P}_{\text{ion}}$ is the dipole moment caused by the polarization of the ions. Moreover, $\mathbf{P}$ can be expressed as a function of the relative displacement

$$\mathbf{P} = \frac{e^* N}{V} \Delta \mathbf{u}. \tag{2.49}$$

Since $\mathbf{P}_{\mathrm{ion}}$ is related to $\kappa_0$ via $\kappa_0 \varepsilon_0 \mathbf{E} = \varepsilon_0 \mathbf{E} + \mathbf{P}_{\mathrm{ion}}$, we find

$$\mathbf{D} = \kappa_0 \varepsilon_0 \mathbf{E} + \mathbf{P} = \kappa_0 \varepsilon_0 \mathbf{E} + \frac{e^* N}{V} \Delta \mathbf{u}. \tag{2.50}$$

The assumption of zero macroscopic, free charge ($\nabla \cdot \mathbf{D} = 0$) and the form of the longitudinal elastic polarization waves $\mathbf{D} = D \mathbf{e_q} \exp(i\mathbf{q} \cdot \mathbf{r})$ implies that $i\mathbf{q} \cdot \mathbf{D} = 0$ and, consequently, $\mathbf{D} = 0$. Hence, we obtain the electric field induced by polar optical phonons

$$\mathbf{E} = -\frac{N}{V} \frac{e^* \Delta \mathbf{u}}{\kappa_0 \varepsilon_0} \tag{2.51}$$

and the associated electrostatic potential $U(\mathbf{r})$ by taking advantage of (2.47):

$$U(\mathbf{r}) = -\int \mathbf{E} \cdot d\mathbf{r} = -i \frac{N}{V} \frac{e^* \Delta u(\mathbf{r})}{q \kappa_0 \varepsilon_0}. \tag{2.52}$$

This result leads immediately to the perturbation potential for polar optical scattering

$$H'_{\mathrm{POP}} = -eU(\mathbf{r}) = \sum_{\mathbf{q}} i\frac{e}{q} \left( \frac{\hbar \omega_0}{2\varepsilon_0 V} \right)^{\frac{1}{2}} \left( \frac{1}{\kappa_\infty} - \frac{1}{\kappa_0} \right)^{\frac{1}{2}} (a_{\mathbf{q}} + a^\dagger_{-\mathbf{q}}) e^{i\mathbf{q} \cdot \mathbf{r}}. \tag{2.53}$$

Inserting this result into (2.30) ends in the matrix elements

$$|\langle \mathbf{k}' | H'_{\mathrm{POP}} | \mathbf{k} \rangle|^2 = \sum_{\mathbf{q}} \frac{e^2 \hbar \omega_0}{2\varepsilon_0 V} \left( \frac{1}{\kappa_\infty} - \frac{1}{\kappa_0} \right) \frac{1}{q^2} \left[ g(\mathbf{q}) + \frac{1}{2} \mp \frac{1}{2} \right] \delta_{\mathbf{k}', \mathbf{k} \pm \mathbf{q}} \tag{2.54}$$

and the transition rate for polar optical (POP) scattering

$$S_{\mathrm{POP}}(\mathbf{k}, \mathbf{k}') = \sum_{\mathbf{q}} \frac{\pi e^2 \omega_0}{\varepsilon_0 V} \left( \frac{1}{\kappa_\infty} - \frac{1}{\kappa_0} \right) \frac{1}{q^2} \left[ g(\mathbf{q}) + \frac{1}{2} \mp \frac{1}{2} \right] \tag{2.55}$$

$$\times \delta_{\mathbf{k}', \mathbf{k} \pm \mathbf{q}} \delta[E(\mathbf{k}') - E(\mathbf{k}) \mp \hbar \omega_0]$$

with the optical phonon distribution function $g(\mathbf{q})$. Integrating this result with respect to $\mathbf{k}'$ yields the scattering rate for polar optical scattering for the equilibrium phonon distribution $g_0$

$$\frac{1}{\tau_{\mathrm{POP}}(k)} = \frac{e^2 \omega_0}{8\pi \varepsilon_0} \left( \frac{1}{\kappa_\infty} - \frac{1}{\kappa_0} \right) \frac{k}{E(k)} \left[ g_0 + \frac{1}{2} \mp \frac{1}{2} \right] \ln \frac{q^\pm_{\mathrm{max}}}{q^\pm_{\mathrm{min}}} \tag{2.56}$$

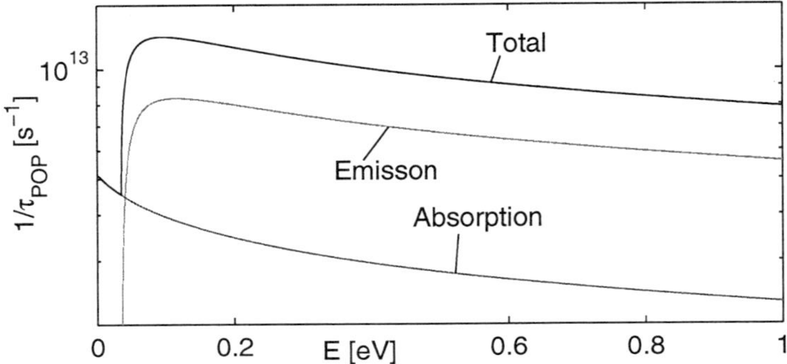


Fig. 2.8   Polar optical scattering rate for electrons in the $\Gamma$ valley of GaAs at 300 K.

with

$$q^{\pm}_{\min} = k\left|1 - \left(1 \pm \frac{\hbar\omega_0}{E(\mathbf{k})}\right)^{\frac{1}{2}}\right|, \quad q^{\pm}_{\max} = k\left|1 + \left(1 \pm \frac{\hbar\omega_0}{E(\mathbf{k})}\right)^{\frac{1}{2}}\right|. \quad (2.57)$$

Figure 2.8 shows the polar optical scattering rate for electrons in the $\Gamma$ valley of GaAs, induced by equilibrium phonons, at 300 K with material parameters as given in Tab. 5.1. This calculations are performed using the parabolic band approximation.

The calculation for the polar acoustic scattering rate is similar to the one presented for the polar optical phonon scattering. In this case, the piezoelectric (PZ) transition rate is given by

$$S_{\mathrm{PZ}}(\mathbf{k}, \mathbf{k}') = \frac{2\pi k_{\mathrm{B}} T_{\mathrm{L}}}{\hbar \rho v_{\mathrm{s}}^2 V} \left(\frac{e e_{\mathrm{PZ}}}{\kappa_0 \varepsilon_0}\right)^2 \frac{\delta[E(\mathbf{k}') - E(\mathbf{k})]}{|\mathbf{k}' - \mathbf{k}|^2} \quad (2.58)$$

with the piezoelectric coupling constant $e_{\mathrm{PZ}}$ and assuming elastic scattering and equipartition of the acoustic phonons (2.19).

### 2.4.3   *Ionized Impurity Scattering*

Carriers in semiconductor devices usually move through highly doped regions, which can be regarded as carrier reservoirs. In these regions, the carrier motion is strongly influenced by scattering by randomly distributed ionized impurities. The electrostatic potential of an impurity charge in the crystal must be regarded as more or less screened depending on the concentration of free carriers. As suggested by Brooks and Herring [Brooks

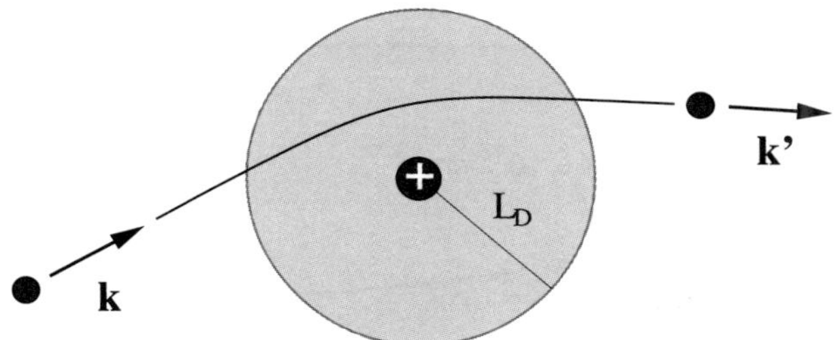


Fig. 2.9   Schematic illustration of ionized impurity scattering in the Brooks-Herring model.

and Herring (1951)], the transition rate for ionized impurity scattering is obtained as follows.

The perturbation potential for ionized impurity scattering is a screened Coulomb potential

$$H'_{\text{imp}} = \frac{ze}{4\pi\kappa_0\epsilon_0 r}e^{-q_{\text{D}}r} \qquad (2.59)$$

with the screening parameter

$$q_{\text{D}} = \sqrt{\frac{e^2 n_0}{\kappa_0\epsilon_0 k_{\text{B}} T_{\text{L}}}}, \qquad (2.60)$$

which is related to the screening length via $L_{\text{D}} = 1/q_{\text{D}}$. In these relations, $ze$ is the charge of a single impurity, $\kappa_0$ is the static dielectric constant and $n_0$ is the equilibrium electron density at temperature $T_{\text{L}}$. Having identified the perturbation potential for the electron scattering, the matrix element for impurity scattering is obtained by inserting (2.59) into (2.30), which leads to

$$\langle \mathbf{k}'|H'_{\text{imp}}|\mathbf{k}\rangle = \frac{1}{V}\frac{ze^2}{4\pi\kappa_0\epsilon_0}\int_V d^3r\, e^{-i\mathbf{k}\cdot\mathbf{r}}\frac{e^{q_{\text{D}}r}}{r}e^{i\mathbf{k}\cdot\mathbf{r}} \qquad (2.61)$$

$$= \frac{ze^2}{V\kappa_0\epsilon_0}\frac{1}{q^2+q_{\text{D}}^2} \qquad (2.62)$$

where $\mathbf{q} = \mathbf{k} - \mathbf{k}'$. With the help of this result and Fermi's golden rule (2.22), we can easily find the transition rate for scattering at a single ionized impurity

$$S_{1\text{imp}}(\mathbf{k},\mathbf{k}') = \frac{2\pi}{\hbar}\left(\frac{ze^2}{V\kappa_0\epsilon_0}\right)^2\frac{\delta[E(\mathbf{k}')-E(\mathbf{k})]}{(q^2+q_{\text{D}}^2)^2}. \qquad (2.63)$$

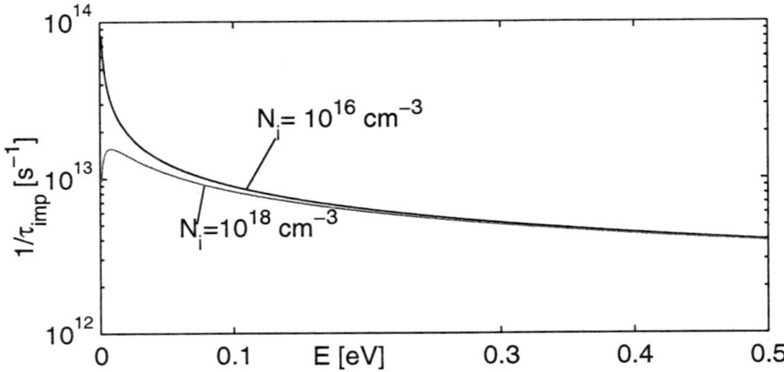


Fig. 2.10    Scattering rate at ionized impurities in GaAs with $n_0 = N_i$.

Assuming $N_i V$ impurities in the volume $V$, we obtain the final result for the transition rate of ionized impurity (IMP) scattering:

$$S_{\text{imp}}(\mathbf{k}, \mathbf{k}') = \frac{2\pi}{\hbar} \frac{N_i z^2 e^4}{V \kappa_0^2 \epsilon_0^2} \frac{\delta[E(\mathbf{k}') - E(\mathbf{k})]}{(q^2 + q_D^2)^2}. \tag{2.64}$$

It should be noted that the electron energy is conserved for impurity scattering.

The scattering rate is calculated by means of (2.31). Some algebra ends in

$$\frac{1}{\tau_{\text{imp}}(k)} = \frac{2\pi N_i z^2 e^4 \mathcal{Z}(E(\mathbf{k}))}{\hbar \kappa_0 \epsilon_0} \frac{1}{q_D^2 (4k^2 + q_D^2)}. \tag{2.65}$$

Figure 2.9 depicts schematically the scattering of an electron in a screened Coulomb potential. Moreover, Fig. 2.10 shows the scattering rate for electrons in the $\Gamma$ valley of GaAs at temperature $T = 300$ K for parabolic band structure and $z = 1$. The material parameters used for this calculation are found in Tab. 5.1.

## 2.5    Semiclassical Dynamics of Electrons

As discussed in Sec. 2.2, electrons in a crystal behave like electrons in free space, except for a change in the electron mass. Because of this fact, the motion of carriers in a crystal may be described by the classical equations of motion. This idea is acceptable if the potential energy felt by the electrons

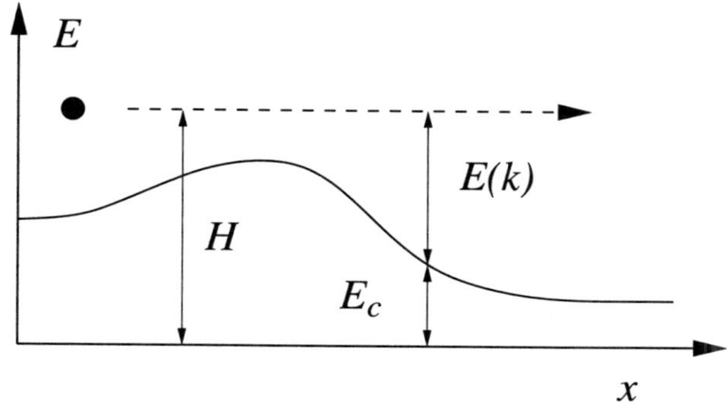


Fig. 2.11  Motion of an electron in a slowly varying potential $E_c$ without scattering. The symbols $E(k)$ and $H$ denote the kinetic energy and the total energy of the electron, respectively.

varies slowly compared to the crystal potential so that quantum effects like reflection and tunneling can be ignored (Wentzel-Kramers-Brillouin approximation [Bohm (1951)]).

The classical motion of a particle is governed by the equations of motion based on the total energy,

$$H(\mathbf{r}, \mathbf{k}) = E(\mathbf{k}) + U(\mathbf{r}), \qquad (2.66)$$

where $E(\mathbf{k})$ is the kinetic energy and $U(\mathbf{r})$ is the potential energy. Considering electrons in the conduction band, $U$ is given by the conduction band minimum $E_c(\mathbf{r})$, so that

$$U(\mathbf{r}) = E_c(\mathbf{r}) = C - \chi(\mathbf{r}) - eV(\mathbf{r}). \qquad (2.67)$$

Here, $\chi(\mathbf{r})$ is the electron affinity, $V(\mathbf{r})$ labels the electrostatic potential and $C$ is related to the reference of the electron energy. If the semiconductor material is compositionally uniform, $\chi(\mathbf{r})$ is constant and can be absorbed in $C$. Figure 2.11 illustrates the case of an electron moving in a slowly varying potential without scattering.

The equations of motion can easily be constructed in analogy to the Hamilton's equations of motion. Thus, we have

$$\dot{\mathbf{r}} = \frac{1}{\hbar}\nabla_{\mathbf{k}}H, \qquad (2.68a)$$

$$\dot{\mathbf{k}} = -\frac{1}{\hbar}\nabla_{\mathbf{r}}H. \qquad (2.68b)$$

The temporal derivative of the position vector $\mathbf{r}$ can be interpreted as the group velocity $\mathbf{v}$ of the electron. Hence, we obtain by combining (2.66) and (2.68a)

$$\mathbf{v}(\mathbf{k}) = \frac{1}{\hbar}\nabla_{\mathbf{k}} E(\mathbf{k}). \tag{2.69}$$

Inserting (2.67) into (2.66) for a spatially homogeneous crystal and taking advantage of (2.68b) leads directly to the pseudo-Newtonian law of the semiclassical motion of electrons

$$\hbar\dot{\mathbf{k}} = e\nabla_{\mathbf{r}} V(\mathbf{r}) = -e\mathbf{E}(\mathbf{r}) \tag{2.70}$$

with the electric field strength $\mathbf{E} = -\nabla_{\mathbf{r}} V(\mathbf{r})$ and the positive elementary charge $e$.

## 2.6   The Bloch-Boltzmann-Peierls Equations

A carrier with charge $Q$ and wave vector $\mathbf{k}$ that is exposed to an external electric field $\mathbf{E}$ changes its state according to

$$\hbar\dot{\mathbf{k}} = Q\mathbf{E}, \tag{2.71}$$

as discussed in the last section. Now, we define the distribution function $f(\mathbf{r}, \mathbf{k}, t)$ of this particle, which has the position vector $\mathbf{r}$ and the wave vector $\mathbf{k}$ at time $t$. First, we assume that the particle is not subject to scattering and that the state is only changed by the external field. This implies that after a time $dt$, the carrier is changed into a new state with the position $\mathbf{r} + \dot{\mathbf{r}}dt$ and the wave vector $\mathbf{k} + \dot{\mathbf{k}}dt$, which lowers $f(\mathbf{r}, \mathbf{k}, t)$. On the other hand, a particle which occupies the state $\mathbf{r} - \dot{\mathbf{r}}dt$ and $\mathbf{k} - \dot{\mathbf{k}}dt$ at $t - dt$ will move into the state $\mathbf{r}$ and $\mathbf{k}$ at time $t$ and increase $f(\mathbf{r}, \mathbf{k}, t)$. Hence, the rate of change in the distribution function by diffusion (changes of $\mathbf{r}$) and drift (changes of $\mathbf{k}$) of particles is given by

$$\left(\frac{df}{dt}\right)_{\mathrm{DD}} = \frac{[f(\mathbf{r} - \dot{\mathbf{r}}dt, \mathbf{k} - \dot{\mathbf{k}}dt, t - dt) - f(\mathbf{r}, \mathbf{k}, t)]}{dt}. \tag{2.72}$$

Since the scattering of particles is not included in the above expression, this rate represents the continuous flow of particles.

The Taylor expansion of the first term on the RHS of (2.72) leads to

$$f(\mathbf{r} - \dot{\mathbf{r}}dt, \mathbf{k} - \dot{\mathbf{k}}dt, t - dt) \tag{2.73}$$

$$= f(\mathbf{r}, \mathbf{k}, t) - \left[ \dot{\mathbf{r}} \cdot \nabla_{\mathbf{r}} f + \dot{\mathbf{k}} \cdot \nabla_{\mathbf{k}} f + \frac{\partial f}{\partial t} \right] + \dots$$

$$= f(\mathbf{r}, \mathbf{k}, t) - \left[ \mathbf{v}(\mathbf{k}) \cdot \nabla_{\mathbf{r}} f + \frac{\mathcal{Q}}{\hbar} \mathbf{E} \cdot \nabla_{\mathbf{k}} f + \frac{\partial f}{\partial t} \right] + \dots,$$

where we used the definition of the group velocity $\mathbf{v}$ (2.69) and the pseudo-Newtonian law (2.71). Inserting this result into (2.72) leads to

$$\left( \frac{df}{dt} \right)_{\mathrm{DD}} = - \left[ \frac{\partial f}{\partial t} + \mathbf{v}(\mathbf{k}) \cdot \nabla_{\mathbf{r}} f + \frac{\mathcal{Q}}{\hbar} \mathbf{E} \cdot \nabla_{\mathbf{k}} f \right]. \tag{2.74}$$

On the other hand, particles change their states by scattering, which can be described by the rate of change in the distribution function due to collisions $\mathcal{C}[f]$. Since the distribution function must satisfy the equilibrium condition (condition of balance), we obtain

$$\left( \frac{df}{dt} \right)_{\mathrm{DD}} + \mathcal{C}[f] = 0. \tag{2.75}$$

By inserting (2.74) into this expression, we get the celebrated Boltzmann transport equation of carriers in semiconductors:

$$\frac{\partial f(\mathbf{r}, \mathbf{k}, t)}{\partial t} + \mathbf{v}(\mathbf{k}) \cdot \nabla_{\mathbf{r}} f(\mathbf{r}, \mathbf{k}, t) + \frac{\mathcal{Q}}{\hbar} \mathbf{E}(\mathbf{r}) \cdot \nabla_{\mathbf{k}} f(\mathbf{r}, \mathbf{k}, t) = \mathcal{C}[f]. \tag{2.76}$$

In this equation, the second and third term on the LHS, $\mathbf{v}(\mathbf{k}) \cdot \nabla_{\mathbf{r}} f$ and $\mathcal{Q}/\hbar\, \mathbf{E} \cdot \nabla_{\mathbf{k}} f$, are called diffusion term and drift term, respectively. The RHS of (2.76) is usually named the collision term, including the changes of the distribution function $f$ due to all the scattering mechanisms taken into account.

Similarly as for carriers, one can deduce a Boltzmann transport equation for phonons with the main difference that the drift term vanishes, since phonons do not interact with the electric field. This equation reads

$$\frac{\partial g(\mathbf{r}, \mathbf{q}, t)}{\partial t} + \mathbf{u}(\mathbf{q}) \cdot \nabla_{\mathbf{r}} g(\mathbf{r}, \mathbf{q}, t) = \mathcal{D}[g]. \tag{2.77}$$

Here, $g$ labels the phonon distribution function depending on the position $\mathbf{r}$, the phonon wave vector $\mathbf{q}$ and time $t$. Moreover, $\mathbf{u}$ denotes the phonon group velocity, defined by $\mathbf{u}(\mathbf{q}) = \nabla_{\mathbf{q}} \omega(\mathbf{q})$ with the phonon frequency $\omega(\mathbf{q})$, and $\mathcal{D}[g]$ is the collision term for phonons.

Next, we consider a semiconductor, containing a mixture of carriers and phonons, which interact. Each type $\nu = 1, 2, \ldots, n_c$ of carriers (e.g., electrons in several valleys) has its own velocity $\mathbf{v}_\nu$ and charge $Q_\nu$ and it is described by the distribution function $f^\nu$. Similarly, the phonon mode (e.g., acoustic and optical phonons), $\gamma = 1, 2, \ldots, n_p$ is distributed according to $g^\gamma$ with its group velocity $\mathbf{u}_\gamma$. The evolution equations for the functions $f^\nu$ and $g^\gamma$ are the so-called Bloch-Boltzmann-Peierls equations [Ferry (1991); Lifschitz and Pitaevskii (1981)],

$$\frac{\partial f^\nu(\mathbf{r}, \mathbf{k}, t)}{\partial t} + \mathbf{v}_\nu(\mathbf{k}) \cdot \nabla_{\mathbf{r}} f^\nu(\mathbf{r}, \mathbf{k}, t) + \frac{Q_\nu}{\hbar} \mathbf{E}(\mathbf{r}) \cdot \nabla_{\mathbf{k}} f^\nu(\mathbf{r}, \mathbf{k}, t)$$

$$= \mathcal{C}^\nu[f^1, \ldots, f^{n_c}, g^1, \ldots, g^{n_p}], \tag{2.78a}$$

$$\frac{\partial g^\gamma(\mathbf{r}, \mathbf{q}, t)}{\partial t} + \mathbf{u}_\gamma(\mathbf{q}) \cdot \nabla_{\mathbf{r}} g^\gamma(\mathbf{r}, \mathbf{q}, t)$$

$$= \mathcal{D}^\gamma[f^1, \ldots, f^{n_c}, g^1, \ldots, g^{n_p}], \tag{2.78b}$$

$\nu = 1, 2, \ldots, n_c$, $\gamma = 1, 2, \ldots, n_p$, which describe the dynamics of the coupled carrier-phonon system in a semiconductor. The BBP equations constitute a set of partial integro-differential equations, which are coupled by the collision terms of carriers $\mathcal{C}^\nu$ and phonons $\mathcal{D}^\gamma$.

These collision terms are heuristically obtained as follows. The rate of change in the distribution function $f^\nu(\mathbf{k})$ due to scattering contains two terms: the rate of increase of $f^\nu(\mathbf{k})$ due to transitions from all possible $\mathbf{k}'$ of carrier type $\mu$ (excluding $\mathbf{k}$) to the state $\mathbf{k}$ and the rate of decrease of $f^\nu(\mathbf{k})$ due to the transitions from the $\mathbf{k}$ state to the other possible states $\mathbf{k}'$ of type $\mu$. With the help of the transition rate $S_\xi^{\nu \to \mu}(\mathbf{k}, \mathbf{k}')$ (cf. Sec. 2.4), which refer to transitions of carriers of type $\nu$, state $\mathbf{k}$ to type $\mu$, state $\mathbf{k}'$ caused by the scattering mechanism $\xi$, the collision term $\mathcal{C}_\xi^\nu[f^\nu]$ for this collision process can be written as

$$\mathcal{C}_\xi^\nu[f^\nu] = \sum_{\mathbf{k}'} \left\{ S_\xi^{\mu \to \nu}(\mathbf{k}', \mathbf{k}) f^\mu(\mathbf{k}')[1 - f^\nu(\mathbf{k})] \right.$$

$$\left. - S_\xi^{\nu \to \mu}(\mathbf{k}, \mathbf{k}') f^\nu(\mathbf{k})[1 - f^\mu(\mathbf{k}')] \right\}, \tag{2.79}$$

where the factor $f^\nu(\mathbf{k})(1 - f^\mu(\mathbf{k}')$ represents the probability of carrier occupation in the initial state $\mathbf{k}$ and of carrier vacancy in the final state $\mathbf{k}'$ according to the Pauli principle. When we replace the summation of $\mathbf{k}'$ by an integral via

$$\sum_{\mathbf{k}'} \rightarrow \frac{V}{8\pi^3} \int d^3k' \qquad (2.80)$$

with the volume of the crystal $V$, we obtain the collision term in the more convenient form

$$C_\xi^\nu[f^\nu] = \frac{V}{8\pi^3} \int d^3k' \left\{ S_\xi^{\mu\rightarrow\nu}(\mathbf{k}',\mathbf{k})f^\mu(\mathbf{k}')[1-f^\nu(\mathbf{k})] \right. \qquad (2.81)$$
$$\left. - S_\xi^{\nu\rightarrow\mu}(\mathbf{k},\mathbf{k}')f^\nu(\mathbf{k})[1-f^\mu(\mathbf{k}')] \right\} .$$

A widely used simplification of (2.81) is the low density approximation. It is valid for cases where the Fermi level lies in the band gap, so that the conditions $f^\mu(\mathbf{k}) \ll 1$ and $f^\mu(\mathbf{k}') \ll 1$ hold. Then, we are allowed to set $1 - f^\nu(\mathbf{k}) \approx 1$ and $1 - f^\mu(\mathbf{k}') \approx 1$, which implies that the collision term simplifies to

$$C_\xi^\nu[f^\nu] = \frac{V}{8\pi^3} \int d^3k' \left[ S_\xi^{\mu\rightarrow\nu}(\mathbf{k}',\mathbf{k})f^\mu(\mathbf{k}') - S_\xi^{\nu\rightarrow\mu}(\mathbf{k},\mathbf{k}')f^\nu(\mathbf{k}) \right]. \qquad (2.82)$$

The collision term of phonons $\mathcal{D}_\eta^\gamma[g^\gamma]$ for the phonon mode $\gamma$ caused by the scattering mechanism $\eta$ (e.g., phonon-phonon, phonon-carrier scattering) reads

$$\mathcal{D}_\eta^\gamma[g^\gamma] = S_\eta^{\gamma,\mathrm{G}}(\mathbf{q})[g^\gamma(\mathbf{q})+1] - S_\eta^{\gamma,\mathrm{L}}(\mathbf{q})g^\gamma(\mathbf{q}). \qquad (2.83)$$

Here, $S_\eta^{\mathrm{G}}(\mathbf{q})$ and $S_\eta^{\mathrm{L}}(\mathbf{q})$ are the rates per unit time for the creation (gain) and annihilation (loss) of phonons. The phonon collision terms cannot be interpreted so illustrative as those of carriers since the appearing of the additional factor 1 in the gain term is a typical quantum effect. However, we can show that there exists a close relation between carrier and phonon collision terms as shown in the following considerations.

We consider a typical intravalley electron-phonon interaction mechanism as discussed in the subsection 2.4.2. This scattering mechanism is supposed to have the transition rate

$$S(\mathbf{k},\mathbf{k}') = \sum_{\mathbf{q}} s(\mathbf{q}) \left[ g(\mathbf{q}) + \frac{1}{2} \mp \frac{1}{2} \right] \delta_{\mathbf{k}',\mathbf{k}\pm\mathbf{q}} \delta[E(\mathbf{k}')-E(\mathbf{k})\mp\hbar\omega(\mathbf{q})], \qquad (2.84)$$

where the scattering function is labeled $s(\mathbf{q})$ and upper/lower signs refer to absorption/emission processes. For this scattering mechanism, the collision term of electrons is obtained by inserting (2.55) into (2.79),

$$\mathcal{C}[f] = \sum_{\mathbf{k}',\mathbf{q}} s(\mathbf{q})\{g(\mathbf{q})f(\mathbf{k}')[1-f(\mathbf{k})]\delta_{\mathbf{k}',\mathbf{k}-\mathbf{q}}\delta[E(\mathbf{k}')-E(\mathbf{k})+\hbar\omega(\mathbf{q})] \quad (2.85)$$

$$+[g(\mathbf{q})+1]f(\mathbf{k}')[1-f(\mathbf{k})]\delta_{\mathbf{k}',\mathbf{k}+\mathbf{q}}\delta[E(\mathbf{k}')-E(\mathbf{k})-\hbar\omega(\mathbf{q})]$$

$$-g(\mathbf{q})f(\mathbf{k})[1-f(\mathbf{k}')]\delta_{\mathbf{k}',\mathbf{k}+\mathbf{q}}\delta[E(\mathbf{k}')-E(\mathbf{k})-\hbar\omega(\mathbf{q})]$$

$$-[g(\mathbf{q})+1]f(\mathbf{k})[1-f(\mathbf{k}')]\delta_{\mathbf{k}',\mathbf{k}-\mathbf{q}}\delta[E(\mathbf{k}')-E(\mathbf{k})+\hbar\omega(\mathbf{q})]\},$$

skipping the indices $\nu$ and $\gamma$ for simplicity. Obviously, the expression

$$s(\mathbf{q})g(\mathbf{q})f(\mathbf{k})[1-f(\mathbf{k}')]\delta_{\mathbf{k}',\mathbf{k}+\mathbf{q}}\delta[E(\mathbf{k}')-E(\mathbf{k})-\hbar\omega(\mathbf{q})]$$

gives the rate of collisions scattering electrons of the state $\mathbf{k}$ to state $\mathbf{k}'$ by absorbing a phonon of the state $\mathbf{q}$. Hence, the total rate of all the possible scattering processes between $\mathbf{k}$ and $\mathbf{k}'$ by absorbing a phonon of wave vector $\mathbf{q}$, which equals the loss term $\mathcal{D}_{\mathrm{L}}[g]$ in the phonon collision term (2.83), is given by

$$\mathcal{D}_{\mathrm{L}}[g] = 2\sum_{\mathbf{k},\mathbf{k}'} s(\mathbf{q})g(\mathbf{q}) \qquad (2.86)$$

$$\times f(\mathbf{k})[1-f(\mathbf{k}')]\delta_{\mathbf{k}',\mathbf{k}+\mathbf{q}}\delta[E(\mathbf{k}')-E(\mathbf{k})-\hbar\omega(\mathbf{q})],$$

when including an additional factor 2 for the electron spin degeneracy. With similar arguments for deducing the gain term of the phonon collision term, we obtain the phonon collision term $\mathcal{D}[g]$ corresponding to the electron collision term $\mathcal{C}[f]$ (2.85)

$$\mathcal{D}[g] = 2\sum_{\mathbf{k},\mathbf{k}'} s(\mathbf{q})f(\mathbf{k})[1-f(\mathbf{k}')]\{[g(\mathbf{q})+1]\delta_{\mathbf{k}',\mathbf{k}-\mathbf{q}} \qquad (2.87)$$

$$\times\delta[E(\mathbf{k}')-E(\mathbf{k})+\hbar\omega(\mathbf{q})]-g(\mathbf{q})\delta_{\mathbf{k}',\mathbf{k}+\mathbf{q}}\delta[E(\mathbf{k}')-E(\mathbf{k})-\hbar\omega(\mathbf{q})]\}.$$

The comparison of this result with the phonon collision term given in (2.83) reveals that the phonon transition rates $S^{\mathrm{G}}(\mathbf{q})$ and $S^{\mathrm{L}}(\mathbf{q})$ associated to the electron transition rate $S(\mathbf{k},\mathbf{k}')$ of the form (2.84) read

$$S^{\mathrm{G}}(\mathbf{q}) = 2\sum_{\mathbf{k},\mathbf{k}'} s(\mathbf{q})f(\mathbf{k})[1-f(\mathbf{k}')]\delta_{\mathbf{k}',\mathbf{k}-\mathbf{q}}\delta[E(\mathbf{k}')-E(\mathbf{k})+\hbar\omega(\mathbf{q})], \quad (2.88\mathrm{a})$$

$$S^{\mathrm{L}}(\mathbf{q}) = 2\sum_{\mathbf{k},\mathbf{k}'} s(\mathbf{q})f(\mathbf{k})[1-f(\mathbf{k}')]\delta_{\mathbf{k}',\mathbf{k}+\mathbf{q}}\delta[E(\mathbf{k}')-E(\mathbf{k})-\hbar\omega(\mathbf{q})]. \quad (2.88\mathrm{b})$$

Having determined the carrier distribution functions $f^\nu$ and the phonon distribution functions $g^\gamma$ from the BBP equations (2.78), we can easily evaluate macroscopic quantities from $f^\nu$ and $g^\gamma$ by forming moments with respect to the wave vectors $\mathbf{k}$ and $\mathbf{q}$. For instance, the carrier density $\langle n^{\mathrm{c},\nu}\rangle$, the momentum density $\langle n\mathbf{k}^{\mathrm{c},\nu}\rangle$, the velocity density $\langle n\mathbf{v}^{\mathrm{c},\nu}\rangle$, the energy density $\langle nE^{\mathrm{c},\nu}\rangle$, the momentum flux $\langle n\mathsf{K}^{\mathrm{c},\nu}\rangle$ and the energy flux $\langle n\mathbf{Q}^{\mathrm{c},\nu}\rangle$ of carriers of the type $\nu$ are given by

$$\langle n^{\mathrm{c},\nu}(\mathbf{r},t)\rangle = \frac{1}{4\pi^3}\int d^3k\, f^\nu(\mathbf{r},\mathbf{k},t), \tag{2.89a}$$

$$\langle n\mathbf{k}^{\mathrm{c},\nu}(\mathbf{r},t)\rangle = \frac{1}{4\pi^3}\int d^3k\, \mathbf{k}\, f^\nu(\mathbf{r},\mathbf{k},t), \tag{2.89b}$$

$$\langle n\mathbf{v}^{\mathrm{c},\nu}(\mathbf{r},t)\rangle = \frac{1}{4\pi^3}\int d^3k\, \mathbf{v}_\nu(\mathbf{k})\, f^\nu(\mathbf{r},\mathbf{k},t), \tag{2.89c}$$

$$\langle nE^{\mathrm{c},\nu}(\mathbf{r},t)\rangle = \frac{1}{4\pi^3}\int d^3k\, E_\nu(\mathbf{k})\, f^\nu(\mathbf{r},\mathbf{k},t), \tag{2.89d}$$

$$\langle n\mathsf{K}^{\mathrm{c},\nu}(\mathbf{r},t)\rangle = \frac{1}{4\pi^3}\int d^3k\, (\mathbf{k}\otimes\mathbf{v}_\nu(\mathbf{k}))\, f^\nu(\mathbf{r},\mathbf{k},t), \tag{2.89e}$$

$$\langle n\mathbf{Q}^{\mathrm{c},\nu}(\mathbf{r},t)\rangle = \frac{1}{4\pi^3}\int d^3k\, E_\nu(\mathbf{k})\mathbf{v}_\nu(\mathbf{k})\, f^\nu(\mathbf{r},\mathbf{k},t). \tag{2.89f}$$

Moreover, the drift velocity $\langle\mathbf{v}^{\mathrm{c},\nu}\rangle$ and the average energy $\langle E^{\mathrm{c},\nu}\rangle$ of carriers of the type $\nu$ are found to equal

$$\langle\mathbf{v}^{\mathrm{c},\nu}(\mathbf{r},t)\rangle = \frac{\langle n\mathbf{v}^{\mathrm{c},\nu}(\mathbf{r},t)\rangle}{\langle n^{\mathrm{c},\nu}(\mathbf{r},t)\rangle}, \tag{2.90a}$$

$$\langle E^{\mathrm{c},\nu}(\mathbf{r},t)\rangle = \frac{\langle nE^{\mathrm{c},\nu}(\mathbf{r},t)\rangle}{\langle n^{\mathrm{c},\nu}(\mathbf{r},t)\rangle}. \tag{2.90b}$$

Turning to the phonons, the phonon density $\langle n^{\mathrm{p},\gamma}\rangle$, the momentum density $\langle n\mathbf{k}^{\mathrm{p},\gamma}\rangle$, the velocity density $\langle n\mathbf{v}^{\mathrm{p},\gamma}\rangle$, the energy density $\langle nE^{\mathrm{p},\gamma}\rangle$, the momentum flux $\langle n\mathsf{K}^{\mathrm{p},\gamma}\rangle$ and the energy flux $\langle n\mathbf{Q}^{\mathrm{p},\gamma}\rangle$ of phonons of the mode $\gamma$ are obtained via

$$\langle n^{\mathrm{p},\gamma}(\mathbf{r},t)\rangle = \frac{1}{8\pi^3} \int d^3q\, g^\gamma(\mathbf{r},\mathbf{q},t), \tag{2.91a}$$

$$\langle n\mathbf{k}^{\mathrm{p},\gamma}(\mathbf{r},t)\rangle = \frac{1}{8\pi^3} \int d^3q\, \mathbf{q}\, g^\gamma(\mathbf{r},\mathbf{q},t), \tag{2.91b}$$

$$\langle n\mathbf{v}^{\mathrm{p},\gamma}(\mathbf{r},t)\rangle = \frac{1}{8\pi^3} \int d^3q\, \mathbf{u}_\gamma(\mathbf{q})\, g^\gamma(\mathbf{r},\mathbf{q},t), \tag{2.91c}$$

$$\langle nE^{\mathrm{p},\gamma}(\mathbf{r},t)\rangle = \frac{1}{8\pi^3} \int d^3q\, \hbar\omega_\gamma(\mathbf{q})\, g^\gamma(\mathbf{r},\mathbf{q},t), \tag{2.91d}$$

$$\langle n\mathsf{K}^{\mathrm{p},\gamma}(\mathbf{r},t)\rangle = \frac{1}{8\pi^3} \int d^3q\, (\mathbf{q}\otimes\mathbf{u}_\gamma(\mathbf{q}))\, g^\gamma(\mathbf{r},\mathbf{q},t), \tag{2.91e}$$

$$\langle n\mathbf{Q}^{\mathrm{p},\gamma}(\mathbf{r},t)\rangle = \frac{1}{8\pi^3} \int d^3q\, \hbar\omega_\gamma(\mathbf{q})\mathbf{u}_\gamma(\mathbf{q})\, g^\gamma(\mathbf{r},\mathbf{q},t), \tag{2.91f}$$

and the relations for the drift velocity $\langle \mathbf{v}^{\mathrm{p},\gamma}\rangle$ and the average energy $\langle E^{\mathrm{p},\gamma}\rangle$ of phonons of the mode $\gamma$ read

$$\langle \mathbf{v}^{\mathrm{p},\gamma}(\mathbf{r},t)\rangle = \frac{\langle n\mathbf{v}^{\mathrm{p},\nu}(\mathbf{r},t)\rangle}{\langle n^{\mathrm{p},\nu}(\mathbf{r},t)\rangle}, \tag{2.92a}$$

$$\langle E^{\mathrm{p},\gamma}(\mathbf{r},t)\rangle = \frac{\langle nE^{\mathrm{p},\nu}(\mathbf{r},t)\rangle}{\langle n^{\mathrm{p},\nu}(\mathbf{r},t)\rangle}. \tag{2.92b}$$

## 2.7   Mathematical Properties of the BBP Equations

In this section, we summarize the most important mathematical properties of the BBP equations in the following theorems. For simplicity, we consider only one type of carriers and one mode of phonons, coupled by one interaction process. Hence, the BBP equations, which we consider, are (2.76) and (2.77) with the collision terms given in (2.85) and (2.87).

**Theorem 1   (Boundedness of the solution)**
*The solution $f(\mathbf{r},\mathbf{k},t)$ and $g(\mathbf{r},\mathbf{q},t)$ of the BBP equations (2.76), (2.77), (2.85) and (2.87) are bounded by*

$$0 \le f(\mathbf{r},\mathbf{k},t) \le 1, \tag{2.93a}$$

$$0 \le g(\mathbf{r},\mathbf{q},t) \tag{2.93b}$$

*for all times $t > t_0$, if the initial distributions fulfill the conditions*

$$0 \le f(\mathbf{r}, \mathbf{k}, t_0) \le 1, \tag{2.94a}$$

$$0 \le g(\mathbf{r}, \mathbf{q}, t_0). \tag{2.94b}$$

The proof of this theorem is found in [Markowich *et al.* (1990)].

**Theorem 2    (Conservation of carrier density)**
*The BBP equations (2.76), (2.77), (2.85) and (2.87) provide the continuity equation*

$$\frac{\partial \langle n^{\mathrm{c}}(\mathbf{r}, t) \rangle}{\partial t} + \nabla_{\mathbf{r}} \cdot \langle n\mathbf{v}^{\mathrm{c}}(\mathbf{r}, t) \rangle = 0 \tag{2.95}$$

*with the density $\langle n^{\mathrm{c}}(\mathbf{r}, t) \rangle$ and the velocity density $\langle n\mathbf{v}^{\mathrm{c}}(\mathbf{r}, t) \rangle$ of carriers defined in (2.89).*

**Theorem 3    (Conservation of total momentum)**
*The BBP equations (2.76), (2.77), (2.85) and (2.87) provide the continuity equation*

$$\frac{\partial \langle n\mathbf{k}^{\mathrm{tot}}(\mathbf{r}, t) \rangle}{\partial t} + \nabla_{\mathbf{r}} \langle n\mathsf{K}^{\mathrm{tot}}(\mathbf{r}, t) \rangle = \mathbf{0} \tag{2.96}$$

*for vanishing electric field $\mathbf{E}$ with the total momentum density $\langle n\mathbf{k}^{\mathrm{tot}} \rangle = \langle n\mathbf{k}^{\mathrm{c}} \rangle + \langle n\mathbf{k}^{\mathrm{p}} \rangle$ and the total momentum flux $\langle n\mathsf{K}^{\mathrm{tot}} \rangle = \langle n\mathsf{K}^{\mathrm{c}} \rangle + \langle n\mathsf{K}^{\mathrm{p}} \rangle$ (cf. (2.89), (2.91)).*

**Theorem 4    (Conservation of total energy density)**
*The BBP equations (2.76), (2.77), (2.85) and (2.87) provide the continuity equation*

$$\frac{\partial \langle nE^{\mathrm{tot}}(\mathbf{r}, t) \rangle}{\partial t} + \nabla_{\mathbf{r}} \cdot \langle n\mathbf{Q}^{\mathrm{tot}}(\mathbf{r}, t) \rangle = 0 \tag{2.97}$$

*for vanishing electric field $\mathbf{E}$ with the total energy density $\langle nE^{\mathrm{tot}} \rangle = \langle nE^{\mathrm{c}} \rangle + \langle nE^{\mathrm{p}} \rangle$ and the total energy flux $\langle n\mathbf{Q}^{\mathrm{tot}} \rangle = \langle n\mathbf{Q}^{\mathrm{c}} \rangle + \langle n\mathbf{Q}^{\mathrm{p}} \rangle$ (cf. (2.89), (2.91)).*

The theorems 2, 3 and 4 can easily be proved by forming the appropriate moments of the BBP equations (2.76), (2.77), (2.85) and (2.87).

**Theorem 5    (H-Theorem)**
*A Lyapounov functional to the BBP equations (2.76), (2.77), (2.85) and*

*(2.87) for the spatially homogeneous case with vanishing electric field is given by*

$$H = \int d^3k\, \mathcal{H}^{\mathrm{c}}[f] + \int d^3q\, \mathcal{H}^{\mathrm{p}}[g], \tag{2.98}$$

*where*

$$\frac{\partial \mathcal{H}^{\mathrm{c}}[f]}{\partial f} = 2\ln \frac{f(\mathbf{k})}{1 - f(\mathbf{k})}, \tag{2.99a}$$

$$\frac{\partial \mathcal{H}^{\mathrm{p}}[g]}{\partial g} = \ln \frac{g(\mathbf{q})}{g(\mathbf{q}) + 1}. \tag{2.99b}$$

For the proof of this theorem, we refer to [Rossani and Kaniadakis (2000); Rossani (2002)].

### Theorem 6    (Equilibrium solution)
*The definition of the equilibrium state of the BBP equations (2.76), (2.77), (2.85) and (2.87), $\mathcal{C}[f] = \mathcal{D}[g] = 0$ is equivalent to*

$$f(\mathbf{k})[1 - f(\mathbf{k}')][g(\mathbf{q}) + 1] = f(\mathbf{k}')[1 - f(\mathbf{k})]g(\mathbf{q}) \tag{2.100}$$

$\forall\, \mathbf{k}, \mathbf{q}$. *For isotropic distributions, (2.100) is equivalent to the Fermi-Dirac distribution $f_{\mathrm{FD}}$ and Bose-Einstein distribution $g_{\mathrm{BE}}$*

$$f_{\mathrm{FD}}(\mathbf{k}) = \left[\exp \frac{E(\mathbf{k}) - E_{\mathrm{F}}}{k_{\mathrm{B}} T_{\mathrm{L}}} + 1\right]^{-1}, \tag{2.101a}$$

$$g_{\mathrm{BE}}(\mathbf{q}) = \left[\exp \frac{\hbar \omega(\mathbf{q})}{k_{\mathrm{B}} T_{\mathrm{L}}} - 1\right]^{-1} \tag{2.101b}$$

*with the Fermi energy $E_{\mathrm{F}}$ and the temperature $T_{\mathrm{L}}$.*

The proof of this theorem is given in [Rossani and Kaniadakis (2000); Rossani (2002)].

### Remark.
It is interesting to note that for the collision terms $\mathcal{C}[f]$ and $\mathcal{D}[g]$ written

in the low density approximation,

$$C[f] = \sum_{\mathbf{k}',\mathbf{q}} s(\mathbf{q})\{g(\mathbf{q})f(\mathbf{k}')\delta_{\mathbf{k}',\mathbf{k}-\mathbf{q}}\delta[E(\mathbf{k}') - E(\mathbf{k}) + \hbar\omega(\mathbf{q})] \quad (2.102)$$
$$+[g(\mathbf{q}) + 1]f(\mathbf{k}')\delta_{\mathbf{k}',\mathbf{k}+\mathbf{q}}\delta[E(\mathbf{k}') - E(\mathbf{k}) - \hbar\omega(\mathbf{q})]$$
$$-g(\mathbf{q})f(\mathbf{k})\delta_{\mathbf{k}',\mathbf{k}+\mathbf{q}}\delta[E(\mathbf{k}') - E(\mathbf{k}) - \hbar\omega(\mathbf{q})]$$
$$-[g(\mathbf{q}) + 1]f(\mathbf{k})\delta_{\mathbf{k}',\mathbf{k}-\mathbf{q}}\delta[E(\mathbf{k}') - E(\mathbf{k}) + \hbar\omega(\mathbf{q})]\},$$

$$\mathcal{D}[g] = 2\sum_{\mathbf{k},\mathbf{k}'} s(\mathbf{q})f(\mathbf{k})\{[g(\mathbf{q}) + 1]\delta_{\mathbf{k}',\mathbf{k}-\mathbf{q}}\delta[E(\mathbf{k}') - E(\mathbf{k}) + \hbar\omega(\mathbf{q})] \quad (2.103)$$
$$-g(\mathbf{q})\delta_{\mathbf{k}',\mathbf{k}+\mathbf{q}}\delta[E(\mathbf{k}') - E(\mathbf{k}) - \hbar\omega(\mathbf{q})]\},$$

the condition (2.100) for the vanishing of the collision terms in theorem 6 becomes

$$f(\mathbf{k})[g(\mathbf{q}) + 1] = f(\mathbf{k}')g(\mathbf{q}) \qquad (2.104)$$

$\forall\, \mathbf{k}, \mathbf{q}$. This implies that for isotropic distributions, the equilibrium solution of the BBP equations (2.76), (2.77), (2.87) with (2.102) are the Bose-Einstein distribution $g_{\mathrm{BE}}$ (2.101b) and the Maxwell distribution

$$f_{\mathrm{M}}(\mathbf{k}) = \exp\frac{E_{\mathrm{F}}}{k_{\mathrm{B}}T_{\mathrm{L}}}\exp-\frac{E(\mathbf{k})}{k_{\mathrm{B}}T_{\mathrm{L}}}. \qquad (2.105)$$

## Chapter 3

# Multigroup Model Equations for Polar Semiconductors

## 3.1 Introduction

The use of III-V compound semiconductor materials for manufacturing semiconductor devices becomes more and more important owing to their excellent velocity characteristics. These features permit the operation of devices at very high frequencies. Especially InP is increasingly used as the bulk material for many electronic devices, whose performance exceed those fabricated of GaAs regarding velocity and frequency response.

Due to the enhanced functional integration of such modern electron devices, the traditional drift diffusion models [Markowich *et al.* (1990)] fail in describing the occurring high-field and sub-micron phenomena. Therefore, the semi-classical Boltzmann transport equations must be applied for dealing with phenomena of hot electrons [Ferry (1991)]. Moreover, non-equilibrium longitudinal-optical phonons have been found to strongly affect the electron distribution function (hot phonon phenomena) in polar semiconductors [Vaissiere *et al.* (1992); Vaissiere *et al.* (1996)]. Consequently, one must also include kinetic equations for the evolution of phonon populations in an accurate description of such materials.

The numerical solution of these BBP equations is not an easy task, because these transport equations are of integro-differential type in six phase space and one time variables. Besides the well-established Monte Carlo techniques [Jacoboni and Lugli (1989); González Sánchez *et al.* (1991); Jungemann and Meinzerhagen (2003)], newly developed deterministic methods are nowadays used for solving the BTEs under transient transport conditions. Here, the publications of Fatemi and Odeh [Fatemi and Odeh (1993)], Niclot *et al.* [Niclot *et al.* (1988)], Majorana and Pidatella [Majorana and Pidatella (2001)], Carrillo *et al.* [Carrillo *et al.* (2002);

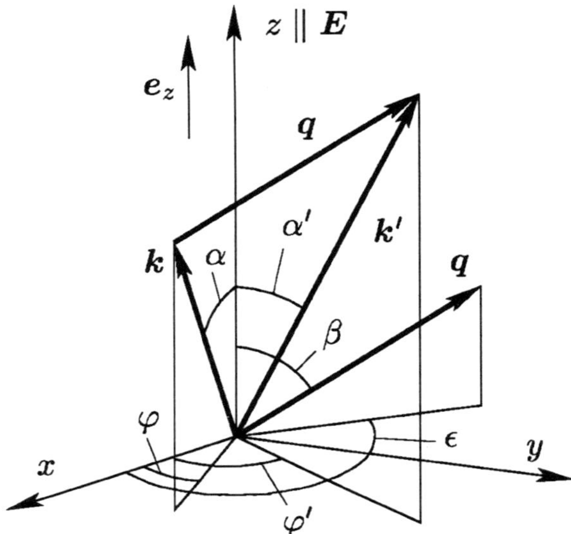



Fig. 3.1   The absorption of a phonon with wave vector **q**, scattering an electron from the initial state **k** to the final state **k′**. (GALLER M. and SCHÜRRER F., *J. Phys. A: Math. Gen.*, **37** (2004), 1479–1497; reproduced with permission from IOP Publishing, www.iop.org/journals/jphysa.)

Carrillo *et al.* (2003)a; Carrillo *et al.* (2003)b] and Ringhofer [Ringhofer (2000); Ringhofer *et al.* (2001)] should be mentioned. These papers present solution methods appropriate for the investigation of the electron transport in silicon, which is actually the most important semiconductor material. On the other hand, deterministic solution methods taking into account the special requirements related to the transport in polar semiconductors are still rare in the literature.

To close this gap, we introduce multigroup model equations to the Bloch-Boltzmann-Peierls equations (2.78) governing the temporal evolution of the coupled hot-electron hot-phonon system in polar semiconductors.

## 3.2   Multigroup Equations to the Bloch-Boltzmann-Peierls Equations

In this section, we present the multigroup model equations (MMEs) describing the transport properties in polar semiconductors. Therefore, we couple the electron BTE with a Boltzmann equation for phonons by taking into account the relevant quantum statistics. Furthermore, we include several valleys in the wave vector space of electrons in our model. In contrast to the above mentioned papers, our approach of discretizing the BTEs is motivated on physical grounds. We divide the wave vector spaces of electrons and phonons into tiny cells and transform the full BTEs into a system of coupled transport equations balancing the particle transfer among these cells. The particle density within each cell is represented by a Dirac distribution, which allows us to perform the collision integrals analytically.

To begin with, we summarize some relations needed below. The dispersion law, which connects the energy $E^\nu$ of an electron, measured from the bottom of the valley $\nu$, with its wave vector $\mathbf{k}$, is given by the non-parabolic band approximation

$$E^\nu(\mathbf{k})\left[1 + \alpha_\nu E^\nu(\mathbf{k})\right] = \frac{\hbar^2 k^2}{2m_\nu^*}. \tag{3.1}$$

Vice versa, we find

$$k^\nu(E) = \frac{1}{\hbar}\left[2m_\nu^* E\left(1 + \alpha_\nu E\right)\right]^{\frac{1}{2}}. \tag{3.2}$$

Here, $k$ denotes the modulus of the wave vector, $\alpha_\nu$ and $m_\nu^*$ are the non-parabolicity factor and the effective mass in the valley $\nu$. This implies that the density of states equals

$$\mathcal{Z}^\nu(E) = \frac{m_\nu^*}{(2\pi\hbar)^3}\left(1 + 2\alpha_\nu E\right)\left[2m_\nu^* E(1 + \alpha_\nu E)\right]^{\frac{1}{2}}. \tag{3.3}$$

For the phonons, we apply the standard approximations for the energy-momentum rule, i.e. the energy $\hbar\omega(\mathbf{q})$ and the wave vector $\mathbf{q}$ are related via $\omega_{\mathrm{ac}}(\mathbf{q}) = v_{\mathrm{s}}|\mathbf{q}|$ and $\omega_{\mathrm{op}}(\mathbf{q}) = \omega_0$ in the cases of acoustic and optical phonons, respectively. The symbol $v_{\mathrm{s}}$ denotes the sound velocity.

The wave vectors $\mathbf{k}$ and $\mathbf{q}$ are represented as

$$\mathbf{k} = (k^\nu(E)\sin\alpha\cos\varphi,\, k^\nu(E)\sin\alpha\sin\varphi,\, k^\nu(E)\cos\alpha), \tag{3.4a}$$

$$\mathbf{q} = (q\sin\beta\cos\epsilon,\, q\sin\beta\sin\epsilon,\, q\cos\beta) \tag{3.4b}$$

in a coordinate system, whose z-axis is parallel to the direction of the external electric field $\mathbf{E}$. Figure 3.1 explains the meaning of the used angles. It illustrates a scattering event from an electron state $\mathbf{k}$ to $\mathbf{k}'$ by absorbing a phonon with wave vector $\mathbf{q}$. For our purposes, it is convenient to introduce the quantities $\mu = \cos\alpha$ and $\chi = \cos\beta$.

Due to the cylindrical symmetry with respect to the electric field $\mathbf{E}$, the electron distribution function (EDF) $f^\nu(\mathbf{k})$ in the valley $\nu$ and the LO phonon distribution function (PDF) $g(\mathbf{q})$ do not depend on the azimuthal angles $\varphi$ and $\epsilon$. In other words, we consider $f^\nu = f^\nu(E, \mu, t)$, where we use the electron energy $E$ as one of the independent variables instead of the modulus of the wave vector, and $g = g(q, \chi, t)$.

### 3.2.1 *The Electron Boltzmann Equation*

The evolution of the electron distribution function $f^\nu$ in bulk semiconductors is governed by the electron BTE (cf. (2.78)):

$$\frac{\partial f^\nu}{\partial t} - \frac{e}{\hbar}\mathbf{E} \cdot \nabla_\mathbf{k} f^\nu = \sum_\xi \mathcal{C}_\xi[f^\nu]. \qquad (3.5)$$

In this equation, $\xi$ represents all the involved scattering mechanisms. The collision terms $\mathcal{C}_\xi[f^\nu]$ typically read, neglecting degeneracy,

$$\mathcal{C}_\xi[f^\nu] = \frac{V}{8\pi^3} \int d^3k' \left[S_\xi(\mathbf{k}' \to \mathbf{k})f^\nu(\mathbf{k}') - S_\xi(\mathbf{k} \to \mathbf{k}')f^\nu(\mathbf{k})\right]. \qquad (3.6)$$

The symbol $S_\xi(\mathbf{k} \to \mathbf{k}')$ is the transition rate from state $\mathbf{k}$ to $\mathbf{k}'$, $V$ the volume of the crystal and $e$ denotes the elementary charge. Here and in all of the other integrals with unspecified integration intervals, the integration is performed over all the states $\mathbf{k}$ associated with the valley $\nu$.

To deal with the electron BTE, we proceed as follows. First, we introduce the new function $F^\nu$, defined by $F^\nu(E, \mu) = \mathcal{Z}^\nu(E)\, f^\nu(E, \mu)$. As a consequence, the integration of the EDF with respect to $\mathbf{k}$ can be written as

$$\frac{1}{8\pi^3} \int d^3k \, f^\nu(\mathbf{k}) = \int dE \int d\mu \int_0^{2\pi} d\varphi \, F^\nu(E, \mu). \qquad (3.7)$$

The second step of constructing the multigroup model BTEs consists of introducing a partition of the $(E, \mu)$ space into cells $C_{ij}^\nu = [E_{i-1/2}^\nu, E_{i+1/2}^\nu] \times [\mu_{j-1/2}^\nu, \mu_{j+1/2}^\nu]$, $i = 1, 2, \ldots, N^\nu$, $j = 1, 2, \ldots, M^\nu$. The boundary values of this partition are set to $\mu_{1/2}^\nu = -1$, $\mu_{M^\nu+1/2}^\nu = 1$, $E_{1/2}^\nu = 0$ and $E_{N^\nu+1/2}^\nu = E_{\max}$, so that $F^\nu(E_{\max}, \mu)$ can be considered negligible.

To obtain an approximative solution to the electron BTE, we express the distribution function $F^\nu$ as the finite sum

$$F^\nu(E,\mu,t) = \sum_{i=1}^{N^\nu} \sum_{j=1}^{M^\nu} n_{ij}^\nu(t)\,\delta(E - E_i^\nu)\,\delta(\mu - \mu_j^\nu). \tag{3.8}$$

The poles of the Dirac distributions must fulfill the conditions $E_i^\nu \in I_i^{E,\nu} = (E_{i-1/2}^\nu, E_{i+1/2}^\nu)$ and $\mu_j^\nu \in I_j^{\mu,\nu} = (\mu_{j-1/2}^\nu, \mu_{j+1/2}^\nu)$. For a physical interpretation of the coefficients $n_{ij}^\nu$ in (3.8), we evaluate the particle density $\langle n^{c,\nu} \rangle_{ij}$, of electrons with energies and directions within the cell $C_{ij}^\nu$ in the valley $\nu$. Following the definitional equation of this quantity and replacing the original EDP by the ansatz made above, we find by taking advantage of (3.7)

$$\langle n_{ij}^{c,\nu} \rangle = \frac{1}{4\pi^3} \int_{C_{ij}^\nu} d^3k\, f^\nu(\mathbf{k}) = 4\pi \sum_{a=1}^{N^\nu} \sum_{b=1}^{M^\nu} \int_{I_i^{E,\nu}} dE$$

$$\times \int_{I_i^{\mu,\nu}} d\mu\, n_{ab}^\nu\, \delta(E - E_a^\nu)\,\delta(\mu - \mu_b^\nu) = 4\pi n_{ij}^\nu. \tag{3.9}$$

In other words, the coefficients $n_{ij}^\nu$ are equal to the electron density in the cell $C_{ij}^\nu$ except for a constant factor. The same procedure allows us to determine other important macroscopic quantities like the total electron density $\langle n^{c,\nu} \rangle$, the drift velocity $\langle v^{c,\nu} \rangle$ in direction of the electric field (unit vector $\mathbf{e}_z$) and the mean energy $\langle E^{c,\nu} \rangle$:

$$\langle n^{c,\nu} \rangle = \frac{1}{4\pi^3} \int d^3k\, f^\nu(\mathbf{k}) = 4\pi \sum_{i=1}^{N^\nu} \sum_{j=1}^{M^\nu} n_{ij}^\nu, \tag{3.10a}$$

$$\langle v^{c,\nu} \rangle = \frac{1}{4\pi^3 \hbar \langle n^{c,\nu} \rangle} \int d^3k\, \frac{\partial E(\mathbf{k})}{\partial \mathbf{k}} \cdot \mathbf{e}_z\, f^\nu(\mathbf{k})$$

$$= \frac{4\pi\hbar}{m_\nu^* \langle n^{c,\nu} \rangle} \sum_{i=1}^{N^\nu} \sum_{j=1}^{M^\nu} \frac{k^\nu(E_i^\nu)}{1 + 2\alpha_\nu E_i^\nu}\, \mu_j^\nu\, n_{ij}^\nu, \tag{3.10b}$$

$$\langle E^{c,\nu} \rangle = \frac{1}{4\pi^3 \langle n^{c,\nu} \rangle} \int d^3k\, E^\nu(\mathbf{k})\, f^\nu(\mathbf{k}) = \frac{4\pi}{\langle n^{c,\nu} \rangle} \sum_{i=1}^{N^\nu} \sum_{j=1}^{M^\nu} E_i^\nu\, n_{ij}^\nu. \tag{3.10c}$$

To construct the equations which govern the evolutions of the coefficients $n_{ij}^\nu$ we follow the method of weighted residuals [Lapidus and Pinder (1982)]. The electron BTE (3.5) is integrated over the cell $C_{ij}^\nu$. By the help of (3.7), the integration variables are transformed into $E$ and $\mu$. Whenever the product $\mathcal{Z}^\nu f^\nu$ appears, it is replaced by the ansatz (3.8). This procedure ends in a set of $N^\nu \times M^\nu$ equations for the $N^\nu \times M^\nu$ unknowns $n_{ij}^\nu$.

Following this strategy, we find for the first term of the electron BTE, containing the temporal derivative,

$$\frac{1}{8\pi^3} \int_{C_{ij}^\nu} d^3k \, \frac{\partial f^\nu}{\partial t} = 2\pi \frac{\partial n_{ij}^\nu}{\partial t}. \tag{3.11}$$

Special effort must be invested in a suitable formulation of the force term $\mathcal{C}_{\mathbf{E}}[f^\nu] = -e\mathbf{E} \cdot \nabla_{\mathbf{k}} f^\nu / \hbar$. Here, our recipe yields, not writing down terms that contain the azimuthal angle $\varphi$ since these terms vanish after the integration with respect to $\varphi$,

$$\frac{1}{8\pi^3} \int_{C_{ij}^\nu} d^3k \, \mathcal{C}_{\mathbf{E}}[f^\nu] = \frac{-e|\mathbf{E}|}{8\pi^3 \hbar} \int_{C_{ij}^\nu} d^3k \left( \mu \frac{\partial f^\nu}{\partial k} + \frac{1 - \mu^2}{k} \frac{\partial f^\nu}{\partial \mu} \right) = -\frac{2\pi \, e|\mathbf{E}|}{\hbar}$$

$$\times \sum_{a=1}^{N^\nu} \sum_{b=1}^{M^\nu} \int_{I_i^{E,\nu}} dE \int_{I_j^{\mu,\nu}} d\mu \left[ \left( \frac{\partial k^\nu(E)}{\partial E} \right)^{-1} \mu \, n_{ab}^\nu \, \delta(\mu - \mu_b^\nu) \frac{\partial}{\partial E} \delta(E - E_a^\nu) \right.$$

$$\left. + \frac{1 - \mu^2}{k^\nu(E)} n_{ab}^\nu \, \delta(E - E_a^\nu) \frac{\partial}{\partial \mu} \delta(\mu - \mu_b^\nu) \right]. \tag{3.12}$$

The application of the standard algebra for evaluating the derivatives of Dirac distributions to this result ends in

$$\frac{1}{8\pi^3} \int_{C_{ij}^\nu} d^3k \, \mathcal{C}_{\mathbf{E}}[f^\nu] \tag{3.13}$$

$$= \frac{2\pi \, e|\mathbf{E}|}{\hbar} \left\{ \mu_j^\nu \int_{I_i^{E,\nu}} dE \, \delta(E - E_i^\nu) \frac{\hbar^2}{m_\nu^*} \frac{\partial}{\partial E} \left( \frac{k^\nu(E)}{1 + 2\alpha_\nu E} \, n_{ij}^\nu \right) \right.$$

$$\left. + \frac{1}{k^\nu(E_i^\nu)} \int_{I_j^{\mu,\nu}} d\mu \, \delta(\mu - \mu_j^\nu) \frac{\partial}{\partial \mu} \left[ (1 - \mu^2) \, n_{ij}^\nu \right] \right\}.$$

Now, we replace the derivatives with respect to $E$ and $\mu$ by

$$\frac{\partial}{\partial E}\left(\frac{k^\nu(E)}{1+2\alpha_\nu E}\,n_{ij}^\nu\right) \tag{3.14a}$$

$$= -\frac{1}{E_{i+\frac{1}{2}}^\nu - E_{i-\frac{1}{2}}^\nu}\left[\frac{k^\nu(E_{i+\frac{1}{2}}^\nu)}{1+2\alpha_\nu E_{i+\frac{1}{2}}^\nu}\,[n_{ij}^\nu]^+ - \frac{k^\nu(E_{i-\frac{1}{2}}^\nu)}{1+2\alpha_\nu E_{i-\frac{1}{2}}^\nu}\,[n_{ij}^\nu]^-\right]$$

$$\frac{\partial}{\partial\mu}\left[(1-\mu^2)\,n_{ij}^\nu\right] \tag{3.14b}$$

$$= -\frac{1}{\mu_{j+\frac{1}{2}}^\nu - \mu_{j-\frac{1}{2}}^\nu}\left[(1-\mu_{j+\frac{1}{2}}^{\nu\,2})\,n_{i,j+1}^\nu - (1-\mu_{j-\frac{1}{2}}^{\nu\,2})\,n_{ij}^\nu\right]$$

with

$$[n_{ij}^\nu]^+ = \begin{cases} n_{i+1,j}^\nu & \text{if } \mu_j^\nu > 0, \\ n_{ij}^\nu & \text{if } \mu_j^\nu < 0, \end{cases} \qquad [n_{ij}^\nu]^- = \begin{cases} n_{ij}^\nu & \text{if } \mu_j^\nu > 0, \\ n_{i-1,j}^\nu & \text{if } \mu_j^\nu < 0. \end{cases} \tag{3.15}$$

These expressions are justified by heuristically gained evolution equations for the particle density $\langle n_{ij}^{c,\nu}\rangle$ in the cell $C_{ij}^\nu$. The detailed derivation of these expressions, which is based on the pseudo-Newtonian law for the electron state $\hbar\dot{\mathbf{k}} = -e\mathbf{E}$, is found in [Ertler and Schürrer (2003)]. Mathematically spoken, we apply an upwind scheme for dealing with the force term.

As a result, we obtain the multigroup version of the force term by inserting (3.14) into (3.13):

$$\frac{1}{8\pi^3}\int_{C_{ij}^\nu} d^3k\, \mathcal{C}_{\mathbf{E}}[f^\nu] = -\frac{2\pi\,e|\mathbf{E}|}{\hbar}\left\{\frac{\hbar^2}{m_\nu^*}\frac{\mu_j^\nu}{E_{i+\frac{1}{2}}^\nu - E_{i-\frac{1}{2}}^\nu}\left[\frac{k^\nu(E_{i+\frac{1}{2}}^\nu)}{1+2\alpha_\nu E_{i+\frac{1}{2}}^\nu}\,[n_{ij}^\nu]^+ \right.\right.$$

$$\left.\left. -\frac{k^\nu(E_{i-\frac{1}{2}}^\nu)}{1+2\alpha_\nu E_{i-\frac{1}{2}}^\nu}\,[n_{ij}^\nu]^-\right] + \frac{(1-\mu_{j+\frac{1}{2}}^{\nu\,2})\,n_{ij+1}^\nu - (1-\mu_{j-\frac{1}{2}}^{\nu\,2})\,n_{ij}^\nu}{k^\nu(E_i^\nu)(\mu_{j+\frac{1}{2}}^\nu - \mu_{j-\frac{1}{2}}^\nu)}\right\}. \tag{3.16}$$

From a physical point of view, the expression (3.16) describes the fluxes between neighboring cells induced by the electric field. The particle conservation with respect to the electric field can be ensured, when terms which describe flows at boundary cells in and out of the chosen finite $(E,\mu)$ space are set to zero.

Now, we turn to the collision terms. As for generality, we consider a scattering mechanism $\xi$ acting between the valleys $\nu$ and $\mu$. With this formalism, we can describe all the scattering mechanisms ($\mu = \nu$ for intravalley scattering) except for the POP interaction (Sec. 3.2.3).

Firstly, the loss terms $L_\xi[f^\nu] = f^\nu(\mathbf{k})\,V \int d^3k'\, S_\xi^{\nu\to\mu}(\mathbf{k}\to\mathbf{k}')/8\pi^3$ for the valley $\nu$ (see (3.6)) are transformed into our multigroup scheme:

$$\frac{1}{8\pi^3}\int_{C_{ij}^\nu} d^3k\, L_\xi[f^\nu] = 2\pi \sum_{a=1}^{N^\mu}\sum_{b=1}^{M^\mu} \langle S_{\xi,ab}^{\nu\to\mu}\rangle_{ij}\, n_{ij}^\nu \tag{3.17}$$

with

$$\langle S_{\xi,ab}^{\mu\to\nu}\rangle_{ij} = \frac{V}{8\pi^3}\int_{I_i^{E,\nu}} dE \int_{I_j^{\mu,\nu}} d\mu \tag{3.18}$$

$$\times \int_{C_{ab}^\mu} d^3k'\, S_\xi^{\nu\to\mu}(\mathbf{k}\to\mathbf{k}')\delta(E-E_i^\nu)\delta(\mu-\mu_j^\nu).$$

Secondly, the gain terms $G_\xi[f^\nu] = V \int d^3k'\, S_\xi^{\mu\to\nu}(\mathbf{k}'\to\mathbf{k})\,f^\mu(\mathbf{k}')/8\pi^3$ are found to equal

$$\frac{1}{8\pi^3}\int_{C_{ij}^\nu} d^3k\, G_\xi[f^\nu] \tag{3.19}$$

$$= \frac{V}{(8\pi^3)^2}\sum_{a=1}^{N^\mu}\sum_{b=1}^{M^\mu}\int_{C_{ij}^\nu} d^3k \int_{C_{ab}^\mu} d^3k'\, S_\xi^{\mu\to\nu}(\mathbf{k}'\to\mathbf{k})\,f^\mu(\mathbf{k}').$$

By exchanging the names of the integration variables $\mathbf{k}$ and $\mathbf{k}'$ and comparing the result with (3.18), it can easily be seen that

$$\frac{1}{8\pi^3}\int_{C_{ij}^\nu} d^3k\, G_\xi[f^\nu] = 2\pi \sum_{a=1}^{N^\mu}\sum_{b=1}^{M^\mu} \langle S_{\xi,ij}^{\mu\to\nu}\rangle_{ab}\, n_{ab}^\mu. \tag{3.20}$$

Finally, we achieve the multigroup formulation of the electron BTE for the valley $\nu$ by adding up the expressions (3.11), (3.16), (3.17) and (3.18). These equations read after canceling a factor $2\pi$ and neglecting obsolete valley indices

$$\frac{\partial n_{ij}^\nu}{\partial t} - \frac{e|\mathbf{E}|}{\hbar} \left\{ \frac{\hbar^2}{m_\nu^*} \frac{\mu_j^\nu}{E_{i+\frac{1}{2}}^\nu - E_{i-\frac{1}{2}}^\nu} \left[ \frac{k^\nu(E_{i+\frac{1}{2}}^\nu)}{1+2\alpha_\nu E_{i+\frac{1}{2}}^\nu} [n_{ij}^\nu]^+ - \frac{k^\nu(E_{i-\frac{1}{2}}^\nu)}{1+2\alpha_\nu E_{i-\frac{1}{2}}^\nu} [n_{ij}^\nu]^- \right] \right.$$

$$\left. + \frac{1}{k^\nu(E_i^\nu)} \frac{1}{\mu_{j+\frac{1}{2}}^\nu - \mu_{j-\frac{1}{2}}^\nu} \left[ (1 - (\mu_{j+\frac{1}{2}}^\nu)^2)\, n_{ij+1}^\nu - (1 - (\mu_{j-\frac{1}{2}}^\nu)^2)\, n_{ij}^\nu \right] \right\}$$

$$= \sum_{\xi,\, \xi \neq \mathrm{IV, POP}} \sum_{a=1}^{N^\nu} \sum_{b=1}^{M^\nu} \left( \langle S_{\xi,ij}^\nu \rangle_{ab}\, n_{ab}^\nu - \langle S_{\xi,ab}^\nu \rangle_{ij}\, n_{ij}^\nu \right) + \frac{1}{16\pi^4} \int_{C_{ij}^\nu} d^3k\, \mathcal{C}_{\mathrm{POP}}[f^\nu]$$

$$+ \sum_{\mu,\, \mu \neq \nu} \sum_{a=1}^{N^\mu} \sum_{b=1}^{M^\mu} \left( Z_\mu\, \langle S_{\mathrm{IV},ij}^{\mu\to\nu} \rangle_{ab}\, n_{ab}^\mu - Z_\nu\, \langle S_{\mathrm{IV},ab}^{\nu\to\mu} \rangle_{ij}\, n_{ij}^\nu \right), \tag{3.21}$$

where $i = 1, 2, \ldots N^\nu$, $j = 1, 2, \ldots, M^\nu$.

### 3.2.2  The LO Phonon Boltzmann Equation

The evolution equation for the distribution function $g(\mathbf{q})$ of longitudinal optical phonons depending on the phonon state $\mathbf{q}$ reads:

$$\frac{\partial g}{\partial t} = \mathcal{D}_{\mathrm{PH-PH}}[g] + \sum_\nu 2\, Z_\nu\, \mathcal{D}_{\mathrm{POP}}^\nu[g]. \tag{3.22}$$

Here, we assume that the PDF can be disturbed by phonon-phonon thermalization processes due to $\mathcal{D}_{\mathrm{PH-PH}}[g]$ and the POP interaction with electrons of the valleys $\nu$ according to $\mathcal{D}_{\mathrm{POP}}^\nu[g]$.

The lattice scattering term in (3.22) is described by means of an appropriate relaxation time $\tau_{\mathrm{L}}$

$$\mathcal{D}_{\mathrm{PH-PH}}[g] = \frac{g_{\mathrm{L}} - g(\mathbf{q})}{\tau_{\mathrm{L}}}, \tag{3.23}$$

where $g_{\mathrm{L}}$ is the Bose-Einstein equilibrium distribution at lattice temperature $T_{\mathrm{L}}$.

As concerns the POP collision term, we find

$$\mathcal{D}_{\mathrm{POP}}^\nu[g] = \frac{V}{8\pi^3} \int d^3k \left\{ S_{\mathrm{POP}}^{\mathrm{em}}(\mathbf{q})[g(\mathbf{q}) + 1] - S_{\mathrm{POP}}^{\mathrm{abs}}(\mathbf{q})\, g(\mathbf{q}) \right\} \tag{3.24}$$

with the transition rates $S_{\mathrm{POP}}^{\mathrm{em}}(\mathbf{q})$ and $S_{\mathrm{POP}}^{\mathrm{abs}}(\mathbf{q})$ for phonon emission and absorption, respectively, which are given in (2.88) by applying the low density approximation.

The LO phonon BTE is treated in the same way as the electron BTE. Hence, we define $G(q, \chi) = q^2 g(q, \chi)/8\pi^3$, which leads to

$$\frac{1}{8\pi^3} \int d^3q\, g(\mathbf{q}) = \int dq \int d\chi \int_0^{2\pi} d\epsilon\, G(q, \chi). \tag{3.25}$$

The wave vector space $(q, \chi)$ is divided into the cells $D_{xy} = [q_{x-1/2}, q_{x+1/2}] \times [\chi_{y-1/2}, \chi_{y+1/2}]$, $x = 1, 2, \ldots, R$, $y = 1, 2, \ldots, S$ with the boundary values $\chi_{1/2} = -1$, $\chi_{S+1/2} = 1$, $q_{1/2} = 0$ and $q_{R+1/2} = q_{\max}$. Here, $q_{\max}$ is chosen so that $g(q_{\max})$ can be considered undisturbed by the POP interaction with electrons.

Similar to (3.8), we represent the distribution function $G$ by the ansatz

$$G(q, \chi) = \sum_{x=1}^{R} \sum_{y=1}^{S} r_{xy}\, \delta(q - q_x)\, \delta(\chi - \chi_y) \tag{3.26}$$

demanding $q_x \in I_x^q = (q_{x-1/2}, q_{x+1/2})$ and $\chi_y \in I_y^\chi = (\chi_{y-1/2}, \chi_{y+1/2})$. Macroscopic quantities for the LO phonons can be evaluated with the help of expressions similar to those in (3.10). As an example, the phonon density $\langle n_{xy}^{\mathrm{p}} \rangle$ of LO phonons with wave vector $\mathbf{q}$ within the cell $D_{xy}$ is given by

$$\langle n_{xy}^{\mathrm{p}} \rangle = \frac{1}{8\pi^3} \int_{D_{xy}} d^3q\, g(\mathbf{q}) = 2\pi r_{xy}. \tag{3.27}$$

The $R \times S$ evolution equations for the coefficients $r_{xy}$ are found by the following strategy. The phonon BTE is integrated over the cell $D_{xy}$. By taking advantage of (3.25), we obtain an evolution equation for the function $G$. Here, we replace the function $G$ by the ansatz (3.26) and carry out all the possible integrations.

Hence, the left hand side of the phonon BTE (3.22) becomes

$$\frac{1}{8\pi^3} \int_{D_{xy}} d^3q\, \frac{\partial g}{\partial t} = 2\pi \frac{\partial r_{xy}}{\partial t}. \tag{3.28}$$

The application of our multigroup formalism to the phonon-phonon interaction term $\mathcal{D}_{\mathrm{PH-PH}}[g]$ yields

$$\frac{1}{8\pi^3} \int_{D_{xy}} d^3q\, \mathcal{D}_{\mathrm{PH-PH}}[g] = \frac{2\pi}{\tau_{\mathrm{L}}} \left( r_{xy}^{\mathrm{equi}} - r_{xy} \right), \tag{3.29}$$

where the equilibrium coefficients $r_{xy}^{\mathrm{equi}}$ are calculated via

$$r_{xy}^{\mathrm{equi}} = \frac{1}{16\pi^4} \int_{D_{xy}} d^3q \left[ \exp\left( \frac{\hbar\omega_{\mathrm{LO}}}{k_{\mathrm{B}}T_{\mathrm{L}}} \right) - 1 \right]^{-1}. \tag{3.30}$$

After summarizing the terms (3.28) and (3.29), the transformation of the phonon BTE into our multigroup scheme ends in

$$\frac{\partial r_{xy}}{\partial t} = \frac{1}{\tau_{\mathrm{L}}} \left( r_{xy}^{\mathrm{equi}} - r_{xy} \right) + \sum_\nu \frac{2\,Z_\nu}{16\pi^4} \int_{D_{xy}} d^3q\, \mathcal{D}_{\mathrm{POP}}^\nu[g], \tag{3.31}$$

$x = 1, 2, \ldots, R$, $y = 1, 2, \ldots, S$, except for the POP interaction term, which is discussed in the next section.

### 3.2.3 *The Coupling POP Interaction Term*

Here, we deal with the most important terms of our multigroup model, the polar optical scattering terms, that couple the electron and the LO phonon Boltzmann equations. On the one hand, these expressions describe how the lattice is modified by the hot electrons. On the other hand, the electron transport properties are essentially influenced by these hot phonons. Consequently, these coupling terms are responsible for a significant deviation of the results from those based on the usual equilibrium phonon calculations.

To begin with, we consider the POP interaction term of the electron BTE $\mathcal{C}_{\mathrm{POP}}[f^\nu]$. In its full form, it reads (cf. Sec. 2.6)

$$\mathcal{C}_{\mathrm{POP}}[f^\nu] = \frac{V}{8\pi^3} \int d^3k' \tag{3.32}$$

$$\times \left\{ \frac{K_{\mathrm{POP}}}{|\mathbf{k} - \mathbf{k}'|^2} f^\nu(\mathbf{k}') \left[ g(\mathbf{k}' - \mathbf{k}) + 1 \right] \delta[E^\nu(\mathbf{k}') - E^\nu(\mathbf{k}) - \hbar\omega_{\mathrm{LO}}] \right.$$

$$+ \frac{K_{\mathrm{POP}}}{|\mathbf{k} - \mathbf{k}'|^2} f^\nu(\mathbf{k}')\, g(\mathbf{k} - \mathbf{k}')\, \delta[E^\nu(\mathbf{k}') - E^\nu(\mathbf{k}) + \hbar\omega_{\mathrm{LO}}]$$

$$- \frac{K_{\mathrm{POP}}}{|\mathbf{k} - \mathbf{k}'|^2} f^\nu(\mathbf{k}) \left[ g(\mathbf{k} - \mathbf{k}') + 1 \right] \delta[E^\nu(\mathbf{k}') - E^\nu(\mathbf{k}) + \hbar\omega_{\mathrm{LO}}]$$

$$\left. - \frac{K_{\mathrm{POP}}}{|\mathbf{k} - \mathbf{k}'|^2} f^\nu(\mathbf{k})\, g(\mathbf{k}' - \mathbf{k})\, \delta[E^\nu(\mathbf{k}') - E^\nu(\mathbf{k}) - \hbar\omega_{\mathrm{LO}}] \right\}$$

with $K_{\mathrm{POP}} = \pi e^2 \omega_{\mathrm{LO}} (1/\kappa_\infty - 1/\kappa_0)/\varepsilon_0 V$. Following our multigroup scheme, we integrate the electron BTE (3.5) over the cell $C_{ij}^\nu$ and replace the distribution function by the ansatz (3.8). For the PDF $g$, we use the expression (3.26) divided by $q^2/8\pi^3$. The additional factor 1 in the phonon emission terms is expanded to $\sum_{x'y'} V_{x'y'}\, \delta(q - q_{x'})\, \delta(\chi - \chi_{y'})/q^2$ with $V_{xy} = (q_{x+1/2}^3 - q_{x-1/2}^3)(\chi_{y+1/2} - \chi_{x-1/2})/3$. This replacement is justified by the fact that these expressions are equal after the integration over the cell $D_{xy}$:

$$\frac{1}{8\pi^3} \int_{I_x^q} dq\, q^2 \int_{I_y^\chi} d\chi \tag{3.33}$$

$$= \frac{1}{8\pi^3} \sum_{x'=1}^{R} \sum_{y'=1}^{S} V_{x'y'} \int_{I_x^q} dq \int_{I_y^\chi} d\chi\, \delta(q - q_{x'})\, \delta(\chi - \chi_{y'}).$$

In this way, we find the loss term $L_{\mathrm{POP}}[f^\nu]$ of the electron BTE due to POP interaction

$$\frac{1}{8\pi^3} \int_{C_{ij}^\nu} d^3k\, L_{\mathrm{POP}}[f^\nu] = 2\pi \sum_{a=1}^{N^\nu} \sum_{b=1}^{M^\nu} \sum_{x=1}^{R} \sum_{y=1}^{S} \tag{3.34}$$

$$\times \left[ \left( \langle \mathcal{E}_{\mathrm{POP},ab}^{\nu,xy} \rangle_{ij}\, r_{xy} + \langle 1_{\mathrm{POP},ab}^{\nu,xy} \rangle_{ij} \right) + \langle \mathcal{A}_{\mathrm{POP},ab}^{\nu,xy} \rangle_{ij}\, r_{xy} \right] n_{ij}^\nu$$

with

$$\langle \mathcal{E}_{\mathrm{POP},ab}^{\nu,xy} \rangle_{ij} = \frac{V\, K_{\mathrm{POP}}}{q_x^3} \int_{I_i^{E,\nu}} dE \int_{I_j^{\mu,\nu}} d\mu \tag{3.35a}$$

$$\times \int_{C_{ab}^\nu} d^3k'\, \delta(E - E_i^\nu)\, \delta(\mu - \mu_j^\nu)\, \delta(|\mathbf{k} - \mathbf{k}'| - q_x)$$

$$\times \delta\left[ (\mathbf{k} - \mathbf{k}') \cdot \mathbf{e}_z - q_x \chi_y \right] \delta(E^\nu(\mathbf{k}') - E + \hbar\omega_{\mathrm{LO}}),$$

$$\langle \mathcal{A}_{\mathrm{POP},ab}^{\nu,xy} \rangle_{ij} = \frac{V\, K_{\mathrm{POP}}}{q_x^3} \int_{I_i^{E,\nu}} dE \int_{I_j^{\mu,\nu}} d\mu \tag{3.35b}$$

$$\times \int_{C_{ab}^\nu} d^3k'\, \delta(E - E_i^\nu)\, \delta(\mu - \mu_j^\nu)\, \delta(|\mathbf{k}' - \mathbf{k}| - q_x)$$

$$\times \delta\left[ (\mathbf{k}' - \mathbf{k}) \cdot \mathbf{e}_z - q_x \chi_y \right] \delta(E^\nu(\mathbf{k}') - E - \hbar\omega_{\mathrm{LO}}),$$

$$\langle 1_{\mathrm{POP},ab}^{\nu,xy} \rangle_{ij} = V_{xy}\, \langle \mathcal{E}_{\mathrm{POP},ab}^{\nu,xy} \rangle_{ij}/8\pi^3 \tag{3.35c}$$

and $\mathbf{e}_z = \mathbf{E}/|\mathbf{E}|$. With the help of similar symmetry arguments as applied in Sec. 3.2.1, the multigroup version of the corresponding gain term can be deduced. Therefore, the collision term for the POP scattering of the electron BTE in our multigroup scheme is given by

$$\frac{1}{8\pi^3} \int_{C^\nu_{ij}} d^3k\, \mathcal{C}_{\text{POP}}[f^\nu] = 2\pi \sum_{a=1}^{N^\nu} \sum_{b=1}^{M^\nu} \sum_{x=1}^{R} \sum_{y=1}^{S} \tag{3.36}$$

$$\left[ \left( \langle \mathcal{E}^{\nu,xy}_{\text{POP},ij} \rangle_{ab}\, r_{xy} + \langle 1^{\nu,xy}_{\text{POP},ij} \rangle_{ab} \right) n^\nu_{ab} + \langle \mathcal{A}^{\nu,xy}_{\text{POP},ij} \rangle_{ab}\, r_{xy}\, n^\nu_{ab} \right.$$

$$\left. - \left( \langle \mathcal{E}^{\nu,xy}_{\text{POP},ab} \rangle_{ij}\, r_{xy} + \langle 1^{\nu,xy}_{\text{POP},ab} \rangle_{ij} + \langle \mathcal{A}^{\nu,xy}_{\text{POP},ab} \rangle_{ij}\, r_{xy} \right) n^\nu_{ij} \right].$$

The absorption collision coefficient $\langle \mathcal{A}^{\nu,xy}_{\text{POP},ab} \rangle_{ij}$, for instance, is evaluated in the next section.

As concerns the POP scattering term $\mathcal{D}^\nu_{\text{POP}}[g]$ in the phonon BTE, we find

$$\mathcal{D}^\nu_{\text{POP}}[g] = \frac{V}{8\pi^3} \int d^3k\, \frac{K_{\text{POP}}}{|\mathbf{q}|^2}\, f^\nu(\mathbf{k})\, \{[g(\mathbf{q}) + 1] \tag{3.37}$$

$$\times \delta[E(\mathbf{k} - \mathbf{q}) - E(\mathbf{k}) + \hbar\omega_{\text{LO}}] - g(\mathbf{q})\, \delta[E(\mathbf{k} + \mathbf{q}) - E(\mathbf{k}) - \hbar\omega_{\text{LO}}]\}$$

by inserting the POP scattering rates into (3.24). After performing our multigroup transformation, we obtain with the help of (3.8), (3.26) and a similar procedure for handling the additive factor 1 for phonon emission

$$\frac{1}{8\pi^3} \int_{D_{xy}} d^3q\, \mathcal{D}^\nu_{\text{POP}}[g] \tag{3.38}$$

$$= 2\pi \sum_{i=1}^{N^\nu} \sum_{j=1}^{M^\nu} \left[ \left( \langle E^{ij}_{\text{POP}} \rangle_{xy} - \langle A^{ij}_{\text{POP}} \rangle_{xy} \right) r_{xy} + \langle 1^{ij}_{\text{POP}} \rangle_{xy} \right] n^\nu_{ij},$$

where

$$\langle E^{ij}_{\text{POP}} \rangle_{xy} = \frac{V\, K_{\text{POP}}}{8\pi^3\, q_x^2} \int_{I_x^q} dq \int_{I_y^\chi} d\chi \tag{3.39a}$$

$$\times \int_{C^\nu_{ij}} d^3k\, \frac{1}{\mathcal{Z}^\nu(E^\nu(\mathbf{k}))} \delta(E - E_i^\nu)\, \delta(\mu - \mu_j^\nu)\, \delta(q - q_x)$$

$$\times \delta(\chi - \chi_y)\, \delta[E^\nu(\mathbf{k} - \mathbf{q}) - E^\nu(\mathbf{k}) + \hbar\omega_{\text{LO}}],$$

$$\langle A^{ij}_{\text{POP}} \rangle_{xy} = \frac{V\, K_{\text{POP}}}{8\pi^3\, q_x^2} \int_{I_x^q} dq \int_{I_y^\chi} d\chi \tag{3.39b}$$

$$\times \int_{C^\nu_{ij}} d^3k\, \frac{1}{\mathcal{Z}^\nu(E^\nu(\mathbf{k}))} \delta(E - E_i^\nu)\, \delta(\mu - \mu_j^\nu)\, \delta(q - q_x)$$

$$\times \delta(\chi - \chi_y)\, \delta[E^\nu(\mathbf{k} + \mathbf{q}) - E^\nu(\mathbf{k}) - \hbar\omega_{\text{LO}}],$$

$$\langle 1^{ij}_{\text{POP}} \rangle_{xy} = V_{xy}\, \langle E^{ij}_{\text{POP}} \rangle_{xy}. \tag{3.39c}$$

Inserting the expressions (3.36) and (3.38) into the multigroup versions of the electron and phonon equations (3.21) and (3.31), respectively, ends in the final set of evolution equations for the coefficients $n_{ij}^{\nu}$ and $r_{xy}$:

$$\frac{\partial n_{ij}^{\nu}}{\partial t} = \mathcal{C}_{\mathbf{E}}[n_{ij}^{\nu}] + \sum_{\xi,\ \xi \neq \mathrm{IV,POP}} \mathcal{C}_{\xi}[n_{ij}^{\nu}] + \mathcal{C}_{\mathrm{IV}}[n_{ij}^{\nu}] + \mathcal{C}_{\mathrm{POP}}[n_{ij}^{\nu}], \qquad (3.40\mathrm{a})$$

$$\frac{\partial r_{xy}}{\partial t} = \mathcal{D}_{\mathrm{PH-PH}}[r_{xy}] + \mathcal{D}_{\mathrm{POP}}[r_{xy}] \qquad (3.40\mathrm{b})$$

for $i = 1, 2, \ldots, N^{\nu}$, $j = 1, 2, \ldots, M^{\nu}$ and $x = 1, 2, \ldots, R$, $y = 1, 2, \ldots, S$ with the electron collision terms

$$\begin{aligned}
\mathcal{C}_{\mathbf{E}}[n_{ij}^{\nu}] = &\frac{e|\mathbf{E}|}{\hbar} \left\{ \frac{\hbar^2}{m_{\nu}^*} \frac{\mu_j^{\nu}}{E_{i+\frac{1}{2}}^{\nu} - E_{i-\frac{1}{2}}^{\nu}} \right. \\
&\times \left[ \frac{k^{\nu}(E_{i+\frac{1}{2}}^{\nu})}{1 + 2\alpha_{\nu} E_{i+\frac{1}{2}}^{\nu}} [n_{ij}^{\nu}]^+ - \frac{k^{\nu}(E_{i-\frac{1}{2}}^{\nu})}{1 + 2\alpha_{\nu} E_{i-\frac{1}{2}}^{\nu}} [n_{ij}^{\nu}]^- \right] \\
&\left. + \frac{1}{k^{\nu}(E_i^{\nu})} \frac{1}{\mu_{j+\frac{1}{2}}^{\nu} - \mu_{j-\frac{1}{2}}^{\nu}} \left[ (1 - (\mu_{j+\frac{1}{2}}^{\nu})^2)\, n_{i,j+1}^{\nu} - (1 - (\mu_{j-\frac{1}{2}}^{\nu})^2)\, n_{ij}^{\nu} \right] \right\},
\end{aligned} \qquad (3.41\mathrm{a})$$

$$\mathcal{C}_{\xi}[n_{ij}^{\nu}] = \sum_{a=1}^{N^{\nu}} \sum_{b=1}^{M^{\nu}} \left( \langle S_{\xi,ij}^{\nu} \rangle_{ab}\, n_{ab}^{\nu} - \langle S_{\xi,ab}^{\nu} \rangle_{ij}\, n_{ij}^{\nu} \right) \qquad (3.41\mathrm{b})$$

for $\xi = \mathrm{ADP, PZ, ODP, IMP}$ and

$$\mathcal{C}_{\mathrm{IV}}[n_{ij}^{\nu}] = \sum_{\mu,\ \mu \neq \nu} \sum_{a=1}^{N^{\mu}} \sum_{b=1}^{M^{\mu}} \left( \frac{Z_{\mu}}{Z_{\nu}} \langle S_{\mathrm{IV},ij}^{\mu \to \nu} \rangle_{ab}\, n_{ab}^{\mu} - \langle S_{\mathrm{IV},ab}^{\nu \to \mu} \rangle_{ij}\, n_{ij}^{\nu} \right), \qquad (3.41\mathrm{c})$$

$$\begin{aligned}
\mathcal{C}_{\mathrm{POP}}[n_{ij}^{\nu}] = &\sum_{a=1}^{N^{\nu}} \sum_{b=1}^{M^{\nu}} \sum_{x=1}^{R} \sum_{y=1}^{S} \left\{ \left[ \left( \langle \mathcal{E}_{\mathrm{POP},ij}^{\nu,xy} \rangle_{ab} + \langle \mathcal{A}_{\mathrm{POP},ij}^{\nu,xy} \rangle_{ab} \right) r_{xy} \right. \right. \\
&\left. + \langle 1_{\mathrm{POP},ij}^{\nu,xy} \rangle_{ab} \right] n_{ab}^{\nu} - \left[ \left( \langle \mathcal{E}_{\mathrm{POP},ab}^{\nu,xy} \rangle_{ij} + \langle \mathcal{A}_{\mathrm{POP},ab}^{\nu,xy} \rangle_{ij} \right) r_{xy} + \langle 1_{\mathrm{POP},ab}^{\nu,xy} \rangle_{ij} \right] n_{ij}^{\nu} \right\}.
\end{aligned} \qquad (3.41\mathrm{d})$$

and the phonon collision terms

$$\mathcal{D}_{\mathrm{PH-PH}}[r_{xy}] = \frac{1}{\tau_L} \left( r_{xy}^{\mathrm{equi}} - r_{xy} \right), \qquad (3.42\mathrm{a})$$

$$\begin{aligned}
\mathcal{D}_{\mathrm{POP}}[r_{xy}] = &\sum_{\nu} 2\, Z_{\nu} \sum_{i=1}^{N^{\nu}} \sum_{j=1}^{M^{\nu}} \left[ -\langle A_{\mathrm{POP}}^{ij} \rangle_{xy}\, r_{xy}\, n_{ij}^{\nu} \right. \\
&\left. + \left( \langle E_{\mathrm{POP}}^{ij} \rangle_{xy}\, r_{xy} + \langle 1_{\mathrm{POP}}^{ij} \rangle_{xy} \right) n_{ij}^{\nu} \right].
\end{aligned} \qquad (3.42\mathrm{b})$$

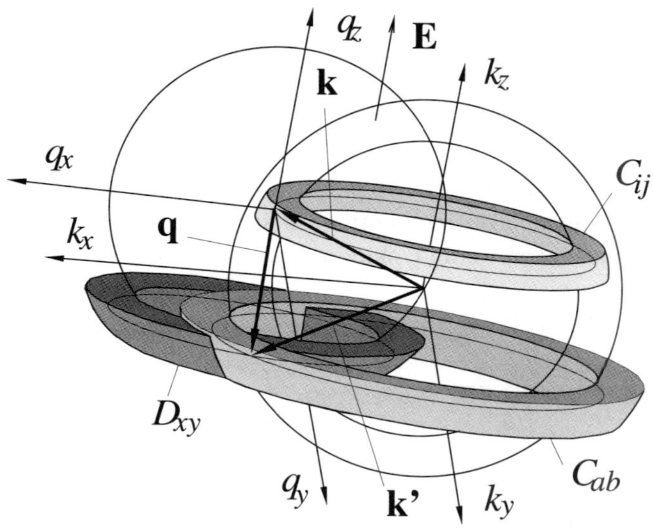


Fig. 3.2 Schematic illustration of the absorption of a phonon with wave vector **q** in the cell $D_{xy}$, which scatters an electron from the initial state **k** in the cell $C_{ij}$ to the final state **k**′ in $C_{ab}$. (GALLER M. and SCHÜRRER F., A multigroup approach to the coupled electron-phonon Boltzmann equations in InP, *Trans. Theo. Stat. Phys.*, **33** (2004), 485–501; reproduced with permission from Taylor & Francis.)

From a physical point of view, the resulting multigroup equations can be interpreted very descriptively. For instance, our model implies that a decrease of the electron density in the cell $C_{ij}$ due to the POP interaction by absorbing LO phonons of the cell $D_{xy}$ simply equals the product of the particle densities within these cells and a mean value of the transition rates with respect to these cells. The loss of LO phonons in the cell $D_{xy}$ because of absorption processes is given by a corresponding expression. Figure 3.2 depicts such an absorption process which annihilates a phonon with wave vector **q** in the cell $D_{xy}$ by scattering an electron from the initial state **k** in the cell $C_{ij}$ to the final state **k**′ in $C_{ab}$.

### 3.2.4  *The Evaluation of the Collision Coefficients*

For demonstrating how to proceed, we have evaluated the collision coefficient $\langle \mathcal{A}^{\nu,xy}_{\mathrm{POP},ab} \rangle_{ij}$ in this section. Therefore, we consider equation (3.35b).

First of all, we recognize that

$$|\mathbf{k}' - \mathbf{k}| = \left\{ k^2 + k'^2 - 2kk' \left[ [(1-\mu^2)(1-\mu'^2)]^{\frac{1}{2}} \cos\varphi' + \mu\mu' \right] \right\}^{\frac{1}{2}}, \quad (3.43a)$$

$$(\mathbf{k}' - \mathbf{k}) \cdot \mathbf{e}_z = k'\mu' - k\mu, \quad (3.43b)$$

as it can easily be verified in the polar representation of $\mathbf{k}$ and $\mathbf{k}'$, setting $\mathbf{k}$ to the (x,z) plane. By representing the integration with respect to $\mathbf{k}'$ in spherical coordinates, we find, skipping the valley index $\nu$,

$$\langle \mathcal{A}^{xy}_{\mathrm{POP},ab} \rangle_{ij} = 8\pi^3 \, V \, \frac{K_{\mathrm{POP}}}{q_x^3} \quad (3.44)$$

$$\times \int_{I_i^E} dE \int_{I_j^\mu} d\mu \int_{I_a^E} dE' \, \mathcal{Z}(E') \int_{I_b^\mu} d\mu' \int_0^{2\pi} d\varphi' \delta(E - E_i)$$

$$\times \delta\left\{ \left\{ k(E)^2 + k(E')^2 - 2k(E)k(E') \left[ [(1-\mu^2)(1-\mu'^2)]^{\frac{1}{2}} \cos\varphi' + \mu\mu' \right] \right\}^{\frac{1}{2}} - q_x \right\}$$

$$\times \delta(\mu - \mu_j) \, \delta\left[ (k(E')\mu' - k(E)\mu) - q_x\chi_y \right] \delta(E' - E + \hbar\omega_{\mathrm{LO}}).$$

Carrying out the integrations with respect to $E$, $E'$ and $\mu$ yields

$$\langle \mathcal{A}^{xy}_{\mathrm{POP},ab} \rangle_{ij} = 8\pi^3 V \frac{K_{\mathrm{POP}}}{q_x^3} \mathcal{Z}(E_i + \hbar\omega_{\mathrm{LO}}) \quad (3.45)$$

$$\times \Theta[E_{a+\frac{1}{2}} - (E_i + \hbar\omega_{\mathrm{LO}})]\Theta(E_i + \hbar\omega_{\mathrm{LO}} - E_{a-\frac{1}{2}}) \int_{I_b^\mu} d\mu' \int_0^{2\pi} d\varphi'$$

$$\times \delta\left[ k(E_i + \hbar\omega_{\mathrm{LO}})\mu' - k(E_i)\mu_j - q_x\chi_y \right] \delta\left\{ \left\{ k(E_i)^2 + k(E_i + \hbar\omega_{\mathrm{LO}})^2 \right. \right.$$

$$\left. \left. - 2k(E_i)k(E_i + \hbar\omega_{\mathrm{LO}}) \left[ [(1 - \mu_j^2)(1 - \mu'^2)]^{\frac{1}{2}} \cos\varphi' + \mu_j\mu' \right] \right\}^{\frac{1}{2}} - q_x \right\}.$$

The symbol $\Theta$ denotes the Heaviside step function. By rewriting the Dirac distribution

$$\delta(|\mathbf{k}' - \mathbf{k}| - q_x) \quad (3.46)$$

$$= \frac{2q_x \, \delta[\varphi' - \arccos \mathcal{N}^x_{ij}(\mu')]}{\left| k(E_i)k(E_i + \hbar\omega_{\mathrm{LO}}) \left[ (1 - \mu_j^2)(1 - \mu'^2)(1 - \mathcal{N}^x_{ij}(\mu')^2) \right]^{\frac{1}{2}} \right|},$$

where

$$\mathcal{N}^x_{ij}(\mu') = \frac{k(E_i)^2 + k(E_i + \hbar\omega_{\mathrm{LO}})^2 - q_x^2 - 2k(E_i)k(E_i + \hbar\omega_{\mathrm{LO}})\mu_j\mu'}{2k(E_i)k(E_i + \hbar\omega_{\mathrm{LO}}) \left[ (1 - \mu_j^2)(1 - \mu'^2) \right]^{\frac{1}{2}}}, \quad (3.47)$$

we achieve

$$\langle \mathcal{A}^{xy}_{\mathrm{POP},ab}\rangle_{ij} = 16\pi^3 V \frac{K_{\mathrm{POP}}}{q_x^2} \tag{3.48}$$

$$\times \; \mathcal{Z}(E_i + \hbar\omega_{\mathrm{LO}})\Theta[E_{a+\frac{1}{2}} - (E_i + \hbar\omega_{\mathrm{LO}})]\,\Theta(E_i + \hbar\omega_{\mathrm{LO}} - E_{a-\frac{1}{2}})$$

$$\times \int_{I_b^\mu} d\mu' \frac{\Theta[\arccos \mathcal{N}^x_{ij}(\mu')]\,\delta\left[k(E_i + \hbar\omega_{\mathrm{LO}})\mu' - k(E_i)\mu_j - q_x\chi_y\right]}{\left| k(E_i)k(E_i + \hbar\omega_{\mathrm{LO}})\left[(1 - \mu_j^2)(1 - \mu'^2)(1 - \mathcal{N}^x_{ij}(\mu')^2)\right]^{\frac{1}{2}} \right|}.$$

Finally, we find

$$\langle \mathcal{A}^{xy}_{\mathrm{POP},ab}\rangle_{ij} = 16\pi^3 \, V \, \frac{K_{\mathrm{POP}}}{q_x^2} \, \frac{\mathcal{Z}(E_i + \hbar\omega_{\mathrm{LO}})}{k(E_i)k(E_i + \hbar\omega_{\mathrm{LO}})} \tag{3.49}$$

$$\times \; \Theta\left[\arccos \hat{\mathcal{N}}^{xy}_{ij}\right]\Theta[E_{a+\frac{1}{2}} - (E_i + \hbar\omega_{\mathrm{LO}})]\Theta(E_i + \hbar\omega_{\mathrm{LO}} - E_{a-\frac{1}{2}})$$

$$\times \; \Theta\left[\mu_{b+\frac{1}{2}} - \frac{k(E_i)\mu_j + q_x\chi_y}{k(E_i + \hbar\omega_{\mathrm{LO}})}\right] H\left[\frac{k(E_i)\mu_j + q_x\chi_y}{k(E_i + \hbar\omega_{\mathrm{LO}})} - \mu_{b-\frac{1}{2}}\right]$$

$$\times \; \left\{ (1 - \mu_j^2)[k(E_i + \hbar\omega_{\mathrm{LO}})^2 - (k(E_i)\mu_j + q_x\chi_y)^2](1 - \hat{\mathcal{N}}^{xy}_{ij}{}^2) \right\}^{-\frac{1}{2}}$$

with

$$\hat{\mathcal{N}}^{xy}_{ij} = \mathcal{N}^x_{ij}\left( \frac{k(E_i)\mu_j + q_x\chi_y}{k(E_i + \hbar\omega_{\mathrm{LO}})} \right). \tag{3.50}$$

As the most important result of this section, we notice that the collision coefficient $\langle \mathcal{A}^{xy}_{\mathrm{POP},ab}\rangle_{ij}$ as well as the other POP interaction coefficients can be calculated analytically.

## 3.3  Conservation Laws

In the following, we deduce the conservation laws for the electron density and the total energy density for our multigroup model in the case of a spatially homogeneous problem. These quantities are conserved when the $(E, \mu)$ space of electrons is discretized in a way that

$$E^\nu_{i+\frac{1}{2}} - E^\nu_{i-\frac{1}{2}} = \hbar\omega_{\mathrm{LO}}/n, \; n \in \mathbb{N}, \; i = 1, 2, 3, \ldots, N^\nu, \tag{3.51a}$$

$$E^\nu_i - E^\nu_{i-1} = \hbar\omega_{\mathrm{LO}}/n, \; n \in \mathbb{N}, \; i = 2, 3, \ldots, N^\nu, \tag{3.51b}$$

$$\mu^\nu_{j+\frac{1}{2}} = \frac{2\,j}{M^\nu} - 1, \; j = 0, 1, \ldots, M^\nu. \tag{3.51c}$$

This partition of the electron **k**-space is preconditioned in all the calculations of this section.

To begin with, we consider the following proposition on the properties of the multigroup collision terms.

**Proposition 1**   *The collision terms $C_\xi[n_{ij}^\nu]$ fulfill the relation*

$$\sum_\nu 4\pi Z_\nu \sum_{i=1}^{N^\nu} \sum_{j=1}^{M^\nu} C_\xi[n_{ij}^\nu] = 0 \qquad (3.52)$$

*with $\xi = \mathbf{E}, \mathrm{ADP}, \mathrm{PZ}, \mathrm{ODP}, \mathrm{IMP}, \mathrm{IV}, \mathrm{POP}$.*

*Proof:* The proof of proposition 1 is very simple in the case of ADP, PZ, ODP and IMP scattering (3.41b). The renaming of appropriate summation indices yields

$$\sum_{i=1}^{N^\nu} \sum_{j=1}^{M^\nu} C_\xi[n_{ij}^\nu] = \sum_{i=1}^{N^\nu} \sum_{j=1}^{M^\nu} \sum_{a=1}^{N^\nu} \sum_{b=1}^{N^\nu} (\langle S_{\xi,ab}^\nu \rangle_{ij} - \langle S_{\xi,ab}^\nu \rangle_{ij}) = 0, \qquad (3.53)$$

$$\xi = \mathrm{ADP}, \mathrm{PZ}, \mathrm{ODP}, \mathrm{IMP}.$$

A similar procedure shows that proposition 1 also holds for the POP interaction (3.41d). As for the intervalley scattering (3.41c), we regard two valleys named $\nu$ and $\mu$. The evaluating of (3.52) ends in

$$4\pi Z_\nu \sum_{i=1}^{N^\nu} \sum_{j=1}^{M^\nu} C_{\mathrm{IV}}[n_{ij}^\nu] + 4\pi Z_\mu \sum_{a=1}^{N^\mu} \sum_{b=1}^{M^\mu} C_{\mathrm{IV}}[n_{ab}^\mu]$$

$$= 4\pi \sum_{i=1}^{N^\nu} \sum_{j=1}^{M^\nu} \sum_{a=1}^{N^\mu} \sum_{b=1}^{N^\mu} \Big[ Z_\mu \langle S_{\mathrm{IV},ij}^{\mu\to\nu} \rangle_{ab}\, n_{ab}^\mu - Z_\nu \langle S_{\mathrm{IV},ab}^{\nu\to\mu} \rangle_{ij}\, n_{ij}^\nu \qquad (3.54)$$

$$+ Z_\nu \langle S_{\mathrm{IV},ab}^{\nu\to\mu} \rangle_{ij}\, n_{ij}^\nu - Z_\mu \langle S_{\mathrm{IV},ij}^{\mu\to\nu} \rangle_{ab}\, n_{ab}^\mu \Big] = 0.$$

The generalization of this derivation to an arbitrary number of involved valleys proves proposition 1 for intervalley scattering. Finally, we turn to the force term (3.41a). Here, we evaluate (3.52) for the first part of $C_{\mathbf{E}}[n_{ij}^\nu]$. Because of (3.51), $\Delta E^\nu = E_{i+1/2}^\nu - E_{i-1/2}^\nu$ is constant for all the indices $i$ and we obtain with the help of (3.15)

$$\frac{\hbar\, e|\mathbf{E}|\, \mu_j^\nu}{m_\nu^* \Delta E^\nu} \sum_{i=1}^{N^\nu} \left[ \frac{k^\nu(E_{i+\frac{1}{2}}^\nu)}{1 + 2\alpha_\nu E_{i+\frac{1}{2}}^\nu} [n_{ij}^\nu]^+ - \frac{k^\nu(E_{i-\frac{1}{2}}^\nu)}{1 + 2\alpha_\nu E_{i-\frac{1}{2}}^\nu} [n_{ij}^\nu]^- \right] \qquad (3.55)$$

$$= \frac{\hbar\, e|\mathbf{E}|\, \mu_j^\nu}{m_\nu^* \Delta E^\nu} \left[ \sum_{i=1}^{N^\nu} \frac{k^\nu(E_{i+\frac{1}{2}}^\nu)}{1 + 2\alpha_\nu E_{i+\frac{1}{2}}^\nu} [n_{ij}^\nu]^+ - \sum_{i=0}^{N^\nu-1} \frac{k^\nu(E_{i+\frac{1}{2}}^\nu)}{1 + 2\alpha_\nu E_{i+\frac{1}{2}}^\nu} [n_{i+1,j}^\nu]^- \right] = 0.$$

Similar calculations for the second part of the force term complete proof of proposition 1.

With the help of proposition 1, we can easily deduce the conservation law of the electron density. For this purpose, we multiply the electron multigroup equations (3.40a) by $4\pi Z_\nu$ and sum the result over all the involved valleys, energies and polar angles. This procedure ends in

$$\frac{\partial}{\partial t} \sum_\nu 4\pi Z_\nu \sum_{i=1}^{N^\nu} \sum_{j=1}^{M^\nu} n_{ij}^\nu = \sum_\xi \sum_\nu 4\pi Z_\nu \sum_{i=1}^{N^\nu} \sum_{j=1}^{M^\nu} \mathcal{C}_\xi[n_{ij}^\nu] = 0. \qquad (3.56)$$

From (3.10), we find that the total electron density $\langle n^{\mathrm{c}} \rangle$ expressed in terms of the coefficients $n_{ij}^\nu$ is given by

$$\langle n^{\mathrm{c}} \rangle = \sum_\nu 4\pi Z_\nu \sum_{i=1}^{N^\nu} \sum_{j=1}^{M^\nu} n_{ij}^\nu. \qquad (3.57)$$

Combining (3.56) and (3.57) proves the following theorem.

**Theorem 7** *The multigroup model equations (3.40) conserve the total electron density $n^{el}$:*

$$\frac{\partial \langle n^{\mathrm{c}}(t) \rangle}{\partial t} = 0.$$

Concerning the total energy density, we only take into account the interaction between electrons and LO phonons, since this type of phonon is the only one in our model which is regarded in a dynamic way. Hence, we consider

$$\frac{\partial n_{ij}^\nu}{\partial t} = \mathcal{C}_{\mathrm{POP}}[n_{ij}^\nu], \qquad\qquad i = 1, 2 \ldots N^\nu,\, j = 1, 2 \ldots M^\nu, \qquad (3.58\mathrm{a})$$

$$\frac{\partial r_{xy}}{\partial t} = \sum_\nu 2 Z_\nu \mathcal{D}_{\mathrm{POP}}[r_{xy}], \qquad x = 1, 2 \ldots R,\, y = 1, 2 \ldots S, \qquad (3.58\mathrm{b})$$

in the case of a vanishing external field $\mathbf{E} = 0$. These equations are multiplied by the corresponding electron energies $4\pi Z_\nu\, E_i^\nu$ and the phonon energy $2\pi \hbar \omega_{\mathrm{LO}}$, respectively. The results are summed over all the involved valleys and the indices $i, j, x, y$. Hence, we find

$$\frac{\partial}{\partial t}\left(\sum_\nu \sum_{i=1}^{N^\nu}\sum_{j=1}^{M^\nu} 4\pi Z_\nu\, E_i^\nu\, n_{ij}^\nu + \sum_{x=1}^{R}\sum_{y=1}^{S} 2\pi\, \hbar\omega_{\mathrm{LO}}\, r_{xy}\right) \tag{3.59}$$

$$= \sum_\nu 4\pi Z_\nu \sum_{i=1}^{N^\nu}\sum_{j=1}^{M^\nu}\sum_{x=1}^{R}\sum_{y=1}^{S}\left\{\hbar\omega_{\mathrm{LO}}\left[\left(\langle E_{\mathrm{POP}}^{ij}\rangle_{xy} - \langle A_{\mathrm{POP}}^{ij}\rangle_{xy}\right)r_{xy} + \langle 1_{\mathrm{POP}}^{ij}\rangle_{xy}\right] n_{ij}^\nu \right.$$

$$+ \sum_{a=1}^{N^\nu}\sum_{b=1}^{M^\nu} E_i^\nu\left[\left(\langle \mathcal{E}_{\mathrm{POP},ij}^{\nu,xy}\rangle_{ab} + \langle \mathcal{A}_{\mathrm{POP},ij}^{\nu,xy}\rangle_{ab}\right)r_{xy} + \langle 1_{\mathrm{POP},ij}^{\nu,xy}\rangle_{ab}\right] n_{ab}^\nu$$

$$\left. - \sum_{a=1}^{N^\nu}\sum_{b=1}^{M^\nu} E_i^\nu\left[\left(\langle \mathcal{E}_{\mathrm{POP},ab}^{\nu,xy}\rangle_{ij} + \langle \mathcal{A}_{\mathrm{POP},ab}^{\nu,xy}\rangle_{ij}\right)r_{xy} + \langle 1_{\mathrm{POP},ab}^{\nu,xy}\rangle_{ij}\right] n_{ij}^\nu\right\}.$$

For the further manipulations of the above expression, we take advantage of the following proposition.

**Proposition 2**    *The POP collision constants fulfill the symmetry relations*

$$\langle E_{\mathrm{POP}}^{ij}\rangle_{xy} = \sum_{a'=1}^{N^\nu}\sum_{b'=1}^{M^\nu}\langle \mathcal{E}_{\mathrm{POP},a'b'}^{\nu,xy}\rangle_{ij}, \tag{3.60a}$$

$$\langle A_{\mathrm{POP}}^{ij}\rangle_{xy} = \sum_{a'=1}^{N^\nu}\sum_{b'=1}^{M^\nu}\langle \mathcal{A}_{\mathrm{POP},a'b'}^{\nu,xy}\rangle_{ij}, \tag{3.60b}$$

$$E_a^\nu\,\langle \mathcal{E}_{\mathrm{POP},ab}^{\nu,xy}\rangle_{ij} = (E_i^\nu - \hbar\omega_{\mathrm{LO}})\,\langle \mathcal{E}_{\mathrm{POP},ab}^{\nu,xy}\rangle_{ij}, \tag{3.60c}$$

$$E_a^\nu\,\langle \mathcal{A}_{\mathrm{POP},ab}^{\nu,xy}\rangle_{ij} = (E_i^\nu + \hbar\omega_{\mathrm{LO}})\,\langle \mathcal{A}_{\mathrm{POP},ab}^{\nu,xy}\rangle_{ij} \tag{3.60d}$$

*for $i, a = 1, 2, \ldots, N^\nu$, $j, b = 1, 2, \ldots, M^\nu$, $x = 1, 2, \ldots, R$ and $y = 1, 2, \ldots, S$.*

*Proof:* Since this proposition deals with an intravalley scattering mechanism, we are allowed to skip the valley index in the following calculations for simplicity.

To deduce (3.60a), we sum the electron collision terms $\langle \mathcal{E}_{\mathrm{POP},ab}^{\nu,xy}\rangle_{ij}$ over all the possible indices $a$ and $b$. The result $I$ reads (cf. (3.32))

$$I = \frac{C_x}{q_x}\int_{I_i^E} dE \int_{I_j^\mu} d\mu \sum_{a=1}^{N}\sum_{b=1}^{M}\int_{C_{ab}} d^3k'\, \delta(E - E_i)\,\delta(\mu - \mu_j) \tag{3.61}$$

$$\times\,\delta(|\mathbf{k} - \mathbf{k}'| - q_x)\,\delta[(\mathbf{k} - \mathbf{k}')\cdot \mathbf{e}_z - q_x\chi_y]\,\delta(E(\mathbf{k}') - E + \hbar\omega_{\mathrm{LO}})$$

with $C_x = \pi e^2 \omega_{\mathrm{LO}} \left(1/\kappa_\infty - 1/\kappa_0\right)/\epsilon_0 \, q_x^2$. The rewriting of $I$ into an integration with respect to $\mathbf{k}$ reveals

$$I = \frac{C_x}{16\pi^4 \, q_x} \int_{C_{ij}} d^3k \, \frac{1}{\mathcal{Z}(E(\mathbf{k}))} \int_{\mathcal{B}} d^3k' \, \delta(E(\mathbf{k}) - E_i) \, \delta(\mu(\mathbf{k}) - \mu_j) \quad (3.62)$$
$$\times \, \delta(|\mathbf{k} - \mathbf{k}'| - q_x) \, \delta\left[(\mathbf{k} - \mathbf{k}') \cdot \mathbf{e}_z - q_x \chi_y\right] \delta(E(\mathbf{k}') - E(\mathbf{k}) + \hbar\omega_{\mathrm{LO}}).$$

Here, we express the integration with respect to $\mathbf{k}'$ as a sum over the Brillouin zone and extend $I$ by $\sum_{\mathbf{q} \in \mathcal{B}} \delta_{\mathbf{k} - \mathbf{k}', \mathbf{q}}$ and find

$$I = \frac{C_x}{2\pi V \, q_x} \int_{C_{ij}} d^3k \, \frac{1}{\mathcal{Z}(E(\mathbf{k}))} \sum_{\mathbf{k}', \mathbf{q} \in \mathcal{B}} \delta_{\mathbf{k} - \mathbf{k}', \mathbf{q}} \, \delta(E(\mathbf{k}) - E_i) \, \delta(\mu(\mathbf{k}) - \mu_j)$$
$$\times \, \delta(|\mathbf{q}| - q_x) \, \delta\left[\mathbf{q} \cdot \mathbf{e}_z - q_x \chi_y\right] \delta(E(\mathbf{k}') - E(\mathbf{k}) + \hbar\omega_{\mathrm{LO}}) \qquad (3.63)$$
$$= \frac{C_x}{2\pi V \, q_x} \int_{C_{ij}} d^3k \, \frac{1}{\mathcal{Z}(E(\mathbf{k}))} \sum_{\mathbf{q} \in \mathcal{B}} \delta(E(\mathbf{k}) - E_i) \, \delta(\mu(\mathbf{k}) - \mu_j)$$
$$\times \, \delta(|\mathbf{q}| - q_x) \, \delta\left[\mathbf{q} \cdot \mathbf{e}_z - q_x \chi_y\right] \delta(E(\mathbf{k} - \mathbf{q}) - E(\mathbf{k}) + \hbar\omega_{\mathrm{LO}}).$$

Replacing the sum over $\mathbf{q}$ by an integral and taking advantage of the fact that the poles of the Dirac distributions $\delta(|\mathbf{q}| - q_x)$ and $\delta\left[\mathbf{q} \cdot \mathbf{e}_z - q_x \chi_y\right]$ lie in the cell $D_{xy}$ yields

$$I = \frac{C_x}{16\pi^4 \, q_x} \int_{C_{ij}} d^3k \, \frac{1}{\mathcal{Z}(E(\mathbf{k}))} \int_{D_{xy}} d^3q \, \delta(E(\mathbf{k}) - E_i) \, \delta(\mu(\mathbf{k}) - \mu_j)$$
$$\times \, \delta(|\mathbf{q}| - q_x) \, \delta\left[\mathbf{q} \cdot \mathbf{e}_z - q_x \chi_y\right] \delta(E(\mathbf{k} - \mathbf{q}) - E(\mathbf{k}) + \hbar\omega_{\mathrm{LO}}) \quad (3.64)$$
$$= \frac{C_x}{8\pi^3} \int_{C_{ij}} d^3k \, \frac{1}{\mathcal{Z}(E(\mathbf{k}))} \int_{I_x^q} dq \int_{I_y^\chi} d\chi \, \delta(E(\mathbf{k}) - E_i) \, \delta(\mu(\mathbf{k}) - \mu_j)$$
$$\times \, \delta(q - q_x) \, \delta\left[\chi - \chi_y\right] \delta(E(\mathbf{k} - \mathbf{q}) - E(\mathbf{k}) + \hbar\omega_{\mathrm{LO}}),$$

which equals the phonon collision coefficient $\langle E_{\mathrm{POP}}^{ij} \rangle_{xy}$ (3.34) as stated in proposition 2. The relation (3.60b) can be deduced in a similar way.

The relation (3.60c) is proved by the following considerations. With the help of (3.32), we find for the POP collision coefficient $\langle \mathcal{E}^{xy}_{\text{POP},ab}\rangle_{ij}$

$$
\langle \mathcal{E}^{\nu,xy}_{\text{POP},ab}\rangle_{ij} = \frac{8\pi^3 C_x}{q_x} \int_{I^E_i} dE \int_{I^\mu_j} d\mu \int_{I^E_a} dE' \, \mathcal{Z}(E') \tag{3.65}
$$

$$
\times \int_{I^\mu_b \times [0,2\pi]} d\boldsymbol{\Omega}' \, \delta(E - E_i)\,\delta(\mu - \mu_j)\,\delta(|\mathbf{k}(E,\mu) - \mathbf{k}(E',\boldsymbol{\Omega}')| - q_x)
$$

$$
\times \delta[(\mathbf{k}(E,\mu) - \mathbf{k}(E',\boldsymbol{\Omega}))\cdot \mathbf{e}_z - q_x \chi_y]\,\delta(E' - E + \hbar\omega_{\text{LO}})
$$

$$
= \frac{8\pi^3 C_x}{q_x}\,\mathcal{Z}(E_i - \hbar\omega_{\text{LO}})\,\Theta[E_{a+\frac12} - (E_i - \hbar\omega_{\text{LO}})]\,\Theta((E_i - \hbar\omega_{\text{LO}}) - E_{a-\frac12})
$$

$$
\times \int_{I^\mu_b \times [0,2\pi]} d\boldsymbol{\Omega}' \, \delta(|\mathbf{k}(E_i,\mu_j) - \mathbf{k}(E_i - \hbar\omega_{\text{LO}},\boldsymbol{\Omega}')| - q_x)
$$

$$
\times \delta[(\mathbf{k}(E_i,\mu_j) - \mathbf{k}(E_i - \hbar\omega_{\text{LO}},\boldsymbol{\Omega}'))\cdot \mathbf{e}_z - q_x \chi_y],
$$

where $\Theta$ denotes the Heaviside step function. This expression can be unequal to zero, only if $E_{a+\frac12} \geq E_i - \hbar\omega_{\text{LO}}$ and $E_i - \hbar\omega_{\text{LO}} \geq E_{a-\frac12}$. In this case $E_i - \hbar\omega_{\text{LO}}$ coincides with the pole $E_a$ of the Dirac's Delta in the interval $I^E_a = [E_{a-\frac12}, E_{a+\frac12}]$ because of the energy partition (3.51), which proves (3.60c). Similarly, (3.60d) can be deduced.

The application of the first part of the above proposition to (3.59) yields

$$
\frac{\partial}{\partial t}\left(\sum_\nu \sum_{i=1}^{N^\nu} \sum_{j=1}^{M^\nu} 4\pi Z_\nu\,E^\nu_i\,n^\nu_{ij} + \sum_{x=1}^{R} \sum_{y=1}^{S} 2\pi\,\hbar\omega_{\text{LO}}\,r_{xy}\right) \tag{3.66}
$$

$$
= \sum_\nu 4\pi Z_\nu \sum_{i=1}^{N^\nu} \sum_{j=1}^{M^\nu} \sum_{a=1}^{N^\nu} \sum_{b=1}^{M^\nu} \sum_{x=1}^{R} \sum_{y=1}^{S}
$$

$$
\left\{ \left[ \left( E^\nu_i\,\langle \mathcal{E}^{\nu,xy}_{\text{POP},ij}\rangle_{ab} + E^\nu_i\,\langle \mathcal{A}^{\nu,xy}_{\text{POP},ij}\rangle_{ab}\right) r_{xy} + E^\nu_i\,\langle 1^{\nu,xy}_{\text{POP},ij}\rangle_{ab}\right] n^\nu_{ab} \right.
$$

$$
- \left[ \left( (E^\nu_i - \hbar\omega_{\text{LO}})\langle \mathcal{E}^{\nu,xy}_{\text{POP},ab}\rangle_{ij} + (E^\nu_i + \hbar\omega_{\text{LO}})\langle \mathcal{A}^{\nu,xy}_{\text{POP},ab}\rangle_{ij}\right) r_{xy} \right.
$$

$$
\left. \left. + (E^\nu_i - \hbar\omega_{\text{LO}})\langle 1^{\nu,xy}_{\text{POP},ab}\rangle_{ij}\right] n^\nu_{ij} \right\}.
$$

After renaming appropriate summation indices, we obtain with the help of the second part of proposition 2

$$\frac{\partial}{\partial t}\left(\sum_\nu \sum_{i=1}^{N^\nu}\sum_{j=1}^{M^\nu} 4\pi Z_\nu\, E_i^\nu\, n_{ij}^\nu + \sum_{x=1}^{R}\sum_{y=1}^{S} 2\pi\,\hbar\omega_{\mathrm{LO}}\, r_{xy}\right) \tag{3.67}$$

$$= \sum_\nu 4\pi Z_\nu \sum_{i=1}^{N^\nu}\sum_{j=1}^{M^\nu}\sum_{a=1}^{N^\nu}\sum_{b=1}^{M^\nu}\sum_{x=1}^{R}\sum_{y=1}^{S}$$

$$\left\{\left[\left(E_a^\nu\,\langle\mathcal{E}_{\mathrm{POP},ab}^{\nu,xy}\rangle_{ij} + E_a^\nu\,\langle\mathcal{A}_{\mathrm{POP},ab}^{\nu,xy}\rangle_{ij}\right) r_{xy} + E_a^\nu\,\langle 1_{\mathrm{POP},ab}^{\nu,xy}\rangle_{ij}\right] n_{ij}^\nu\right.$$

$$-\left[\left((E_i^\nu - \hbar\omega_{\mathrm{LO}})\langle\mathcal{E}_{\mathrm{POP},ab}^{\nu,xy}\rangle_{ij} + (E_i^\nu + \hbar\omega_{\mathrm{LO}})\langle\mathcal{A}_{\mathrm{POP},ab}^{\nu,xy}\rangle_{ij}\right) r_{xy}\right.$$

$$\left.\left.+ (E_i^\nu - \hbar\omega_{\mathrm{LO}})\langle 1_{\mathrm{POP},ab}^{\nu,xy}\rangle_{ij}\right] n_{ij}^\nu\right\} = 0.$$

From (3.10), we find that the total energy density of the electron-LO phonon system equals

$$\langle nE^{\mathrm{tot}}\rangle = \sum_\nu \sum_{i=1}^{N^\nu}\sum_{j=1}^{M^\nu} 4\pi Z_\nu\, E_i^\nu\, n_{ij}^\nu + \sum_{x=1}^{R}\sum_{y=1}^{S} 2\pi\,\hbar\omega_{\mathrm{LO}}\, r_{xy}. \tag{3.68}$$

Inserting this expression into (3.67) completes the proof of the following theorem.

**Theorem 8** *The multigroup model equations (3.58) conserve the total energy density $E^{tot}$ of the coupled electron-phonon system:*

$$\frac{\partial\langle nE^{\mathrm{tot}}(t)\rangle}{\partial t} = 0.$$

Chapter 4

# Particle Transport in Indium Phosphide

## 4.1 Introduction

In this chapter, two types of models for polar semiconductors are applied for investigating the particle transport in indium phosphide: a two-valley model and a three-valley model. Besides the number of the considered energy bands, these models differ mainly in the material parameters used for InP, which vary especially in the energy separation between the $\Gamma$ and the $L$ valleys. In the reference for the material data of the two-valley model [Vaissiere *et al.* (1992)], we find $\Delta_{\Gamma L} = 0.61$ eV, while [Maloney and Frey (1977)] tells us that $\Delta_{\Gamma L} = 0.86$ eV as it is used in the three-valley model. The differences in the resulting transport quantities obtained by the two models are mainly a consequence of the differences in the used input parameter.

## 4.2 Two-valley Model

In this section, we use the multigroup model equations (3.40) for simulating the particle transport in indium phosphide. Therefore, we consider n-type InP with a doping concentration of $N_{\mathrm{D}} = 10^{17} \mathrm{cm}^{-3}$ at the temperature $T_{\mathrm{L}} = 300$ K. Therefore, we are allowed to neglect holes in the valence band. The conduction band is approximated by the $\Gamma$ valley centered at $\langle 0, 0, 0 \rangle$ and four equivalent $L$ valleys along $\langle 1, 1, 1 \rangle$. These valleys are assumed to be spherical and non-parabolic. The $X$ valleys are neglected in our calculations. The material parameters, we use in our calculations are found in table 4.1 and correspond to the values found in [Vaissiere *et al.* (1992)].

As concerns the collision mechanisms (cf. Sec. 2.4), we consider acous-

Table 4.1   Material Parameters for InP: Two-valley Model.

General Characteristics:

| Quantity | Symbol | Unit | Value |
|---|---|---|---|
| Mass density | $\rho$ | $\mathrm{kgm}^{-3}$ | 4830 |
| Sound velocity | $v_\mathrm{s}$ | $\mathrm{ms}^{-1}$ | 5160 |
| HF relative dielectric constant | $\kappa_\infty$ | | 9.56 |
| Static relative dielectric constant | $\kappa_0$ | | 12.3 |
| LO phonon relaxation time | $\tau_\mathrm{L}$ | ps | 5.8 |

Band parameters:

| Quantity | Symbol | Unit | Valley $\Gamma$ | Valley $L$ |
|---|---|---|---|---|
| Relative effective mass | $m^*/m_0$ | | 0.08 | 0.4 |
| Non-parabolicity factor | $\alpha^*$ | $(\mathrm{eV})^{-1}$ | 0.627 | 0.621 |
| Gap referred to $\Gamma$ minimum | $\Delta_{\nu\mu}$ | eV | | 0.610 |
| Number of equivalent valleys | $Z_\mu$ | | 1 | 4 |

Intravalley scattering parameters:

| Quantity | Symbol | Unit | Valley $\Gamma$ | Valley $L$ |
|---|---|---|---|---|
| Acoustic deformation potential | $D_\mathrm{A}$ | $\mathrm{eVm}^{-1}$ | 7 | 12 |
| Piezoelectric constant | $D_\mathrm{PZ}$ | $\mathrm{Cm}^{-2}$ | 0.0131 | 0.0131 |
| TO Phonons: | | | | |
|    Deformation potential | $D_\mathrm{TO}$ | $\mathrm{eVm}^{-1}$ | | $6.7 \times 10^{10}$ |
|    Energy | $\hbar\omega_\mathrm{TO}$ | meV | | 43 |
| LO Phonons: | | | | |
|    Energy | $\hbar\omega_\mathrm{LO}$ | meV | 43.20 | 43.20 |

Intervalley scattering parameters:

Intervalley phonon 1:

| Energy $\hbar\omega_{\nu\mu}$ (meV) | | | Deformation potential $D_{\nu\mu}$ $(10^9\,\mathrm{eVm}^{-1})$ | | |
|---|---|---|---|---|---|
| | $\Gamma$ | $L$ | | $\Gamma$ | $L$ |
| $\Gamma$ | | 33.7 | $\Gamma$ | | 137 |
| $L$ | 33.7 | 33.7 | $L$ | 137 | 56 |

Intervalley phonon 2:

| Energy $\hbar\omega_{\nu\mu}$ (meV) | | | Deformation potential $D_{\nu\mu}$ $(10^9\,\mathrm{eVm}^{-1})$ | | |
|---|---|---|---|---|---|
| | $\Gamma$ | $L$ | | $\Gamma$ | $L$ |
| $\Gamma$ | | 6.8 | $\Gamma$ | | 14 |
| $L$ | 6.8 | | $L$ | 14 | |

tic deformation potential and acoustic piezoelectric scattering, the polar optical, optical deformation potential scattering in $L$ valleys and impurity scattering, which are all intravalley processes, and non-polar optical intervalley scattering. Since ADP and PZ scattering are not efficient at 300 K, these mechanisms are regarded as elastic collisions. For the intervalley transfer, we take into account two types of zone-boundary phonons. The ionized impurity scattering is described with the help of the Brook-Herring model and by assuming equal electron and donor concentrations. In [Vaissiere *et al.* (1992); Vaissiere *et al.* (1996)], the proper choice of the

Table 4.2  The computation times $t_{CPU}$ for the solution of the multigroup equations (3.40a) and (3.31) with $N^\Gamma=44$, $N^L=16$, $R=40$, $M^\Gamma=M^L=S=12$ up to 20 ps after the onset of the electric field pulse for several electric field strengths $|\mathbf{E}|$ and relative accuracies $\varepsilon_r$.

| | Equilibrium phonons | | Hot phonons | |
| | $\varepsilon_r = 10^{-3}$ | $\varepsilon_r = 10^{-5}$ | $\varepsilon_r = 10^{-3}$ | $\varepsilon_r = 10^{-5}$ |
| $|\mathbf{E}|$ [kVcm$^{-1}$] | $t_{CPU}$ [s] | $t_{CPU}$ [s] | $t_{CPU}$ [s] | $t_{CPU}$ [s] |
|---|---|---|---|---|
| 1 | 6.1 | 13.6 | 10.7 | 15.8 |
| 10 | 9.3 | 16.4 | 11.9 | 19.8 |
| 50 | 9.8 | 17.9 | 34.2 | 36.3 |

LO phonon relaxation time $\tau_L$ is discussed. We follow this argumentation leading to the value for $\tau_L$ proposed in the Tab. 4.3.

The application of our multigroup model to the coupled electron LO phonon regime in response to a step-like high dc electric field pulse is demonstrated in this section. All the relevant material parameters of InP are given in Tab. 4.3. In our calculations, the moduli of the wave vectors of electrons and phonons are restricted up to $k_{max} = q_{max} = 1.7 \times 10^9 \mathrm{m}^{-1}$. This implies that the maximal energies in the $\Gamma$ and $L$ valleys are $E_{max}^\Gamma = 0.86\,\mathrm{eV}$ and $E_{max}^L = 0.26\,\mathrm{eV}$, respectively. We assume the $(E,\mu)$ and the $(q,\chi)$ space to be equidistantly partitioned. The poles of the Dirac distributions in (8.15) and (3.26) are simply set to

$$E_i^\nu = \frac{1}{2}(E_{i+\frac{1}{2}}^\nu + E_{i-\frac{1}{2}}^\nu), \qquad i = 1,2,\ldots,N^\nu, \qquad (4.1a)$$

$$\mu_j^\nu = \frac{1}{2}(\mu_{j+\frac{1}{2}}^\nu + \mu_{j-\frac{1}{2}}^\nu), \qquad j = 1,2,\ldots,M^\nu, \qquad (4.1b)$$

$$q_x = \frac{1}{2}(q_{x+\frac{1}{2}} + q_{x-\frac{1}{2}}), \qquad x = 1,2,\ldots,R, \qquad (4.1c)$$

$$\chi_y = \frac{1}{2}(\chi_{y+\frac{1}{2}} + \chi_{y-\frac{1}{2}}), \qquad y = 1,2,\ldots,S. \qquad (4.1d)$$

Several simulations have shown that with an increasing number of intervals, the influence of this choice as well as those of the specific partitions of the $(E,\mu)$ and the $(q,\chi)$ space become negligible.

Before the electric field pulse begins, the electron-phonon system is supposed to be in thermal equilibrium. Therefore, the initial values of the coefficients $n_{ij}^\nu$ and $r_{xy}$ are obtained from the corresponding equilibrium distributions at lattice temperature $T_L$ and electron density $N_D$ via (3.9) and (3.27).

The multigroup equations (3.40) are solved with the help of an explicit Euler scheme with adaptive step-size control demanding the relative accu-

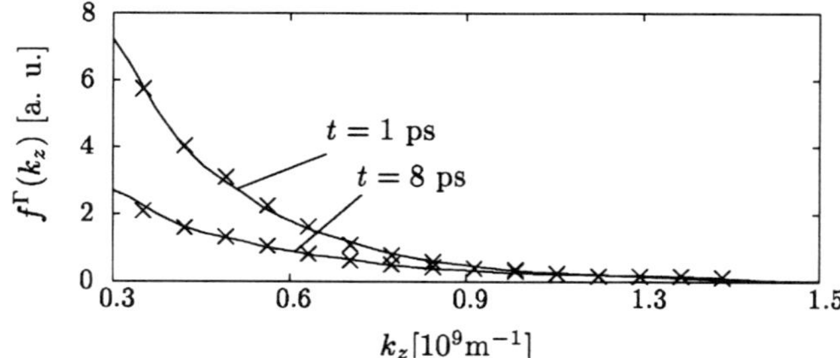


Fig. 4.1 The electron distribution function $f^\Gamma(k_z)$ in the $\Gamma$ valley versus the component of the wave vector $k_z$ in the direction of the electric field with $E = 10$ kVcm$^{-1}$ for several times after the beginning of the electric field pulse. The solid lines refer to calculations performed with our multigroup approach; the crosses refer to the matrix method from Vaissiere *et al.* (GALLER M. and SCHÜRRER F., *J. Phys. A: Math. Gen.*, **37** (2004), 1479–1497; reproduced with permission from IOP Publishing, www.iop.org/journals/jphysa.)

racy $\varepsilon_r$. The calculations are performed with an AMD Athlon MP 2000+ processor, 1666 MHz, 2000 MB RAM. Table 4.2 presents the CPU times needed for integrating the multigroup equations up to 20 ps after the onset of the electric field pulse for several electric field strengths and accuracies. The partitions of the wave vector spaces in these calculations are the same as used for deducing the current-field characteristics displayed in Fig. 4.2. The computation times increase for higher electric fields and the required accuracies and when solving the coupled electron phonon system instead of taking into account only equilibrium phonons. The higher demand on computational power for including hot phonon effects in the transport model is justified by the fact that the influence of the disturbed LO phonon distribution on macroscopic transport quantities cannot be neglected in the presented case (cf. Fig. 4.2). To achieve an accelerated solution technique, which is most welcome in the application to space dependent problems for a realistic device simulation, our multigroup approach can be combined with more advanced differential equation solution routines.

### 4.2.1 *Validation of the Method*

To check the validity of the numerical procedure illustrated in the previous section, we compare the results gained with the help of our method to those calculated by using a matrix method [Vaissiere *et al.* (1996)].

To this end, we study the temporal evolution of the EDF $f^\Gamma$ of the $\Gamma$ valley along the external electric field. According to the above considerations, the main quantities in our model are not the distribution function itself but the particle densities $n^\nu_{ij}$ of electrons with energies and polar angles within the cell $C^\nu_{ij}$. Hence, it is not possible to compare the results of both methods directly. However, we obtain an approximative expression for the distribution function $f^\nu[k^\nu(E^\nu_i)\mu_j]$ via

$$f^\nu[k^\nu(E^\nu_i)\mu_j] = n^\nu_{ij}\left[Z^\nu(E^\nu_i)\,(E_{i+\frac{1}{2}} - E_{i-\frac{1}{2}})(\mu_{j+\frac{1}{2}} - \mu_{j-\frac{1}{2}})\right]^{-1} \quad (4.2)$$

which can be derived from the mean value theorem of integral calculus. It should be noted that the above approximation fails for small energies $E^\nu_i$ because of our equidistant energy partition. Therefore, comparisons can only be performed for sufficiently high energies.

Figure 4.1 displays the distribution function $f(k_z)$ versus the z-component of the wave vector in direction of the electric field at two times after the beginning of the electric field pulse. Unfortunately, the results in [Vaissiere *et al.* (1996)] are given in arbitrary units, and we must introduce one scaling factor. Despite this uncertainty, Fig. 4.1 allows us to state that the results for the EDF in the $\Gamma$ valley as well as its temporal evolution gained with the help of the matrix method and by our model coincide very well.

Additionally, we have evaluated the two transport parameters, the average drift velocity in direction of the electric field and the average electron energy for several values of the applied electric field, by means of the formulae (3.10), which are reported in Fig. 4.2. Our results are compared with those of the stationary iterative method [Vaissiere *et al.* (1992)] and the matrix method [Vaissiere *et al.* (1996)]. Moreover, the results of Monte Carlo calculations [Maloney and Frey (1977)], which are only available without hot phonon effects for InP, are displayed in this figure. It is evident that the characteristics determined by means of our multigroup approach and by the mentioned other methods agree very well in the whole range of the electric field strengths. As a consequence of the nonequilibrium-phonon induced perturbation of the electron distribution, the electron transport parameters are significantly modified. Especially for medium electric field strengths around the threshold field at the onset of the negative differential resistivity, the values of electron drift velocity and the average energy, obtained by taking into account hot phonons, differ notably from those calculated by assuming equilibrium phonons. Hence, the consideration of

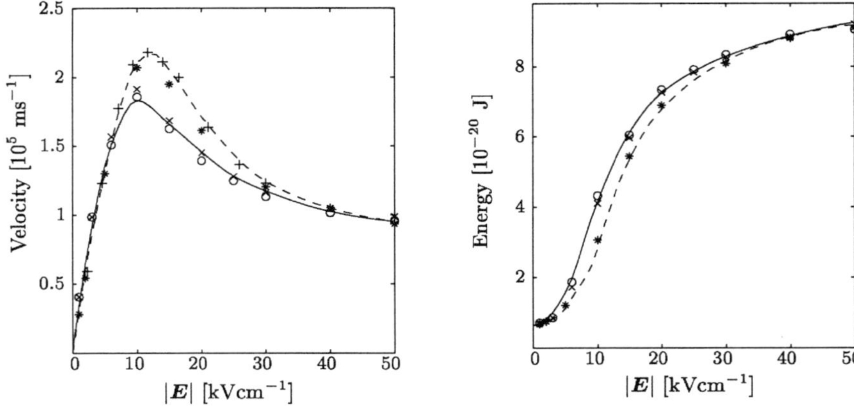

Fig. 4.2   Average drift velocity and average electron energy as a function of the applied electric field, computed at 20 ps from the beginning of the dc field pulse. (——): Presented multigroup approach for hot phonons; (− − −): Presented multigroup approach for equilibrium phonons; ($\circ$): Stationary iterative method for hot phonons; ($\times$): Matrix method for hot phonons; (*): Stationary iterative method for equilibrium phonons; (+): Monte Carlo calculations for equilibrium phonons. (GALLER M. and SCHÜRRER F., *J. Phys. A: Math. Gen.*, **37** (2004), 1479–1497; reproduced with permission from IOP Publishing, www.iop.org/journals/jphysa.)

hot phonon effects is of essential importance for an accurate description of polar semiconductors with high doping concentrations.

According to the comparisons displayed in Figs. 4.1 and 4.2, we find that the investigation of the transient transport regime by means of our multigroup model leads to results which are equivalent to those of the matrix method presented in [Vaissiere *et al.* (1996)].

### 4.2.2   *Electron Distribution Function*

For studying the properties of the EDF in the considered valleys, we display the electron densities $\langle n_{ij}^{c,\nu} \rangle$ versus the cells $C_{ij}^{\nu}$. Figure 4.3 depicts the electron densities $\langle n_{ij}^{c,\Gamma} \rangle$ in the cells $C_{ij}^{\Gamma}$ in the $\Gamma$ valley at $t = 8$ ps after the onset of the electric pulse with $E = 10$ kVcm$^{-1}$. Typically, we find a rapid decrease of the electron density at the energy 0.6 eV related to the bottom of the $L$ valleys. Moreover, we call attention to the strong asymmetry of the displayed particle density with respect to the electric field which reflects the low effective mass of $\Gamma$ electrons. This fact allows us to argue that

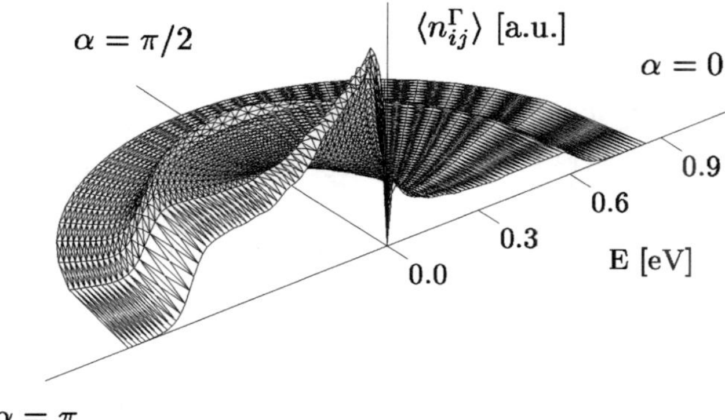

Fig. 4.3   The electron densities $\langle n_{ij}^{\mathrm{c},\Gamma} \rangle$ in the cells $C_{ij}^{\Gamma}$ in the $\Gamma$ valley at $t = 8$ ps after the onset of the electric pulse with $E = 10$ kVcm$^{-1}$. (GALLER M. and SCHÜRRER F., *J. Phys. A: Math. Gen.*, **37** (2004), 1479–1497; reproduced with permission from IOP Publishing, www.iop.org/journals/jphysa.)

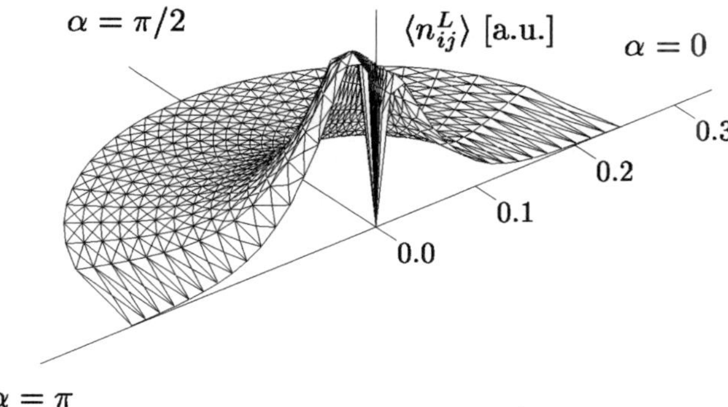

Fig. 4.4   The electron densities $\langle n_{ij}^{\mathrm{c},L} \rangle$ versus the cells $C_{ij}^{L}$ in the $L$ valley, evaluated at $t = 8$ ps after the onset of the electric pulse with $E = 10$ kVcm$^{-1}$. (GALLER M. and SCHÜRRER F., *J. Phys. A: Math. Gen.*, **37** (2004), 1479–1497; reproduced with permission from IOP Publishing, www.iop.org/journals/jphysa.)

the classical drift diffusion models with their restriction to close to equilibrium states would hardly yield reliable results for the considered physical situation. Therefore, a mesoscopic model like our multigroup approach is certainly a good choice for a careful investigation of this problem.

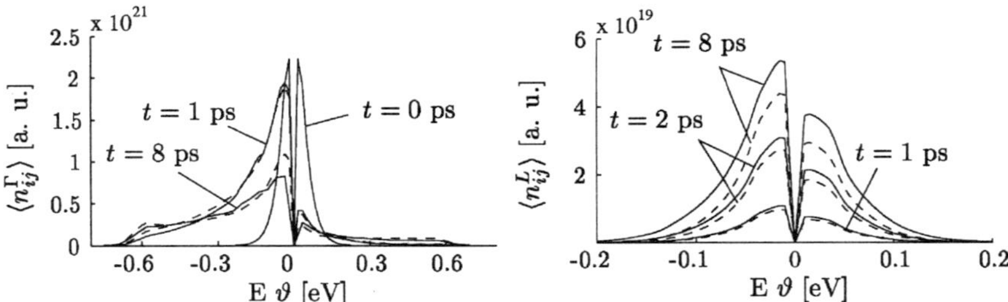

Fig. 4.5  Cuts of electron densities $\langle n_{ij}^{c,\Gamma} \rangle$ and $\langle n_{ij}^{c,L} \rangle$ in the $\Gamma$ and $L$ valleys in direction of the electric field versus the product of energy and cosine of the polar angle, $E\mu$, at several times after the beginning of the electric field pulse with $E = 10$ kVcm$^{-1}$. The solid lines refer to calculations taking into account non-equilibrium phonons; the dashed lines refer to those assuming phonons to be at thermal equilibrium. (GALLER M. and SCHÜRRER F., *J. Phys. A: Math. Gen.*, **37** (2004), 1479–1497; reproduced with permission from IOP Publishing, www.iop.org/journals/jphysa.)

In Fig. 4.4, we display the electron densities $\langle n_{ij}^{c,L} \rangle$ versus the cells $C_{ij}^{L}$ in the $L$ valley, evaluated at $t = 8$ ps after the onset of the electric pulse with $E = 10$ kVcm$^{-1}$. Here, the deviations of the results from the equilibrium distribution without an external electric field are small.

Finally, Fig. 4.5 illustrates cuts of the polar diagram representations of electron densities $\langle n_{ij}^{c,\Gamma} \rangle$ and $\langle n_{ij}^{c,L} \rangle$ in the $\Gamma$ and $L$ valleys in direction of the electric field versus the product of energy and cosine of the polar angle, $E\mu$, at several times after the beginning of the electric field pulse with $E = 10$ kVcm$^{-1}$. Here, we compare the results obtained by taking into account hot electrons with those assuming phonons to be in thermal equilibrium. The differences between the results lead to significantly different results for macroscopic quantities like valley population, drift velocity and electron energy for calculations with and without hot phonons, as discussed in Sec. 4.2.4. The detailed explanation of the physical effects which cause these differences is given in [Vaissiere *et al.* (1996)].

### 4.2.3  *Phonon Distribution Function*

The dynamics of the LO phonon distribution function is discussed in this section. Figure 4.6a illustrates the LO phonon densities $\langle n_{xy}^{p} \rangle$ in the cells $D_{xy}$ at $t = 8$ ps after the onset of the electric pulse with $E = 10$ kVcm$^{-1}$. Here we find that the phonon density is undisturbed at low and high moduli

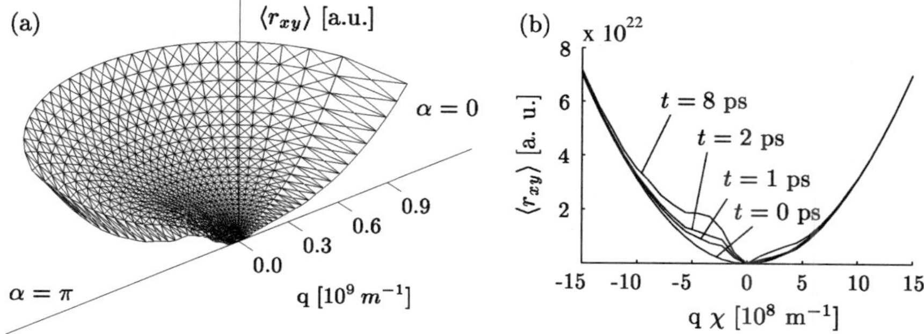



Fig. 4.6   The LO phonon densities $\langle n^{\mathrm{P}}_{xy} \rangle$ in the cells $D_{xy}$ at $t = 8$ ps after the onset of the electric pulse (a) and cuts of the LO phonon densities $\langle n^{\mathrm{P}}_{xy} \rangle$ in direction of the electric field versus the z-component of the wave vector $q\chi$ at several times after the onset of the field pulse (b) with $E = 10$ kVcm$^{-1}$. (GALLER M. and SCHÜRRER F., *J. Phys. A: Math. Gen.*, **37** (2004), 1479–1497; reproduced with permission from IOP Publishing, `www.iop.org/journals/jphysa`.)

of the wave vector, while it is enlarged for immediate $q$. Additionally, an asymmetry of the $\langle n^{\mathrm{P}}_{xy} \rangle$ is observed, which corresponds to the asymmetry of the electron distribution in the $\Gamma$ valley.

In Fig. 4.6b, cuts of the LO phonon densities $\langle n^{\mathrm{P}}_{xy} \rangle$ in direction of the electric field versus the z-component of the wave vector $q\chi$ at several times after the beginning of the electric field pulse with $E = 10$ kVcm$^{-1}$ are displayed. The physical background, how the amplified reabsorption of the hot phonons affects the EDF, is found in [Vaissiere *et al.* (1996)].

### 4.2.4   *Transport Parameters*

To investigate the effect of non-equilibrium phonons on transport parameters in the transient regime, we compare the drift velocities and energies of electrons in the $\Gamma$ and $L$ valleys as functions of time obtained with phonons at thermal equilibrium to those obtained by taking into account phonon disturbance.

Figure 4.7 reports the results for the drift velocity in the $\Gamma$ and $L$ valleys for electric fields of 5 and 20 kVcm$^{-1}$. For times $t \leq 0.4$ ps, there is practically no difference between the drift velocities obtained with and without hot phonons. Thereafter, a modification of the temporal evolution of these quantities is observed, which finally leads to a significant difference between the values for the final stationary drift velocities. In Fig. 4.8, we display the

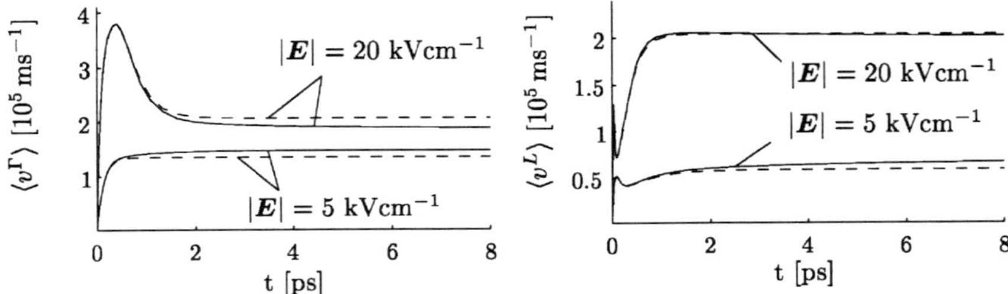

Fig. 4.7    Average drift velocity $\langle v^{c,\Gamma} \rangle$ and $\langle v^{c,L} \rangle$ in the $\Gamma$ and $L$ valleys in direction of the electric field versus the time $t$ and the reported electric field. The solid lines refer to calculations taking into account non-equilibrium phonons; the dashed lines refer to those assuming phonons to be in thermal equilibrium. (GALLER M. and SCHÜRRER F., *J. Phys. A: Math. Gen.*, **37** (2004), 1479–1497; reproduced with permission from IOP Publishing, `www.iop.org/journals/jphysa`.)

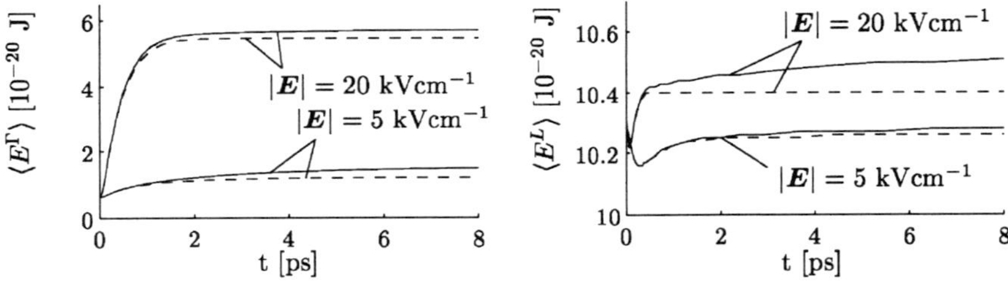

Fig. 4.8    Average electron energy $\langle E^{c,\Gamma} \rangle$ and $\langle E^{c,L} \rangle$ in the $\Gamma$ and $L$ valleys versus the time $t$ and the reported electric field. The solid lines refer to calculations taking into account non-equilibrium phonons; the dashed lines refer to those assuming phonons to be in thermal equilibrium. (GALLER M. and SCHÜRRER F., *J. Phys. A: Math. Gen.*, **37** (2004), 1479–1497; reproduced with permission from IOP Publishing, `www.iop.org/journals/jphysa`.)

average electron energy $\langle E^{c,\Gamma} \rangle$ and $\langle E^{c,L} \rangle$ in the $\Gamma$ and $L$ valleys versus time $t$ for electric field strengths of $E = 5$ kVcm$^{-1}$ and $E = 20$ kVcm$^{-1}$. Here again, we find that hot phonons are responsible for a notable modification of the values of the final electron energy. Consequently, we state that in the cases of high donor densities, hot phonon effects cannot be neglected in an accurate description of the transport parameters in polar semiconductors.

Table 4.3   Material Parameters for InP: Three-valley Model.

General Characteristics:

| Quantity | Symbol | Unit | Value |
|---|---|---|---|
| Mass density | $\rho$ | $\mathrm{kgm^{-3}}$ | 4790 |
| Sound velocity | $v_{\mathrm{s}}$ | $\mathrm{ms^{-1}}$ | 5130 |
| HF dielectric constant | $\kappa_\infty$ | | 9.56 |
| Static dielectric constant | $\kappa_0$ | | 12.40 |
| Phonon relaxation time | $\tau_{\mathrm{L}}$ | ps | 5.8 |

Band parameters:

| Quantity | Symbol | Unit | $\Gamma$ | $L$ | $X$ |
|---|---|---|---|---|---|
| Relative effective mass | $m^*/m_0$ | | 0.078 | 0.260 | 0.325 |
| Non-parabolicity factor | $\alpha^*$ | $\mathrm{(eV)^{-1}}$ | 0.83 | 0.23 | 0.38 |
| Gap to $\Gamma$ minimum | $\Delta_{\nu\mu}$ | eV | | 0.86 | 0.96 |
| Number of valleys | $Z_\mu$ | | 1 | 4 | 3 |

Intravalley scattering parameters:

| Quantity | Symbol | Unit | $\Gamma$ | $L$ | $X$ |
|---|---|---|---|---|---|
| Acoustic def. potential | $D_{\mathrm{A}}$ | $\mathrm{eVm^{-1}}$ | 6.5 | 6.5 | 6.5 |
| Piezoelectric constant | $e_{\mathrm{PZ}}$ | $\mathrm{Cm^{-2}}$ | 0.035 | 0.035 | 0.035 |
| TO phonons: | | | | | |
|    Deformation potential | $D_{\mathrm{TO}}$ | $10^{10}\,\mathrm{eVm^{-1}}$ | | 6.7 | |
|    Energy | $\hbar\omega_{\mathrm{TO}}$ | meV | | 43.2 | |
| LO phonon energy | $\hbar\omega_{\mathrm{LO}}$ | meV | 42.20 | 42.20 | 42.20 |

Intervalley scattering parameters:

| | | | $\Gamma$ | $L$ | $X$ |
|---|---|---|---|---|---|
| Phonon energy | | $\Gamma$ | | 27.8 | 29.9 |
| $\hbar\omega_{\nu\mu}$ [meV] | | $L$ | 27.8 | 29.0 | 29.3 |
| | | $X$ | 29.9 | 29.3 | 29.9 |
| Deformation potential | | $\Gamma$ | | 10 | 10 |
| $D_{\nu\mu}$ $[10^{10}\,\mathrm{eVm^{-1}}]$ | | $L$ | 10 | 10 | 9 |
| | | $X$ | 10 | 9 | 9 |

## 4.3   Three-valley Model

Now, we study the transport properties of n-type InP at the temperature $T_{\mathrm{L}} = 300$ K with the help of a three valley model. Here again, the holes are completely neglected in the considerations. As in the previous section, the conduction band is approximated by the central $\Gamma$ valley and four equivalent $L$ valleys, but we also regard three equivalent $X$ valleys along $\langle 1,0,0\rangle$ [Cohen and Bergstresser (1966)]. These valleys are assumed to be spherical and non-parabolic (cf. (3.1)). In our calculations, we refer to the same collision mechanisms as before. The numerical values for the material parameters used in our calculations are found in Tab. 4.3. They correspond to the values applied in the MC calculations in [González Sánchez *et al.* (1991)].

Figure 4.9 displays the stationary state drift velocity in InP versus the

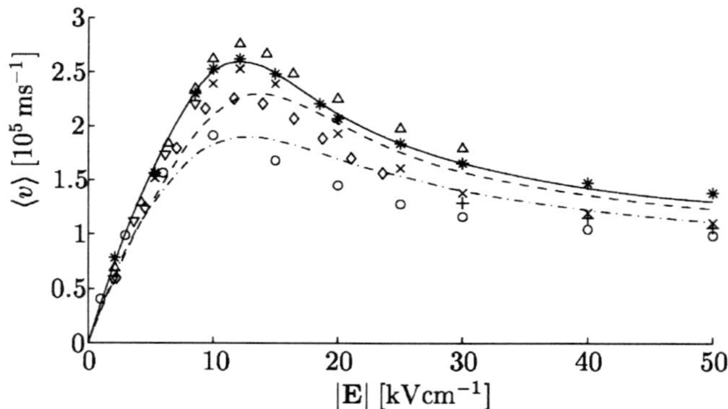

Fig. 4.9 The average drift velocity $\langle v \rangle$ in InP as a function of the electric field at the lattice temperature $T_\mathrm{L} = 300$ K for several electron densities $n_\mathrm{el}$. solid line: present results with $n_\mathrm{el} = 5 \times 10^{14}\,\mathrm{cm}^{-3}$, hot phonons; dashed line: present results with $n_\mathrm{el} = 10^{17}\,\mathrm{cm}^{-3}$, equilibrium phonons; dash-dot line: present results with $n_\mathrm{el} = 10^{17}\,\mathrm{cm}^{-3}$, hot phonons; *: MC calculations, González Sánchez *et al.*; $\triangle$: experiment, Glover, $n_\mathrm{el} = 10^{14}\,\mathrm{cm}^{-3}$; $\times$: experiment, Nielsen, $n_\mathrm{el} = 10^{15}\,\mathrm{cm}^{-3}$; +: experiment, Windhorn *et al.*, $n_\mathrm{el} = 10^{15}\,\mathrm{cm}^{-3}$; $\triangledown$: experiment, Kobayashi *et al.*, $n_\mathrm{el} = 6.7 \times 10^{15}\,\mathrm{cm}^{-3}$; $\Diamond$: MC calculations, Maloney *et al.*, $n_\mathrm{el} = 10^{17}\,\mathrm{cm}^{-3}$, equilibrium phonons; o: matrix method, Vaissiere *et al.*, $n_\mathrm{el} = 10^{17}\,\mathrm{cm}^{-3}$, hot phonons. (GALLER M. and SCHÜRRER F., A multigroup approach to the coupled electron-phonon Boltzmann equations in InP, *Trans. Theo. Stat. Phys.*, **33** (2004), 485–501; reproduced with permission from Taylor & Francis.)

external electric field for several total electron densities at the lattice temperature $T_\mathrm{L} = 300$ K. These steady state values are approximated by the drift velocities at the time $t = 20$ ps after the onset of the electric field. For the low total electron density $n_\mathrm{el} = 5 \times 10^{14}\,\mathrm{cm}^{-3}$, our results agree very well with several theoretical investigations [González Sánchez *et al.* (1991); Maloney and Frey (1977); Vaissiere *et al.* (1996)] and experimental measurements [Glover (1972); Nielsen (1972); Windhorn *et al.* (1982); Kobayashi *et al.* (1978)]. Especially the results of the MC calculations of Gonzáles Sánchez, who used the same material parameter as we do, coincide with ours. The slight difference between the curves for high field strengths might be caused by the fact that two further valleys of the conduction band are included in the MC model, which are neglected in our calculations.

As for the higher electron density $n_\mathrm{el} = 10^{17}\,\mathrm{cm}^{-3}$, we plot results, which are calculated by taking into account hot phonons and by assuming equilibrium LO phonons. We observe that the influence of hot phonons on

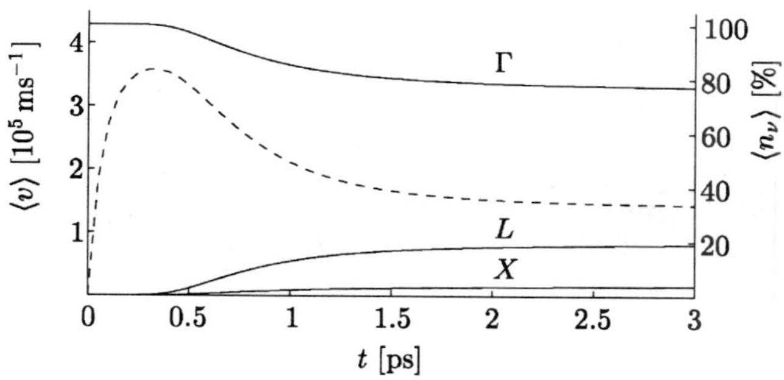

Fig. 4.10   The average drift velocity $\langle v \rangle$ (dashed line) and the populations $\langle n_\nu \rangle$, $\nu = \Gamma, L, X$ of the considered valleys (solid lines) versus the time $t$ after the onset of the electric field with $\mathbf{E} = 25\,\mathrm{kVcm}^{-1}$. Calculations are performed for InP at the lattice temperature $T_\mathrm{L} = 300$ K with the total electron density $n_\mathrm{el} = 10^{17}\,\mathrm{cm}^{-3}$. (GALLER M. and SCHÜRRER F., A multigroup approach to the coupled electron-phonon Boltzmann equations in InP, *Trans. Theo. Stat. Phys.*, **33** (2004), 485–501; reproduced with permission from Taylor & Francis.)

the average drift velocity cannot be neglected for this high electron density. The comparison of our results with some theoretical data shows a similar functional dependency of the drift velocity on the electric field strength with the maximum of the drift velocity shifted towards higher fields. This difference is mainly determined by the fact that Maloney *et al.* [Maloney and Frey (1977)] and Vaissiere *et al.* [Vaissiere *et al.* (1996)] used another set of material parameters in their calculations. The energy separation of $\Delta_{\Gamma L} = 0.86$ eV (calculated in [Zollner *et al.* (1990); Massidda *et al.* (1990)]) between the $\Gamma$ and the $L$ valleys used in our model leads to a higher value of the threshold field than it was obtained with the previously calculated $\Delta_{\Gamma L} = 0.61$ eV applied in the mentioned publications. We adopted the new value, thus achieving a better fit with the experimental results for low fields. Nevertheless, our results comply very well with those of Vaissiere, when the same material data are used as illustrated above.

In Fig. 4.10, the average drift velocity and the populations of the $\Gamma$, $L$ and $X$ valleys versus the time after the onset of the electric field pulse with $\mathbf{E} = 25\,\mathrm{kVcm}^{-1}$ are displayed. Typically, we observe a high velocity overshot just after the onset of the electric field.

For a better understanding of the reasons for this behavior, we consider Fig. 4.11. Here, the temporal evolution of electron densities $\langle n_{ij}^{\mathrm{c},\Gamma} \rangle$ in the cells $C_{ij}^{\Gamma}$ in the $\Gamma$ valley is shown in a polar diagram representation.

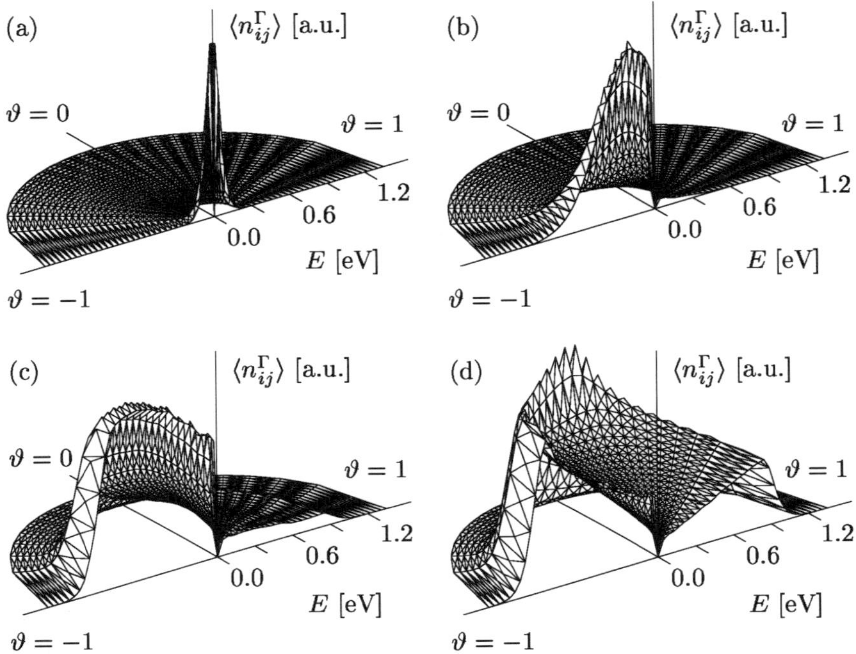

Fig. 4.11   The electron densities $\langle n_{ij}^{c,\Gamma} \rangle$ in the cells $C_{ij}^{\Gamma}$ in the $\Gamma$ valley at (a) $t = 0$ ps, (b) $t = 0.3$ ps, (c) $t = 0.6$ ps and (d) $t = 6$ ps after the onset of the electric field with $|\mathbf{E}| = 25\,\mathrm{kVcm}^{-1}$. Calculations are performed for InP at the lattice temperature $T_{\mathrm{L}} = 300$ K with the total electron density $n_{\mathrm{el}} = 10^{17}$ cm$^{-3}$. (GALLER M. and SCHÜRRER F., A multigroup approach to the coupled electron-phonon Boltzmann equations in InP, *Trans. Theo. Stat. Phys.*, **33** (2004), 485–501; reproduced with permission from Taylor & Francis.)

It should be noted that different scaling factors are used for the subplots 4.11 (a)-(d). The equilibrium distribution of the electrons at $t = 0$ ps is shifted in the electric field as suggested by the concept of ballistic transport, which results in a maximum of the drift velocity around $t = 0.3$ ps. With increasing efficiency of phonon scattering, a redistribution of the electron momenta takes place. This ends in the lower value of the drift velocity associated with the stationary state distribution, which is almost reached at $t = 3$ ps.

In addition, we find in Fig. 4.10 that the energetically high $L$ and $X$ valleys are notably populated in the steady state for $\mathbf{E} = 25\,\mathrm{kVcm}^{-1}$. Due to the high effective masses in and the high equivalent intervalley scattering

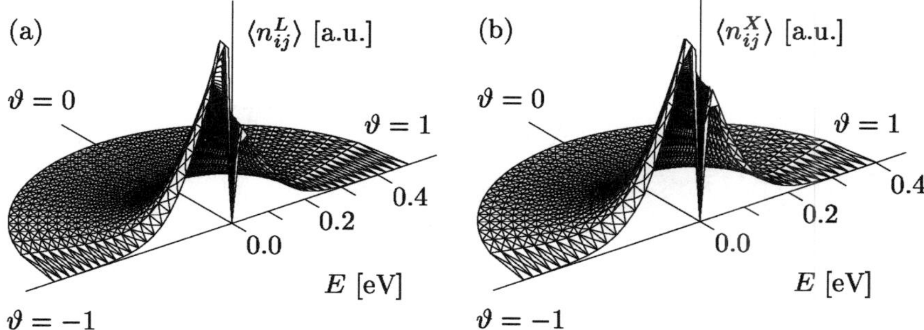



Fig. 4.12　The electron densities $\langle n_{ij}^{L} \rangle$ in the cells $C_{ij}^{L}$ in the $L$ valley (a) and the electron densities $\langle n_{ij}^{c,X} \rangle$ in the cells $C_{ij}^{X}$ in the $X$ valley (b) evaluated at $t = 6$ ps after the onset of the electric field with $|\mathbf{E}| = 25\,\mathrm{kVcm}^{-1}$. Calculations are performed for InP at the temperature $T_{\mathrm{L}} = 300$ K with the electron density $n_{\mathrm{el}} = 10^{17}\,\mathrm{cm}^{-3}$. (GALLER M. and SCHÜRRER F., A multigroup approach to the coupled electron-phonon Boltzmann equations in InP, *Trans. Theo. Stat. Phys.*, **33** (2004), 485–501; reproduced with permission from Taylor & Francis.)

rates among these valleys, the electron distributions in these valleys are hardly disturbed from their equilibrium distributions, as depicted in Fig. 4.12. The high amount of electrons in these valleys with their low mobility is the reason for the negative differential resistivity, in other words, the reason for the decreasing drift velocities with increasing electric field strengths for high fields (cf. Fig. 4.9) observed in polar semiconductors.

Finally, Fig. 4.13 illustrates the LO distribution function $g(\mathbf{q})$ versus the cells $D_{xy}$, which is approximated by dividing the densities of LO phonons in the cells by the volume of $D_{xy}$. This figure clarifies that there are two effects on how non-equilibrium LO phonons affect the electron distribution. At $t = 0.5$ ps after the onset of the electric field, we observe a highly anisotropic rise of the phonon distribution function. Such a peak appears when the electron distribution is slightly shifted in an electric field. This takes place for low electric fields or just after the onset of a high electric field as in our case and results in an increase of the electron drift velocity by the amplified reabsorption of these phonons. When the higher valleys are well populated, the electron densities in the $\Gamma$ valley feature a sharp drop at the energy $\Delta_{\Gamma L}$ as it is found in Fig. 4.11d. This edge is reflected in the LO distribution function by a more symmetrical increase of this function (cf. Fig. 4.13b) as discussed in [Vaissiere *et al.* (1992); Vaissiere *et al.* (1996)]. These hot phonons cause a lower electron drift velocity because

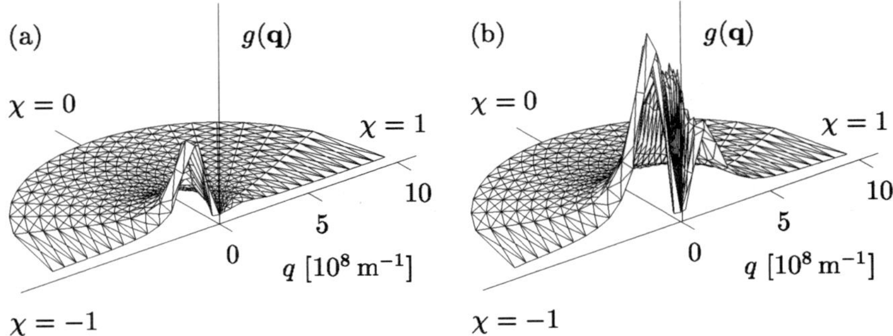



Fig. 4.13    The LO distribution function $g(\mathbf{q})$ versus the cells $D_{xy}$ evaluated at (a) $t = 0.5$ ps and (b) $t = 6$ ps after the onset of the electric field pulse with $|\mathbf{E}| = 25\,\mathrm{kVcm}^{-1}$. Calculations are performed for InP at the lattice temperature $T_\mathrm{L} = 300$ K with the total electron density $n_\mathrm{el} = 10^{17}\,\mathrm{cm}^{-3}$. (GALLER M. and SCHÜRRER F., A multigroup approach to the coupled electron-phonon Boltzmann equations in InP, *Trans. Theo. Stat. Phys.*, **33** (2004), 485–501; reproduced with permission from Taylor & Francis.)

of the higher phonon scattering rate in comparison to the interaction with equilibrium phonons. The interplay of these hot phonon effects results in the notable difference between the results for the drift velocity calculated with and without hot phonons as displayed in Fig. 4.9 for the high electron density. While the usual assumption of equilibrium phonons is certainly justified for low electron densities, a realistic description of the transport in III-V compound semiconductors must take into account non-equilibrium phonons for high electron densities.

Chapter 5

# Particle Transport in Gallium Arsenide

## 5.1 Introduction

This chapter deals with the investigation of the particle transport in gallium arsenide with the help of the multigroup model equations (3.40). The calculations are performed for n-type GaAs at the temperature $T_{\mathrm{L}} = 300\,\mathrm{K}$. The influence of holes in the valence band is completely neglected. The conduction band is approximated by the $\Gamma$ valley centered at $\langle 0, 0, 0 \rangle$, four equivalent $L$ valleys along $\langle 1, 1, 1 \rangle$ and three equivalent $X$ valleys along $\langle 1, 0, 0 \rangle$. All of these valleys are assumed to be spherical and non-parabolic [Conwell and Vassel (1968)].

Concerning the scattering mechanisms, we consider acoustic deformation potential and acoustic piezoelectric scattering. Furthermore, we take into account polar optical, optical deformation potential scattering in the $L$ valleys and impurity scattering, which are all intravalley processes, as well as non-polar optical intervalley scattering. Since the acoustic deformation potential and the piezoelectric scattering are not efficient at $300\,\mathrm{K}$ [Conwell and Vassel (1968); Fawcett $et\ al.$ (1979)], these mechanisms are regarded as being elastic. The ionized impurity scattering is described with the help of the Brooks-Herring model [Brooks and Herring (1951)] and by assuming equal electron and donor concentrations. As for the LO phonons, we include the polar optical interaction and phonon-phonon thermalization processes in their kinetic equations. The material parameters used in our calculations are found in Tab. 4.3 and are taken from [Klemens (1966); Constant (1985); von der Linde $et\ al.$ (1980)]. The maximal electron energy considered in these calculations is set to $E_{\mathrm{max}} = 0.9\,\mathrm{eV}$; the phonon momentums is restricted up to $q_{\mathrm{max}} = 1.5 \times 10^9\ \mathrm{m}^{-1}$.

Table 5.1    Material Parameters of GaAs

| Quantity | Symbol | Unit | Value | | |
|---|---|---|---|---|---|
| Mass density | $\rho$ | kgm$^{-3}$ | 5360 | | |
| Longitudinal sound velocity | $v_s$ | ms$^{-1}$ | 5240 | | |
| Static relative dielectric constant | $\kappa_0$ | | 12.90 | | |
| HF relative dielectric constant | $\kappa_\infty$ | | 10.92 | | |
| LO phonon energy | $\hbar\omega_{LO}$ | meV | 35.36 | | |
| LO phonon relaxation time | $\tau_L$ | ps | 5.6 | | |
| Piezoelectric constant | $e_{PZ}$ | Cm$^{-2}$ | 0.16 | | |
| Quantity | Symbol | Unit | $\Gamma$ | $L$ | $X$ |
| Number of equivalent valleys | $Z_\nu$ | | 1 | 4 | 3 |
| Effective mass ratio | $m_\nu^*/m_0$ | | 0.063 | 0.222 | 0.58 |
| Non-parabolicity factor | $\alpha_\nu$ | (eV)$^{-1}$ | 0.610 | 0.461 | 0.204 |
| Gap referred to $\Gamma$ minimum | $\Delta_{\Gamma\mu}$ | eV | 0 | 0.330 | 0.522 |
| Acoustic deformation potential | $D_A$ | eV | 7.0 | 9.2 | 9.27 |
| Optical deformation potential | $D_{TO}$ | eVm$^{-1}$ | 0 | $3\times10^{10}$ | 0 |
| Optical-phonon energy | $\hbar\omega_{TO}$ | meV | 0 | 34.3 | 0 |

| Quantity | Symbol | Unit | | $\Gamma$ | $L$ | $X$ |
|---|---|---|---|---|---|---|
| Intervalley deformation | | | $\Gamma$ | 0 | $10^{11}$ | $10^{11}$ |
| potential $\nu\rightarrow\mu$ | $D_{\nu\mu}$ | eVm$^{-1}$ | $L$ | $10^{11}$ | $10^{11}$ | $5\times10^{10}$ |
| | | | $X$ | $10^{11}$ | $5\times10^{10}$ | $7\times10^{10}$ |
| Intervalley phonon | | | $\Gamma$ | 0 | 27.8 | 29.9 |
| energy $\nu\rightarrow\mu$ | $\hbar\omega_{\nu\mu}$ | meV | $L$ | 27.8 | 29.0 | 29.3 |
| | | | $X$ | 29.9 | 29.3 | 29.9 |

## 5.2    Transport in a Time-dependent Electric Field

We study the transport properties of GaAs in response to a time-dependent electric field and report the results concerning overshoot phenomena as well as additional effects usually referred to as undershoot phenomena and the Rees effect [Fawcett and Rees (1969)]. In this simulation, an external electric field strength of $|\mathbf{E}| = 2$ kVcm$^{-1}$ is enlarged to $|\mathbf{E}| = 20$ kVcm$^{-1}$ between 2 and 4 ps as illustrated in the subplot of Fig. 5.1. This figure shows the temporal evolution of the drift velocity and the average electron energy in GaAs at the lattice temperature $T_L = 300$ K and with a doping concentration of $n_{el} = 2\times10^{14}$ cm$^{-3}$ in response to the applied electric field. The solid lines refer to the results obtained by means of our multigroup model, which agree very well with those of Monte Carlo calculations from [Constant (1985)] indicated by dashed lines.

The observed results can be understood as follows. Up to $t = 2$ ps, most electrons remain in the $\Gamma$ valley and the mobility is high due to the low effective mass in this valley. After the onset of the high electric field pulse, a typical velocity overshoot is observed [Ruch (1972)]. The electrons

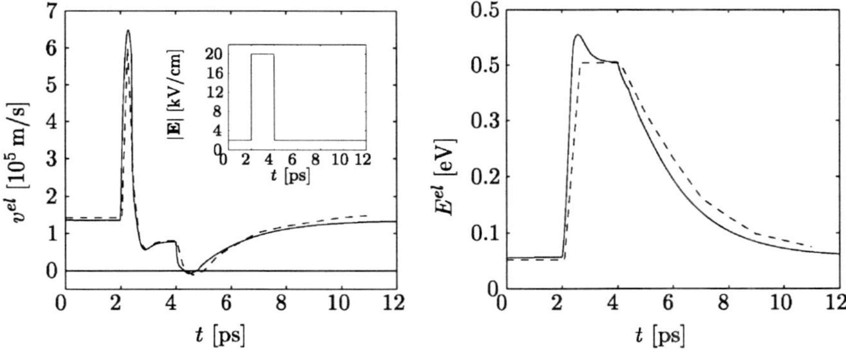

Fig. 5.1   The drift velocity $v_{\mathrm{el}}$ and the average energy $E_{\mathrm{el}}$ of electrons versus time $t$ in GaAs in response to a time pulse configuration of the electric field $|\mathbf{E}|$ as illustrated in the subplot. The lattice temperature $T_{\mathrm{L}}$ equals 300 K; the electron density is set to $n_{\mathrm{el}} = 2 \times 10^{14}\,\mathrm{cm}^{-3}$. The solid lines refer to the results obtained with the multigroup model; dashed lines refer to Monte Carlo calculations. (GALLER M. and SCHÜRRER F., *Comp. Methods Appl. Mech. Engrg.*, **194** (2005), 2806–2818; reproduced with permission from Elsevier.)

are freely accelerated in the electric field (ballistic transport [Shur (1976)]) until the increasing efficiency of scattering mechanisms reduces the drift velocity to a new stationary state value just before $t = 4\,\mathrm{ps}$. A great amount of the electrons is in the $L$ and $X$ valleys. Thus, the mobility is very low because of the high effective masses in these valleys and the high scattering rates due to intervalley transitions. This results in an average drift velocity, which is even lower than it is at $t = 2\,\mathrm{ps}$ in spite of the high electric field strength. After $t = 4\,\mathrm{ps}$, when the electric field is lowered to its original value of $|\mathbf{E}| = 2\ \mathrm{kVcm}^{-1}$, a drastic decrease in the drift velocity is observed, which is known as velocity undershoot phenomenon. It occurs when the instantaneous average energy exceeds its steady-state value corresponding to the instantaneous value of the electric field, which leads to a mobility lower than the stationary state value. However, at increasing times the electrons lose energy by phonon scattering and return to the $\Gamma$ valley. The mobility and the drift velocity increase and reach steady-state values identical to those of the initial state before the onset of the high electric field pulse.

Some explanations are required for the dips in the velocity transient around $t = 3\,\mathrm{ps}$ and $t = 5\,\mathrm{ps}$, where the undershoot phenomenon is amplified in a way that the drift velocity actually becomes negative. These low

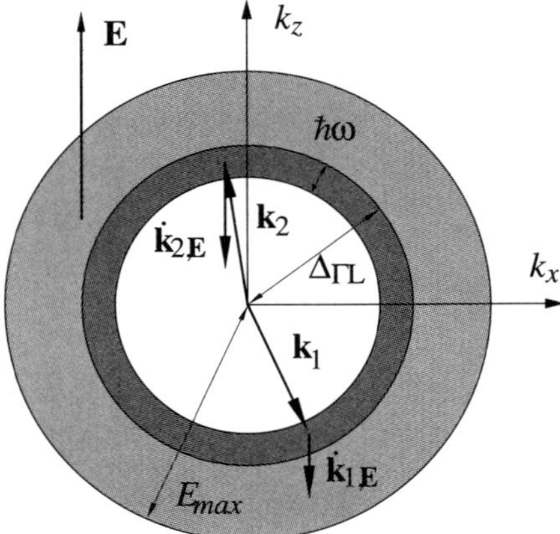


Fig. 5.2   Schematic illustration of the **k**-space of electrons in the $\Gamma$ valley for explaining the Rees effect. The maximum electron energy is labeled by $E_{\mathrm{max}}$, the symbols $\Delta_{\Gamma L}$ and $\hbar\omega_{\mathrm{IV}}$ denote the energy gap between $\Gamma$ and $L$ valleys and the energy of phonons causing intervalley scattering, respectively. The typical electron states $\mathbf{k}_m$, $m = 1, 2$, explaining the Rees effect, and their changes in the electric field $\dot{\mathbf{k}}_{m,\mathbf{E}}$ are displayed. (GALLER M. and SCHÜRRER F., *Comp. Methods Appl. Mech. Engrg.*, **194** (2005), 2806–2818; reproduced with permission from Elsevier.)

values of the drift velocity, the so-called Rees effect, can be explained by the following mechanism, as first suggested by Fawcett and Rees [Fawcett and Rees (1969)]. Figure 5.2 gives a schematic representation of the **k**-space of electrons in the $\Gamma$ valley. It should be noted that electrons with wave vectors **k** in the filled area $A$, in other words, electrons with energies above the energy gap $\Delta_{\Gamma L}$ between the $\Gamma$ and $L$ valleys, are likely to be scattered into the $L$ valleys. When non-stationary intervalley scattering occurs, the fate of the two typical electrons labeled by 1 and 2, just now transferred into the $\Gamma$ valley, is quite different. The electron 1 with a wave vector component parallel to the electric field leaves the area of possible intervalley scattering because of the acceleration in the electric field $\dot{\mathbf{k}}_{1,\mathbf{E}}$ and remains in the $\Gamma$ valley. The electron 2 with a **k** component anti-parallel to **E** remains in the area $A$ and will be rapidly re-transferred to the original valley. Consequently, when intervalley transfer occurs, the life time of type 1 electrons in the central valley, which cause negative contributions to the drift velocity, is much longer than the life time of type 2 electrons. Hence, the drift velocity is reduced in comparison to the final steady-state value.

Figure 5.3 reveals the influence of the electron density $n_{\mathrm{el}}$ on the tem-

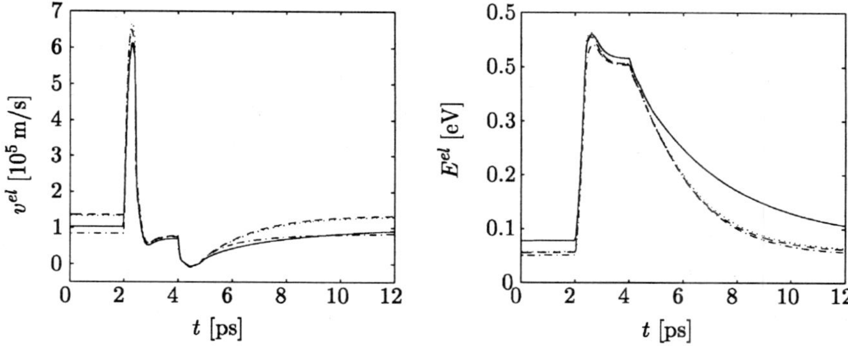

Fig. 5.3   The drift velocity $v_{el}$ and the average energy $E_{el}$ of electrons versus time $t$ in GaAs in response to an electric field pulse as illustrated in Fig 5.1. The calculations are performed for: $n_{el} = 2 \times 10^{14}\,\mathrm{cm}^{-3}$, equilibrium-phonons $(\cdots)$; $n_{el} = 2 \times 10^{14}\,\mathrm{cm}^{-3}$, hot-phonons $(---)$; $n_{el} = 2 \times 10^{17}\,\mathrm{cm}^{-3}$, equilibrium-phonons $(-\cdot-)$; $n_{el} = 2 \times 10^{17}\,\mathrm{cm}^{-3}$, hot-phonons $(\text{———})$. (GALLER M. and SCHÜRRER F., *Comp. Methods Appl. Mech. Engrg.*, **194** (2005), 2806–2818; reproduced with permission from Elsevier.)

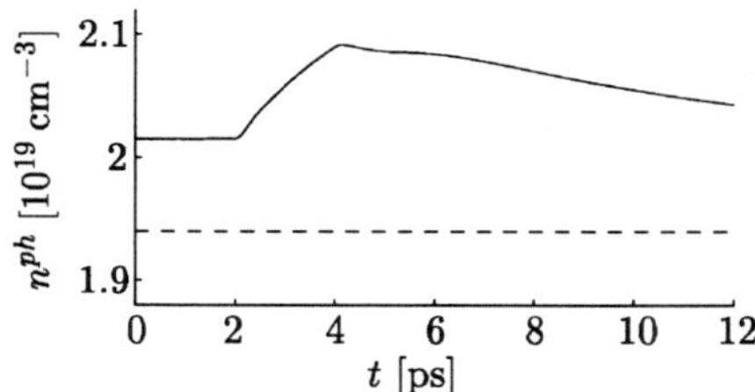

Fig. 5.4   Temporal evolution of the phonon density $n_{ph}$ of LO phonon with moduli of the wave vector up to $q_{max} = 1.5 \times 10^9\,\mathrm{m}^{-1}$ in GaAs at the temperature $T_L = 300\,\mathrm{K}$ in response to the electric field as illustrated in Fig. 5.1. The electron density $n_{el}$ equals $2 \times 10^{17}\,\mathrm{cm}^{-3}$. The dashed line refers to the equilibrium value of $n_{ph}$. (GALLER M. and SCHÜRRER F., *Comp. Methods Appl. Mech. Engrg.*, **194** (2005), 2806–2818; reproduced with permission from Elsevier.)

poral evolution of the drift velocity and the average energy in GaAs at the temperature $T_L = 300\,\mathrm{K}$ in response to the high field pulse of Fig. 5.1. Moreover, the results for the coupled electron-phonon system are compared with those assuming equilibrium-state LO phonons. The results obtained by taking into account non-equilibrium phonons coincide very well with those assuming an equilibrium distribution of LO phonons when the electron density is low ($n_{el} = 2 \times 10^{14}\,\mathrm{cm}^{-3}$). Obviously, the low number of electrons and, consequently, the low number of electron-phonon interaction processes does not cause a notable deviation of the LO phonon distribution from equilibrium. In contrast, the influence of hot phonons cannot be neglected at high electron densities. Here, the low drift velocity at low electric field strength is related to the increased influence of the impurity scattering. Moreover, we observe that the steady-state drift velocity for $|\mathbf{E}| = 2$ kVcm$^{-1}$ is higher when non-equilibrium phonons are taken into account. This result seems to contradict Fig. 5.4, which displays the time-dependence of the total LO phonon density $n_{ph}$ with moduli of the wave vectors up to $q_{max}$ for this physical situation. The increased number of phonons in comparison to equilibrium should end in a decreased drift velocity. However, as discussed in the previous section, most of these additional LO phonons occupy states in the direction of the drift velocity and their amplified reabsorption raises the electron drift velocity.

Within the time interval between 2 and 4 ps, where the electric field is high, all the displayed results are rather similar. The transient regime is mainly determined by intervalley processes, and the electrons are only slightly influenced by the POP interaction. However, many LO phonons are created, which strongly affect the relaxation towards the steady-state values of the drift velocity and the average energy after $t = 4$ ps. Since the ratio of the electron scattering rate for absorbing and emitting phonons is mainly determined by $g(\mathbf{q})/[g(\mathbf{q}) + 1]$, the enlarged phonon density shifts this ratio towards absorption processes. Hence, the average electron energy decreases slower when hot phonons are taken into account (cf. Fig. 5.3).

In summary, we state that our multigroup model describes overshoot and undershoot phenomena as well as the Rees effect in good agreement with the related MC calculations. Moreover, it copes with hot-phonon effects, which cannot be neglected for sufficiently high electron densities.

## 5.3    The Stationary-state Electron Distribution

As a further test case, we use our multigroup model for investigating the steady-state electron distributions of GaAs at the lattice temperature $T_L = 300$ K. The electron density $n_{el}$ is assumed to be $2 \times 10^{14}\,\mathrm{cm}^{-3}$, and the electric field $|\mathbf{E}|$ is fixed to 8 kVcm$^{-1}$. In Tab. 5.2, the stationary state values for the most important macroscopic quantities are summarized.

Figure 5.5 depicts a polar diagram representation of the steady-state

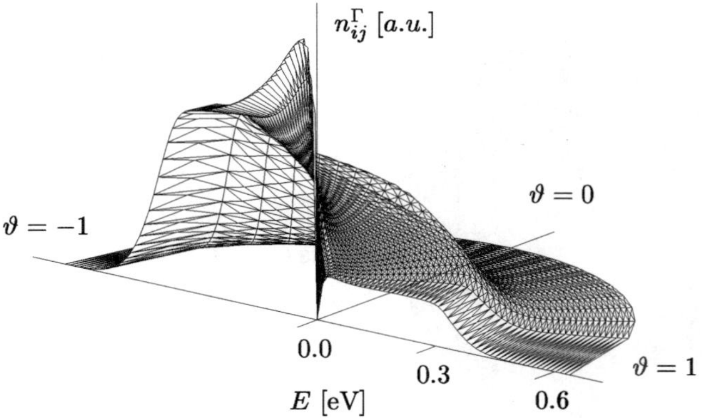

Fig. 5.5   The stationary state electron densities $\langle n_{ij}^{c,\Gamma} \rangle$ in the cells $C_{ij}^{\Gamma}$ in the $\Gamma$ valley at the lattice temperature $T_{\mathrm{L}} = 300\,\mathrm{K}$ for the total electron density $n_{\mathrm{el}} = 2 \times 10^{14}\,\mathrm{cm}^{-3}$ and the electric field $|\mathbf{E}| = 8\,\mathrm{kVcm}^{-1}$. (GALLER M. and SCHÜRRER F., *Comp. Methods Appl. Mech. Engrg.*, **194** (2005), 2806–2818; reproduced with permission from Elsevier.)

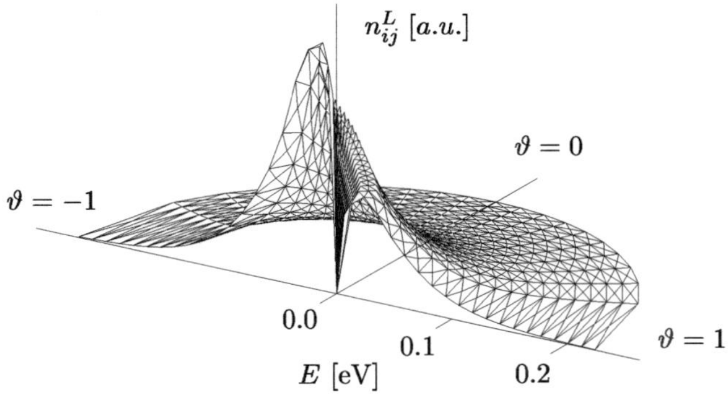

Fig. 5.6   The stationary state electron densities $\langle n_{ij}^{c,L} \rangle$ in the cells $C_{ij}^{L}$ in the $L$ valley at the lattice temperature $T_{\mathrm{L}} = 300\,\mathrm{K}$ for the total electron density $n_{\mathrm{el}} = 2 \times 10^{14}\,\mathrm{cm}^{-3}$ and the electric field $|\mathbf{E}| = 8\,\mathrm{kVcm}^{-1}$. (GALLER M. and SCHÜRRER F., *Comp. Methods Appl. Mech. Engrg.*, **194** (2005), 2806–2818; reproduced with permission from Elsevier.)

electron densities $\langle n_{ij}^{c,\Gamma} \rangle$ versus the cells $C_{ij}^{\Gamma}$. In the $\Gamma$ valley, in Fig. 5.6, the $L$-electron densities $\langle n_{ij}^{c,L} \rangle$ are displayed versus the cells $C_{ij}^{L}$. We observe that the $L$-electron distribution is only slightly disturbed from the equilibrium distribution, which reflects the high effective mass $m_L^*$ and the high total scattering rate due to equivalent intervalley scattering processes

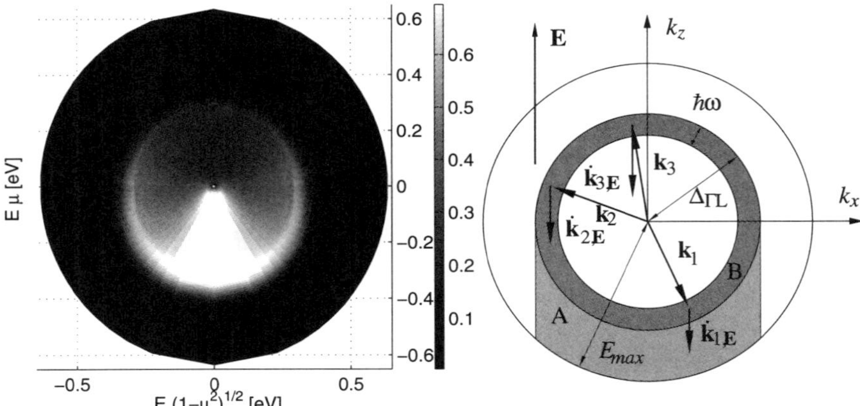

Fig. 5.7    The stationary state electron densities $\langle n_{ij}^{c,\Gamma} \rangle$ in the cells $C_{ij}^{\Gamma}$ in the $\Gamma$ valley at the lattice temperature $T_L = 300\,\mathrm{K}$ for the total electron density $n_{\mathrm{el}} = 2 \times 10^{14}\,\mathrm{cm}^{-3}$ and the electric field $|\mathbf{E}| = 8\,\mathrm{kVcm}^{-1}$ (a) and a schematic illustration of the $\mathbf{k}$-space of the $\Gamma$ valley (b). The maximum electron energy is labeled by $E_{\max}$, $\Delta_{\Gamma L}$ and $\hbar\omega_{IV}$ are the energy gap between $\Gamma$ and $L$ valley and the energy of phonons causing intervalley scattering, respectively. The typical electron states $\mathbf{k}_m$ and their changes in the electric field $\dot{\mathbf{k}}_{m,\mathbf{E}}$, $m = 1, 2, 3$ are displayed.

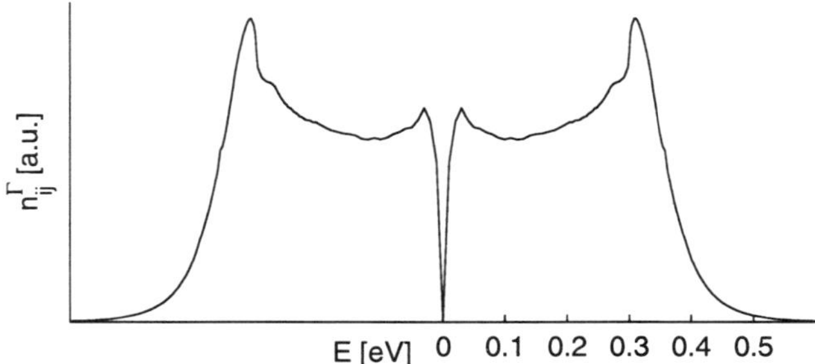

Fig. 5.8    Cut of the stationary state $\Gamma$ electron densities $\langle n_{ij}^{c,\Gamma} \rangle$ in the cells $C_{ij}^{\Gamma}$ along $\mu = 0$ normal to the electric field at the lattice temperature $T_L = 300\,\mathrm{K}$ for the total electron density $n_{\mathrm{el}} = 2 \times 10^{14}\,\mathrm{cm}^{-3}$ and the electric field $|\mathbf{E}| = 8\,\mathrm{kVcm}^{-1}$. (GALLER M. and SCHÜRRER F., *Comp. Methods Appl. Mech. Engrg.*, **194** (2005), 2806–2818; reproduced with permission from Elsevier.)

among these valleys. In contrast, the deviation of the distribution of the $\Gamma$-electrons from their equilibrium states is significant. We find a sharp drop

Table 5.2 The steady-state values of macroscopic quantities in GaAs at the temperature $T_{\mathrm{L}} = 300\,\mathrm{K}$ with the electron density $n_{\mathrm{el}} = 2 \times 10^{14}\,\mathrm{cm}^{-3}$ and the electric field $|\mathbf{E}| = 8\,\mathrm{kVcm}^{-1}$.

| Quantity | $\Gamma$ | $L$ | $X$ | total |
|---|---|---|---|---|
| Valley population [%] | 57.75 | 41.85 | 0.40 | |
| Drift velocity [$10^4$ ms$^{-1}$] | 23.62 | 3.43 | 0.24 | 15.08 |
| Average energy [meV] | 197 | 380 | 563 | 275 |

of the electron densities $\langle n_{ij}^{c,\Gamma} \rangle$ at the energy $E = 0.3\,\mathrm{eV}$, which corresponds to the bottom of the $L$ valleys. Moreover, we find a strong asymmetry of the $\Gamma$-electron distribution, which finds its explanation in the low effective mass $m_{\Gamma}^*$ in this valley.

Special attention should be called to the increased electron densities at energies just below $\Delta_{\Gamma L}$. This wall in the $\Gamma$ electron distribution is almost insignificant for $\mathbf{k}$ vectors parallel to the electric field, while it has its maximum height for wave vectors normal to $\mathbf{E}$ for $\mu = 0$. Figure 5.7a shows the electron densities $\langle n_{ij}^{c,\Gamma} \rangle$ versus the corresponding cells from another point of view. Here, brighter colors refer to higher electron densities. The mentioned wall of increased electron densities is notably visible. For the explanation of this effect we consider Fig. 5.7b. It depicts a schematic representation of the $\mathbf{k}$-space of the $\Gamma$ valley up to the maximal energy $E_{\mathrm{max}}$. Moreover, the energy gap between the bottoms of the $\Gamma$ and $L$ valleys as well as the energy of phonons, which cause the intervalley scattering, are drawn in. In the case of steady-state transport, the electrons are accelerated in the electric field into the area $A$. From $A$, most of them are scattered into the $L$ valleys due to the high intervalley scattering rate. Thereafter, these electrons lose their high energies by several scattering mechanisms, which results in high electron densities $\langle n_{ij}^{c,L} \rangle$ in the $L$ valleys for low energies (cf. Fig. 5.6). Hence, a fraction of electrons, which are isotropically retransferred into the $\Gamma$ valley, is scattered into the energy region $B$, which has the width $\hbar\omega_{\mathrm{IV}}$ just below the energy $\Delta_{\Gamma L}$. Since the converse processes (the scattering events from $B$ into the $L$ valleys) are rare because of the low density of $L$-states for energies just above the band minimum, a wall of enlarged electron densities develops in $B$. Besides the flattening effect of the isotropic phonon scattering mechanisms, the electric field affects the form of this wall. While electrons with states almost parallel to the electric fields ($\mathbf{k}_1$ and $\mathbf{k}_3$ in Fig. 5.7b) are transferred out of $B$, which results in a diminished height of the wall in these directions, electrons with states normal to $\mathbf{E}$ (cf. $\mathbf{k}_2$ in Fig. 5.7b) stay within this region and the wall reaches its greatest height.

In Fig. 5.8 we display the cut of the steady-state $\Gamma$ electron densities along $\mu = 0$ (cf. Fig. 5.5), which is the direction normal to the applied electric field. Here, we observe another interesting detail in the electron distribution. Besides the peak in the electron densities around the energy $\Delta_{\Gamma L}$ as explained above, there are further peaks shifted from the maximum by $\hbar\omega_0$. Such peaked structures are typically observed in the relaxation of laser excited electrons [Lugli *et al.* (1989); Auer *et al.* (2004)], where the emission of optical phonons results in copies of an initial peak in the electron distribution functions, just shifted by $\hbar\omega_0$. The same explanation is true for the emergence of additional peaks in Fig. 5.8. However, we find this structure not only in the transients as for photo-excited electrons, but also in the stationary state, since the original peak at $\Delta_{\Gamma L}$ cannot be broken down completely by LO phonon emission.

Chapter 6

# Multigroup Equations for Degenerated Carrier Gases

## 6.1 Introduction

In this chapter, we introduce deterministic multigroup model equations to the BBP equations, which are based on a general carrier dispersion law and contain the full quantum statistics of both, the carriers and the phonons. We study the spatially homogeneous case with vanishing electric field, but we construct the multigroup equations in a way that they allow the description of particle distributions of arbitrary anisotropy with respect to a main direction imposed, for example, by an electric field. Hence, the transport model can easily be extend to describe field dependent, spatially inhomogeneous problems by including multigroup formulations of the advection terms in the BBP equations following the receipts given in chapter 3.

Since the reliability of a deterministic transport model increases with the amount of mathematical properties, which the model equations and the original BBP equations possess in common grounds, it is the main goal of this paper to investigate the most important features of our multigroup transport model. Therefore, we show the boundedness of the distribution coefficients, which reflects the Pauli principle as it is found in the BBP equations [Markowich *et al.* (1990)]. The conservational properties of the multigroup model are discussed and a Boltzmann H-theorem for the obtained evolution equations is proved, similar to that of the BBP equations [Rossani (2002)]. We consider the equilibrium distribution of our multigroup equations and find that it is given by a set of discretized Fermi-Dirac and Bose-Einstein distributions for non-drifting particles, which corresponds to the features of the original BBP equations.

87

## 6.2 The Bloch-Boltzmann-Peierls Equations

To begin with, we summarize the Bloch-Boltzmann-Peierls (BBP) equations governing the transport of carriers and phonons in semiconductors. Here, we restrict ourselves to considering only two carrier scattering mechanisms for keeping the proofs of the mathematical features of the multigroup equations concise. However, the formalism can easily be extended to a more complicated physical model by including more scattering mechanisms without the theorems mentioned in this chapter losing their validities.

In the semiclassical kinetic theory of transport in semiconductors, the evolution equations for the distribution functions $f = f(\mathbf{r}, \mathbf{k}, t)$ and $g = g(\mathbf{r}, \mathbf{q}, t)$ of carriers and phonons are the BBP equations, which read in the spatially homogeneous case for vanishing electric fields

$$\frac{\partial}{\partial t} f(\mathbf{k}, t) = \mathcal{C}^1[f, g] + \mathcal{C}^2[f], \quad \frac{\partial}{\partial t} g(\mathbf{q}, t) = \mathcal{D}[f, g]. \tag{6.1}$$

These equations include the collision terms of carriers $\mathcal{C}^1$ and optical phonons $\mathcal{D}$ for an inelastic carrier-optical phonon collision process, which is, for example, the polar optical interaction. Additionally, we consider an elastic scattering mechanism (e.g. impurity scattering) by the carrier collision term $\mathcal{C}^2$. The collision terms $\mathcal{C}^1$, $\mathcal{C}^2$ and $\mathcal{D}$ read, containing the full quantum statistics and coupling the functions $f$ and $g$,

$$\mathcal{C}^1[f, g] = \sum_{\mathbf{k}' \in \mathcal{B}} \sum_{\mathbf{q} \in \mathcal{B}} c_1(\mathbf{q}) \tag{6.2a}$$

$$\times \{f'(1-f)[g\, \delta_{\mathbf{k}',\mathbf{k}-\mathbf{q}} \delta(E'-E+\hbar\omega_0) + (g+1)\delta_{\mathbf{k}',\mathbf{k}+\mathbf{q}} \delta(E'-E'-\hbar\omega_0)$$

$$-f(1-f')[g\, \delta_{\mathbf{k}',\mathbf{k}+\mathbf{q}} \delta(E'-E-\hbar\omega_0) - (g+1)\delta_{\mathbf{k}',\mathbf{k}-\mathbf{q}} \delta(E'-E+\hbar\omega_0)\},$$

$$\mathcal{C}^2[f] = \sum_{\mathbf{k}' \in \mathcal{B}} [c_2(\mathbf{k}', \mathbf{k})f'(1-f) - c_2(\mathbf{k}, \mathbf{k}')f(1-f')]\delta(E'-E) \tag{6.2b}$$

$$\mathcal{D}[f, g] = 2 \sum_{\mathbf{k}' \in \mathcal{B}} \sum_{\mathbf{q} \in \mathcal{B}} c_1(\mathbf{q})f(1-f') \tag{6.2c}$$

$$\times [(g+1)\delta_{\mathbf{k}',\mathbf{k}-\mathbf{q}} \delta(E'-E+\hbar\omega_0) - g\, \delta_{\mathbf{k}',\mathbf{k}+\mathbf{q}} \delta(E'-E-\hbar\omega_0)],$$

with the abbreviations $f = f(\mathbf{k}, t)$, $f' = f(\mathbf{k}', t)$, $g = g(\mathbf{q}, t)$. The carrier energies are related to their wave vectors by the dispersion law via $E = E(\mathbf{k})$ and $E' = E(\mathbf{k}')$, while the optical phonon energies $\hbar\omega_0$ are constant according to the Einstein approximation [Ziman (2001)]. The functions $c_1$ and $c_2$ are related to the microscopic scattering cross sections for the considered scattering mechanism. For more details, we refer to chapter 2.

## 6.3 The Multigroup Model Equations

In this section, we present the multigroup transport equations to the coupled system of carriers and phonons and sketch the way, how to deduce them from the BBP equations (6.1). To begin with, we state the quite general assumptions, which our numerical scheme is based on:

**A1.**: The carrier energy $E$ only depends on the modulus of the carrier wave vector $k = |\mathbf{k}|$ by the invertible, differentiable function $E = E(k)$. Hence, $k = k(E)$ relates $k$ and $E$ uniquely. For instance, the well-known non-parabolic dispersion law for electrons [Lundstrom (2000)]

$$E(1 + \alpha^* E) = \frac{\hbar^2 k^2}{2m^*}, \quad k(E) = \frac{1}{\hbar}\left[2m^* E(1 + \alpha^* E)\right]^{\frac{1}{2}} \tag{6.3}$$

with the effective mass $m^*$ and the non-parabolicity factor $\alpha^*$ satisfies A1.

**A2.**: The function $c_1$ in (6.2a) depends only on the modulus of the phonon wave vector $q = |\mathbf{q}|$. This is true for the most important electron-optical phonon interaction processes, i.e., the polar optical scattering and the intervalley scattering [Lundstrom (2000); Ziman (2001)].

**A3.**: The carrier distribution function $f$ and the optical phonon distribution function $g$ are rotationally symmetric with respect to a main direction $\mathbf{e}_z$, for example, imposed by the electric field.

Based on these assumptions, we proceed as follows. According to A1, we can express $\mathbf{k}$ and $\mathbf{q}$ via

$$\mathbf{k} = k(E)(\sqrt{1 - \mu^2}\cos\varphi, \sqrt{1 - \mu^2}\sin\varphi, \mu), \tag{6.4a}$$

$$\mathbf{q} = q(\sqrt{1 - \chi^2}\cos\varepsilon, \sqrt{1 - \chi^2}\sin\varepsilon, \chi) \tag{6.4b}$$

in spherical coordinates, where $\mu$ is the cosine of the angle between $\mathbf{k}$ and $\mathbf{e}_z$ and $\varphi$ is the corresponding azimuth angle. In addition, $\chi$ denotes the cosine of the angle between $\mathbf{q}$ and $\mathbf{e}_z$ and $\varepsilon$ means the azimuth angle of $\mathbf{q}$. From A3, we conclude that $f = f(E, \mu, t)$ and $g = g(q, \chi, t)$. The independent variables $E$, $\mu$, $q$ and $\chi$ are discretized according to

$$E_i = (i - 1/2)\Delta E, \; i = 1, 2, \ldots, N, \quad E_{i+1/2} = i\Delta E, \; i = 0, 1, \ldots, N, \tag{6.5a}$$

$$q_i = (i - 1/2)\Delta q, \; i = 1, 2, \ldots, R, \quad q_{i+1/2} = i\Delta q, \; i = 0, 1, \ldots, R \tag{6.5b}$$

and

$$\mu_i = (i - 1/2)\Delta\mu - 1, \ i = 1, 2, \ldots, M, \tag{6.5c}$$

$$\mu_{i+1/2} = i\Delta\mu - 1, \ i = 0, 1, \ldots, M, \tag{6.5d}$$

$$\chi_{i+1/2} = i\Delta\chi - 1, \ i = 0, 1, \ldots, S, \tag{6.5e}$$

where $\Delta E = \hbar\omega_0/n_{\mathrm{mul}}$, $n_{\mathrm{mul}} \in \mathbb{N}$, $\Delta q = q_{\max}/R$, $\Delta\mu = 2/M$ and $\Delta\chi = 2/S$. Here, $q_{\max}$ and $E_{\max} = N\Delta E$ are chosen in a way that $f(E_{\max})$ is negligible for all $\mu$ and $t$ and $g(q_{\max})$ is undisturbed by the carrier-phonon interaction for all $\chi$ and $t$. The carrier and the optical phonon distribution functions are approximated as the finite sums

$$\mathcal{Z}^{\mathrm{c}}(E)f(E,\mu,t) \approx \sum_{i=1}^{N}\sum_{j=1}^{M} n_{ij}(t)\delta(E - E_i)\delta(\mu - \mu_j), \tag{6.6a}$$

$$\mathcal{Z}^{\mathrm{c}}(E)[1 - f(E,\mu,t)] \approx \sum_{i=1}^{N}\sum_{j=1}^{M} \left[1_{ij}^{\mathrm{c}} - n_{ij}(t)\right]\lambda_i^{\mathrm{E}}(E)\delta(\mu - \mu_j), \tag{6.6b}$$

$$\mathcal{Z}^{\mathrm{P}}(q)g(q,\chi,t) \approx \sum_{x=1}^{R}\sum_{y=1}^{S} r_{xy}(t)\delta(q - q_x)\lambda_y^{\mathrm{X}}(\chi), \tag{6.6c}$$

$$\mathcal{Z}^{\mathrm{P}}(q)[g(q,\chi,t) + 1] \approx \sum_{x=1}^{R}\sum_{y=1}^{S} \left[r_{xy}(t) + 1_{xy}^{\mathrm{P}}\right]\delta(q - q_x)\lambda_y^{\mathrm{X}}(\chi) \tag{6.6d}$$

containing the $N \times M$ new unknown carrier coefficients $n_{ij}$ and the $R \times S$ new unknown phonon coefficients $r_{xy}$. In these expressions, the carrier density of states $\mathcal{Z}^{\mathrm{c}}(E)$ and the phonon density of states $\mathcal{Z}^{\mathrm{P}}(q)$ equal

$$\mathcal{Z}^{\mathrm{c}}(E) = \frac{k(E)^2}{8\pi^3}\frac{\partial k(E)}{\partial E}, \quad \mathcal{Z}^{\mathrm{P}}(q) = \frac{q^2}{8\pi^3}. \tag{6.7}$$

Additionally, we use the abbreviations

$$1_{ij}^{\mathrm{c}} = (24\pi^3)^{-1}[k(E_{i+1/2})^3 - k(E_{i+1/2})^3](\mu_{j+1/2} - \mu_{j-1/2}), \tag{6.8a}$$

$$1_{xy}^{\mathrm{P}} = (24\pi^3)^{-1}(q_{x+1/2}^3 - q_{x-1/2}^3)(\chi_{y+1/2} - \chi_{y-1/2}), \tag{6.8b}$$

$$\lambda_i^{\mathrm{E}}(E) = \begin{cases} \Delta E^{-1}, & E \in [E_{i-1/2}, E_{i+1/2}], \\ 0, & E \notin [E_{i-1/2}, E_{i+1/2}], \end{cases} \tag{6.8c}$$

$$\lambda_i^{\mathrm{X}}(\chi) = \begin{cases} \Delta\chi^{-1}, & \chi \in [\chi_{i-1/2}, \chi_{i+1/2}], \\ 0, & \chi \notin [\chi_{i-1/2}, \chi_{i+1/2}]. \end{cases} \tag{6.8d}$$

The evolution equations for the coefficients $n_{ij}$ and $r_{xy}$ are constructed as suggested by the method of weighted residuals [Lapidus and Pinder

(1982)]. The BBP equations (6.1) are multiplied by $\mathcal{Z}^c$ and $\mathcal{Z}^P$, respectively. Inserting ansatz (6.6) into the result and integrating it over the intervals $[E_{i-1/2}, E_{i+1/2}] \times [\mu_{j-1/2}, \mu_{j+1/2}] \times [0, 2\pi]$ and $[q_{x-1/2}, q_{x+1/2}] \times [\chi_{y-1/2}, \chi_{y+1/2}] \times [0, 2\pi]$ leads to a set of coupled differential equations for the $n_{ij}$ and $r_{xy}$. By defining the vectors of coefficients $\mathbf{n}$ and $\mathbf{r}$, the vectors of collision terms $\mathbf{C}$ and $\mathbf{D}$ as well as $\mathbf{1}^c$ and $\mathbf{1}^P$,

$$\mathbf{n} = (\ldots, n_{ij}, \ldots), \qquad\qquad \mathbf{r} = (\ldots, r_{xy}, \ldots), \tag{6.9a}$$

$$\mathbf{1}^c = (\ldots, 1^c_{ij}, \ldots), \qquad\qquad \mathbf{1}^P = (\ldots, 1^P_{xy}, \ldots), \tag{6.9b}$$

$$\mathbf{C} = (\ldots, \mathcal{C}^1_{ij} + \mathcal{C}^2_{ij}, \ldots), \qquad\qquad \mathbf{D} = (\ldots, \mathcal{D}_{xy}, \ldots), \tag{6.9c}$$

the resulting system of multigroup equations can be written in the very compact form

$$\frac{\partial \mathbf{n}}{\partial t} = \mathbf{C}, \tag{6.10a}$$

$$\frac{\partial \mathbf{r}}{\partial t} = \mathbf{D}. \tag{6.10b}$$

The multigroup version of the collision terms $\mathcal{C}^1_{ij}$, $\mathcal{C}^2_{ij}$ and $\mathcal{D}_{xy}$ are given by

$$\mathcal{C}^1_{ij} = \sum_{a=1}^{N}\sum_{b=1}^{M}\sum_{x=1}^{R}\sum_{y=1}^{S}\left\{ n_{ab}(1^c_{ij} - n_{ij})\left[\langle \mathcal{A}^{xy}_{ab\to ij}\rangle r_{xy} + \langle \mathcal{E}^{xy}_{ab\to ij}\rangle(r_{xy} + 1^P_{xy})\right]\right.$$

$$\left. - n_{ij}(1^c_{ab} - n_{ab})\left[\langle \mathcal{A}^{xy}_{ij\to ab}\rangle r_{xy} + \langle \mathcal{E}^{xy}_{ij\to ab}\rangle(r_{xy} + 1^P_{xy})\right]\right\}, \tag{6.11a}$$

$$\mathcal{C}^2_{ij} = \sum_{a=1}^{N}\sum_{b=1}^{M}\left[\langle \mathcal{C}_{ab\to ij}\rangle n_{ab}(1^c_{ij} - n_{ij}) - \langle \mathcal{C}_{ij\to ab}\rangle n_{ij}(1^c_{ab} - n_{ab})\right], \tag{6.11b}$$

$$\mathcal{D}_{xy} = 2\sum_{a=1}^{N}\sum_{b=1}^{M}\sum_{i=1}^{N}\sum_{j=1}^{M} n_{ij}(1^c_{ab} - n_{ab})$$

$$\times \left[\langle \mathcal{E}^{xy}_{ij\to ab}\rangle(r_{xy} + 1^P_{xy}) - \langle \mathcal{A}^{xy}_{ij\to ab}\rangle r_{xy}\right] \tag{6.11c}$$

with the collision coefficients $\langle \mathcal{A}^{xy}_{ij\to ab}\rangle$ and $\langle \mathcal{E}^{xy}_{ij\to ab}\rangle$ for inelastic carrier-optical phonon interactions by absorbing and emitting phonons, respectively,

$$\langle \mathcal{A}^{xy}_{ij\to ab}\rangle = \frac{1}{2\pi}\int\limits_{E_{i-1/2}}^{E_{i+1/2}} dE \int\limits_{\mu_{j-1/2}}^{\mu_{j+1/2}} d\mu \int\limits_0^{2\pi} d\varphi \int\limits_{E_{a-1/2}}^{E_{a+1/2}} dE' \int\limits_{\mu_{b-1/2}}^{\mu_{b+1/2}} d\mu' \tag{6.12a}$$

$$\times \int\limits_0^{2\pi} d\varphi'\, \frac{c_1(q_x)}{\mathcal{Z}^{\mathrm{P}}(q_x)}\lambda_y^{\mathrm{X}}\left[\frac{(\mathbf{k}'-\mathbf{k})\cdot \mathbf{e}_z}{q_x}\right]\delta(E-E_i)\lambda_a^{\mathrm{E}}(E')$$

$$\times\, \delta(\mu-\mu_j)\delta(\mu'-\mu_b)\delta(|\mathbf{k}'-\mathbf{k}|-q_x)\delta(E'-E-\hbar\omega_0),$$

$$\langle \mathcal{E}^{xy}_{ij\to ab}\rangle = \frac{1}{2\pi}\int\limits_{E_{i-1/2}}^{E_{i+1/2}} dE \int\limits_{\mu_{j-1/2}}^{\mu_{j+1/2}} d\mu \int\limits_0^{2\pi} d\varphi \int\limits_{E_{a-1/2}}^{E_{a+1/2}} dE' \int\limits_{\mu_{b-1/2}}^{\mu_{b+1/2}} d\mu' \tag{6.12b}$$

$$\times \int\limits_0^{2\pi} d\varphi'\, \frac{c_1(q_x)}{\mathcal{Z}^{\mathrm{P}}(q_x)}\lambda_y^{\mathrm{X}}\left[\frac{(\mathbf{k}-\mathbf{k}')\cdot \mathbf{e}_z}{q_x}\right]\delta(E-E_i)\lambda_a^{\mathrm{E}}(E')$$

$$\times\, \delta(\mu-\mu_j)\delta(\mu'-\mu_b)\delta(|\mathbf{k}-\mathbf{k}'|-q_x)\delta(E'-E+\hbar\omega_0)$$

and the collision coefficients for elastic interactions $\langle \mathcal{C}_{ij\to ab}\rangle$:

$$\langle \mathcal{C}_{ij\to ab}\rangle = \frac{1}{2\pi}\int\limits_{E_{i-1/2}}^{E_{i+1/2}} dE \int\limits_{\mu_{j-1/2}}^{\mu_{j+1/2}} d\mu \int\limits_0^{2\pi} d\varphi \int\limits_{E_{a-1/2}}^{E_{a+1/2}} dE' \int\limits_{\mu_{b-1/2}}^{\mu_{b+1/2}} d\mu' \tag{6.12c}$$

$$\times \int\limits_0^{2\pi} d\varphi'\, c_2(\mathbf{k},\mathbf{k}')\delta(E-E_i)\lambda_a^{\mathrm{E}}(E')\delta(\mu-\mu_j)\delta(\mu'-\mu_b)\delta(E'-E)$$

with $\mathbf{k} = \mathbf{k}(E,\mu,\varphi)$ and $\mathbf{k}' = \mathbf{k}(E',\mu',\varphi')$, $\forall(i,j)\in\mathcal{N}\times\mathcal{M}$, $\forall(a,b)\in\mathcal{N}\times\mathcal{M}$ and $\forall(x,y)\in\mathcal{R}\times\mathcal{S}$, $\mathcal{N}=\{1,2,\dots,N\}$, $\mathcal{M}=\{1,2,\dots,M\}$, $\mathcal{R}=\{1,2,\dots,R\}$, $\mathcal{S}=\{1,2,\dots,S\}$.

Expressions for macroscopic quantities in the multigroup formulation are deduced by forming moments of the ansatz (6.6). Defining

$$\mathbf{n}^{\mathrm{c}} = (\dots,1,\dots), \qquad\qquad \mathbf{n}^{\mathrm{P}} = (\dots,,\dots), \tag{6.13a}$$

$$\mathbf{k}^{\mathrm{c}} = (\dots,k(E_i)\mu_j,\dots), \quad \mathbf{k}^{\mathrm{P}} = (\dots,q_x(\chi_{y-1/2}+\chi_{y+1/2})/2,\dots), \tag{6.13b}$$

$$\mathbf{E}^{\mathrm{c}} = (\dots,E_i,\dots), \qquad\qquad \mathbf{E}^{\mathrm{P}} = (\dots,\hbar\omega_0,\dots), \tag{6.13c}$$

the carrier density $\langle n^{\mathrm{c}}\rangle$, the optical phonon density $\langle n^{\mathrm{P}}\rangle$, the momenta of carriers, phonons and of the whole system in main direction $\langle k^{\mathrm{c}}\rangle$, $\langle k^{\mathrm{P}}\rangle$ and $\langle k^{\mathrm{tot}}\rangle$, as well as carrier, phonon and total energy density $\langle E^{\mathrm{c}}\rangle$, $\langle E^{\mathrm{P}}\rangle$ and

$\langle E^{\mathrm{tot}} \rangle$ read in terms of $\mathbf{n}$ and $\mathbf{r}$

$$\langle n^{\mathrm{c}} \rangle = 4\pi \mathbf{n}^{\mathrm{c}} \cdot \mathbf{n}, \qquad \langle n^{\mathrm{p}} \rangle = 2\pi \mathbf{n}^{\mathrm{p}} \cdot \mathbf{r}, \tag{6.14a}$$

$$\langle k^{\mathrm{c}} \rangle = 4\pi \mathbf{k}^{\mathrm{c}} \cdot \mathbf{n}, \qquad \langle k^{\mathrm{p}} \rangle = 2\pi \mathbf{p}^{\mathrm{p}} \cdot \mathbf{r}, \qquad \langle k^{\mathrm{tot}} \rangle = \langle k^{\mathrm{c}} \rangle + \langle k^{\mathrm{p}} \rangle, \tag{6.14b}$$

$$\langle E^{\mathrm{c}} \rangle = 4\pi \mathbf{E}^{\mathrm{c}} \cdot \mathbf{n}, \qquad \langle E^{\mathrm{p}} \rangle = 2\pi \mathbf{E}^{\mathrm{p}} \cdot \mathbf{r}, \qquad \langle E^{\mathrm{tot}} \rangle = \langle E^{\mathrm{c}} \rangle + \langle E^{\mathrm{p}} \rangle \tag{6.14c}$$

by applying the natural scalar product. Moreover, these expressions allow a simple interpretation of the physical meaning of the unknowns. The $n_{ij}$ equal the density of carriers with wave vectors $\mathbf{k} \in [k(E_{i-1/2}), k(E_{i+1/2})] \times [\mu_{j-1/2}, \mu_{j+1/2}] \times [0, 2\pi]$ except for a constant factor. Similar statements are true for the phonon coefficients $r_{xy}$.

## 6.4 Mathematical Aspects of the Multigroup Model Equations

In this section we investigate the most important mathematical properties of the multigroup model equations (6.10) and compare them with the features of the original BBP equations (6.1).

### 6.4.1 *Boundedness of the Solution*

The solutions of the BBP equations (6.1) $f$ and $g$ are bounded by $0 \le f \le 1$ and $0 \le g$ for all times $t$, if the initial distribution functions are bounded [Markowich *et al.* (1990)]. Analogously, the solution of our transport model obeys the following theorem.

**Theorem 9** *The coefficients $\mathbf{n}(t)$ and $\mathbf{r}(t)$ are bounded by*

$$0 \le \mathbf{n}(t) \le \mathbf{1}^{\mathrm{c}}, \quad 0 \le \mathbf{r}(t) \tag{6.15}$$

*for all times $t > t_0$, if the initial values $\mathbf{n}(t_0)$ and $\mathbf{r}(t_0)$ at time $t_0$ fulfill the relation*

$$0 \le \mathbf{n}(t_0) \le \mathbf{1}^{\mathrm{c}}, \quad 0 \le \mathbf{r}(t_0). \tag{6.16}$$

*Proof:* We consider the discrete time series for the variables $\mathbf{n}^t$ and $\mathbf{r}^t$ given by

$$\mathbf{n}^{t+1} = \mathbf{n}^t + \Delta \tau^t \left[ \mathbf{C}_{\mathrm{G}}^t (\mathbf{1}^{\mathrm{c}} - \mathbf{n}^t) - \mathbf{C}_{\mathrm{L}}^t \, \mathbf{n}^t \right], \tag{6.17a}$$

$$\mathbf{r}^{t+1} = \mathbf{r}^t + \Delta \tau^t \left[ \mathbf{D}_{\mathrm{G}}^t (\mathbf{r}^t + \mathbf{1}^{\mathrm{p}}) - \mathbf{D}_{\mathrm{L}}^t \, \mathbf{r}^t \right] \tag{6.17b}$$

with $\mathbf{n}^0$ and $\mathbf{r}^0$ equal to the initial values of the coefficients $\mathbf{n}(t_0)$ and $\mathbf{r}(t_0)$ of the multigroup model fulfilling (6.16) and

$$\mathbf{C}_{\mathrm{G}}^t = (\ldots, \mathcal{C}_{\mathrm{G},ij}^t, \ldots,), \qquad\qquad \mathbf{C}_{\mathrm{L}}^t = (\ldots, \mathcal{C}_{\mathrm{L},ij}^t, \ldots,), \qquad (6.18a)$$

$$\mathbf{D}_{\mathrm{G}}^t = (\ldots, \mathcal{D}_{\mathrm{G},xy}^t, \ldots,), \qquad\qquad \mathbf{D}_{\mathrm{L}}^t = (\ldots, \mathcal{D}_{\mathrm{L},xy}^t, \ldots,). \qquad (6.18b)$$

The used abbreviations mean

$$\mathcal{C}_{\mathrm{G},ij}^t = \sum_{a=1}^{N} \sum_{b=1}^{M} n_{ab}^t \Big\{ \langle \mathcal{C}_{ab \to ij} \rangle \qquad\qquad\qquad\qquad (6.19a)$$

$$+ \sum_{x=1}^{R} \sum_{y=1}^{S} \Big[ \langle \mathcal{A}_{ab \to ij}^{xy} \rangle r_{xy}^t + \langle \mathcal{E}_{ab \to ij}^{xy} \rangle (r_{xy}^t + 1_{xy}^{\mathrm{p}}) \Big] \Big\},$$

$$\mathcal{C}_{\mathrm{L},ij}^t = \sum_{a=1}^{N} \sum_{b=1}^{M} (1_{ab}^{\mathrm{c}} - n_{ab}^t) \Big\{ \langle \mathcal{C}_{ij \to ab} \rangle \qquad\qquad\qquad (6.19b)$$

$$+ \sum_{x=1}^{R} \sum_{y=1}^{S} \Big[ \langle \mathcal{A}_{ij \to ab}^{xy} \rangle r_{xy}^t + \langle \mathcal{E}_{ij \to ab}^{xy} \rangle (r_{xy}^t + 1_{xy}^{\mathrm{p}}) \Big] \Big\},$$

$$\mathcal{D}_{\mathrm{G},xy}^t = 2 \sum_{a=1}^{N} \sum_{b=1}^{M} \sum_{i=1}^{N} \sum_{j=1}^{M} \langle \mathcal{E}_{ij \to ab}^{xy} \rangle n_{ij}^t (1_{ab}^{\mathrm{c}} - n_{ab}^t), \qquad (6.19c)$$

$$\mathcal{D}_{\mathrm{L},xy}^t = 2 \sum_{a=1}^{N} \sum_{b=1}^{M} \sum_{i=1}^{N} \sum_{j=1}^{M} \langle \mathcal{A}_{ij \to ab}^{xy} \rangle n_{ij}^t (1_{ab}^{\mathrm{c}} - n_{ab}^t). \qquad (6.19d)$$

Every time step $\Delta\tau^t$ is chosen according to $\Delta\tau^t \leq \min(1/\mathbf{C}_{\mathrm{G}}^t, 1/\mathbf{C}_{\mathrm{L}}^t, 1/\mathbf{D}_{\mathrm{L}}^t)$. We note that $\Delta\tau^t$ is finite due to the conservational properties of the multigroup equations (cf. theorem 10). At time $t'$, we demand $0 \leq \mathbf{n}^{t'} \leq 1^{\mathrm{c}}$ and $0 \leq \mathbf{r}^{t'}$. Hence, we find $\mathbf{C}_{\mathrm{G}}^{t'} \geq 0$, $\mathbf{C}_{\mathrm{L}}^{t'} \geq 0$ and $\mathbf{D}_{\mathrm{L}}^{t'} \geq 0$. Moreover, we can estimate

$$\mathbf{n}^{t'+1} \leq \mathbf{n}^{t'} + \Delta\tau^{t'} \, \mathbf{C}_{\mathrm{G}}^{t'} (1^{\mathrm{c}} - \mathbf{n}^{t'}) \leq \mathbf{n}^{t'} + (1^{\mathrm{c}} - \mathbf{n}^{t'}) \leq 1^{\mathrm{c}}. \qquad (6.20)$$

Together with similar arguments for the lower boundary of $\mathbf{n}^{t'}$ and $\mathbf{r}^{t'}$, we conclude that $0 \leq \mathbf{n}^{t'} \leq 1^{\mathrm{c}}$ and $0 \leq \mathbf{r}^{t'}$ leads to $0 \leq \mathbf{n}^{t'+1} \leq 1^{\mathrm{c}}$ and $0 \leq \mathbf{r}^{t'+1}$. This implies that the $\mathbf{n}^t$ and $\mathbf{r}^t$ are bounded by $0 \leq \mathbf{n}^t \leq 1^{\mathrm{c}}$ and $0 \leq \mathbf{r}^t$ for all times $t$ because of the choice of the initial coefficients $\mathbf{n}^0$ and $\mathbf{r}^0$. Since the system (6.17a) converges to the multigroup model equations (6.10) for $\Delta\tau^t \to 0$, theorem 9 is proved.

## 6.4.2 Conservation Laws

For studying the conservational properties of our multigroup equations, we need information on the symmetry properties of the collision coefficients defined in (6.12). Hence, we state:

**Proposition 3** *The collision coefficients* $\langle \mathcal{C}_{ij\to ab} \rangle$, $\langle \mathcal{A}^{xy}_{ij\to ab} \rangle$ *and* $\langle \mathcal{E}^{xy}_{ij\to ab} \rangle$ *fulfill the symmetry relations*

$$E_a \langle \mathcal{C}_{ij\to ab} \rangle = E_i \langle \mathcal{C}_{ij\to ab} \rangle, \tag{6.21a}$$

$$E_a \langle \mathcal{A}^{xy}_{ij\to ab} \rangle = (E_i + \hbar\omega_0)\langle \mathcal{A}^{xy}_{ij\to ab} \rangle, \tag{6.21b}$$

$$E_a \langle \mathcal{E}^{xy}_{ij\to ab} \rangle = (E_i - \hbar\omega_0)\langle \mathcal{E}^{xy}_{ij\to ab} \rangle \tag{6.21c}$$

$\forall (i,j) \in \mathcal{N} \times \mathcal{M}$, $\forall (a,b) \in \mathcal{N} \times \mathcal{M}$ *and* $\forall (x,y) \in \mathcal{R} \times \mathcal{S}$.

*Proof:* The validity of (6.21a) follows immediately from (6.12c), since the coefficients $\langle \mathcal{C}_{ij\to ab} \rangle \neq 0$ only if $i = a$. Concerning (6.21b), we find that $\langle \mathcal{A}^{xy}_{ij\to ab} \rangle \neq 0$ only if $E_i + \hbar\omega_0 \in [E_{a-1/2}, E_{a+1/2}]$, according to (6.12a). In this case, $E_i + \hbar\omega_0 = E_a$ because of our discretization of the carrier energy (6.5a). Similar arguments applied to (6.21c) complete the proof of proposition 3. Based on this result, we can easily prove the following theorem on the conservation of carrier density and total energy density.

**Theorem 10** *The multigroup model equations (6.10) conserve the carrier density* $\langle n^{\mathrm{c}} \rangle$ *and the total energy density* $\langle E^{\mathrm{tot}} \rangle$ *(cf. (3.10)) of the coupled carrier phonon system via*

$$\frac{\partial}{\partial t}\langle n^{\mathrm{c}} \rangle = 0, \tag{6.22a}$$

$$\frac{\partial}{\partial t}\langle E^{\mathrm{tot}} \rangle = 0. \tag{6.22b}$$

*Proof:* We multiply the multigroup equations (6.10a) by $4\pi\mathbf{n}^{\mathrm{c}}$. Renaming of some summation indices in the collision terms and inserting the multigroup definition of the carrier density (6.14a) into the result proves (6.22a). For the proof of (6.22b), we multiply (6.10a) and (6.10b) by $4\pi\mathbf{E}^{\mathrm{c}}$ and $4\pi\mathbf{E}^{\mathrm{p}}$, respectively, and sum the results. In this way, we obtain by appropriately changing the names of some indices

$$\frac{\partial}{\partial t}\left(4\pi \mathbf{E}^{\mathrm{c}}\cdot\mathbf{n} + 2\pi \mathbf{E}^{\mathrm{P}}\cdot\mathbf{r}\right) = 4\pi \mathbf{E}^{\mathrm{c}}\cdot\mathbf{C} + 2\pi \mathbf{E}^{\mathrm{P}}\cdot\mathbf{D} \tag{6.23}$$

$$
\begin{aligned}
=&4\pi \sum_{i=1}^{N}\sum_{j=1}^{M}\sum_{a=1}^{N}\sum_{b=1}^{M}\sum_{x=1}^{R}\sum_{y=1}^{S}\Big\{-n_{ij}(1^{\mathrm{c}}_{ab}-n_{ab})\Big[\langle\mathcal{C}_{ij\to ab}\rangle E_i \\
&+ \langle\mathcal{E}^{xy}_{ij\to ab}\rangle(E_i+\hbar\omega_0)(r_{xy}+1^{\mathrm{P}}_{xy}) + \langle\mathcal{A}^{xy}_{ij\to ab}\rangle(E_i-\hbar\omega_0)r_{xy}\Big] \\
&+ n_{ab}(1^{\mathrm{c}}_{ij}-n_{ij})\Big[\langle\mathcal{A}^{xy}_{ab\to ij}\rangle E_i r_{xy}+\langle\mathcal{E}^{xy}_{ab\to ij}\rangle E_i(r_{xy}+1^{\mathrm{P}}_{xy})+\langle\mathcal{C}_{ab\to ij}\rangle E_i\Big]\Big\} \\
=&4\pi \sum_{i=1}^{N}\sum_{j=1}^{M}\sum_{a=1}^{N}\sum_{b=1}^{M}\sum_{x=1}^{R}\sum_{y=1}^{S}n_{ij}(1^{\mathrm{c}}_{ab}-n_{ab})\Big[\langle\mathcal{C}_{ij\to ab}\rangle(E_i-E_a) \\
&+ \langle\mathcal{A}^{xy}_{ij\to ab}\rangle(E_a-E_i+\hbar\omega_0)r_{xy} + \langle\mathcal{E}^{xy}_{ij\to ab}\rangle(E_a-E_i-\hbar\omega_0)(r_{xy}+1^{\mathrm{P}}_{xy})\Big],
\end{aligned}
$$

which vanishes identically according to proposition 3. Taking advantage of the definition of the total energy density completes the proof of theorem 10.

### 6.4.3   *H-theorem*

**Proposition 4**   *The collision coefficients* $\langle\mathcal{C}_{ij\to ab}\rangle$, $\langle\mathcal{A}^{xy}_{ij\to ab}\rangle$ *and* $\langle\mathcal{E}^{xy}_{ij\to ab}\rangle$ *fulfill the symmetry relations*

$$\langle\mathcal{C}_{ab\to ij}\rangle = \langle\mathcal{C}_{ij\to ab}\rangle, \tag{6.24a}$$

$$\langle\mathcal{A}^{xy}_{ab\to ij}\rangle = \langle\mathcal{E}^{xy}_{ij\to ab}\rangle \tag{6.24b}$$

$\forall(i,j)\in\mathcal{N}\times\mathcal{M}$, $\forall(a,b)\in\mathcal{N}\times\mathcal{M}$ *and* $\forall(x,y)\in\mathcal{R}\times\mathcal{S}$.

*Proof:* Taking advantage of the fact that $i=a$ for every non-vanishing coefficient $\langle\mathcal{C}_{ij\to ab}\rangle$, (6.24a) can be easily deduced from (6.12c). The relation (6.24b) is obtained by exchanging the names of the integration variables $E$, $E'$, $\mu$, $\mu'$ and $\varphi$, $\varphi'$ in the collision coefficient $\langle\mathcal{A}^{xy}_{ab\to if}\rangle$ (6.12a), respectively.

This yields

$$\langle \mathcal{A}^{xy}_{ab \to ij} \rangle = \frac{1}{2\pi} \int\limits_{E_{i-1/2}}^{E_{i+1/2}} dE \int\limits_{\mu_{j-1/2}}^{\mu_{j+1/2}} d\mu \int\limits_{0}^{2\pi} d\varphi \int\limits_{E_{a-1/2}}^{E_{a+1/2}} dE' \int\limits_{\mu_{b-1/2}}^{\mu_{b+1/2}} d\mu' \quad (6.25)$$

$$\times \int\limits_{0}^{2\pi} d\varphi' \, \frac{S(q_x)}{\mathcal{Z}^{\mathrm{P}}(q_x)} \lambda^{\mathrm{X}}_{y} \left[ \frac{(\mathbf{k} - \mathbf{k}') \cdot \mathbf{e}_z}{q_x} \right] \delta(E' - E_a) \lambda^{\mathrm{E}}_{i}(E)$$

$$\times \delta(\mu - \mu_j)\delta(\mu' - \mu_b)\delta(|\mathbf{k} - \mathbf{k}'| - q_x)\delta(E' - E + \hbar\omega_0).$$

Considering a non-vanishing term $\langle \mathcal{A}^{xy}_{ij \to ab} \rangle$ with $E_i = E_a + \hbar\omega_0$, we find that

$$\delta(E' - E_a)\delta(E' - E + \hbar\omega_0)\lambda^{\mathrm{E}}_{i}(E) = \delta(E' - E_a)\delta(E_a - E + \hbar\omega_0)/\Delta E \quad (6.26)$$
$$= \delta(E' - E_i - \hbar\omega_0)\delta(E - E_i)/\Delta E = \delta(E' - E - \hbar\omega_0)\delta(E - E_i)\lambda^{\mathrm{E}}_{a}(E').$$

The connection of this result with (6.25) and the comparison with the definition of the collision coefficient $\langle \mathcal{E}^{xy}_{ij \to ab} \rangle$ (6.12b) proves proposition 4.

**Theorem 11** *A Lyapounov functional $H[\mathbf{n}, \mathbf{r}]$ to the considered model equations (6.10) is given by*

$$H[\mathbf{n}, \mathbf{r}] = 2\,\mathbf{n} \cdot \ln \mathbf{n} + (\mathbf{1}^{\mathrm{c}} - \mathbf{n}) \cdot \ln(\mathbf{1}^{\mathrm{c}} - \mathbf{n}) + \mathbf{r} \cdot \ln \mathbf{r} - (\mathbf{r} + \mathbf{1}^{\mathrm{P}}) \cdot \ln(\mathbf{r} + \mathbf{1}^{\mathrm{P}}). \quad (6.27)$$

*Proof:* Some algebra yields that $\partial_t H$ can be written in the form

$$\frac{\partial H}{\partial t} = 2\mathbf{h}_{\mathrm{n}} \cdot \frac{\partial \mathbf{n}}{\partial t} + \mathbf{h}_{\mathrm{r}} \cdot \frac{\partial \mathbf{r}}{\partial t} = 2\mathbf{h}_{\mathrm{n}} \cdot \mathbf{C} + \mathbf{h}_{\mathrm{r}} \cdot \mathbf{D}. \quad (6.28)$$

with the functionals $\mathbf{h}_{\mathrm{n}}$ and $\mathbf{h}_{\mathrm{r}}$ defined by

$$\mathbf{h}_{\mathrm{n}} = (\ldots, \ln \frac{n_{ij}}{1^{\mathrm{c}}_{ij} - n_{ij}}, \ldots), \quad \mathbf{h}_{\mathrm{r}} = (\ldots, \ln \frac{r_{xy}}{r_{xy} + 1^{\mathrm{P}}_{xy}}, \ldots). \quad (6.29)$$

As a consequence of proposition 4, the quantity $2\mathbf{h}_{\mathrm{n}} \cdot \mathbf{C} + \mathbf{h}_{\mathrm{r}} \cdot \mathbf{D}$ may be written in the form

$$2\mathbf{h}_{\mathrm{n}} \cdot \mathbf{C} + \mathbf{h}_{\mathrm{r}} \cdot \mathbf{D} = 2 \sum_{i=1}^{N} \sum_{j=1}^{M} \sum_{a=1}^{N} \sum_{b=1}^{M} \sum_{x=1}^{R} \sum_{y=1}^{S} \langle \mathcal{A}^{xy}_{ij \to ab} \rangle n_{ij}(1^{\mathrm{c}}_{ab} - n_{ab}) r_{xy}$$

$$\times \left[ 1 - \frac{(1^{\mathrm{c}}_{ij} - n_{ij})n_{ab}(r_{xy} + 1^{\mathrm{P}}_{xy})}{n_{ij}(1^{\mathrm{c}}_{ab} - n_{ab})r_{xy}} \right] \ln \frac{(1^{\mathrm{c}}_{ij} - n_{ij})n_{ab}(r_{xy} + 1^{\mathrm{P}}_{xy})}{n_{ij}(1^{\mathrm{c}}_{ab} - n_{ab})r_{xy}} \quad (6.30)$$

$$+ 2 \sum_{i=1}^{N} \sum_{j=1}^{M} \sum_{a=1}^{N} \sum_{b=1}^{M} \langle \mathcal{C}_{ij \to ab} \rangle n_{ij}(1^{\mathrm{c}}_{ab} - n_{ab}) \left[ 1 - \frac{(1^{\mathrm{c}}_{ij} - n_{ij})n_{ab}}{n_{ij}(1^{\mathrm{c}}_{ab} - n_{ab})} \right] \ln \frac{(1^{\mathrm{c}}_{ij} - n_{ij})n_{ab}}{n_{ij}(1^{\mathrm{c}}_{ab} - n_{ab})}.$$

The collision coefficients $\langle \mathcal{C}_{ij\to ab}\rangle$ and $\langle \mathcal{A}^{xy}_{ij\to ab}\rangle$ are all non-negative as it follows immediately from their definitions (6.11). The expressions $n_{ij}(1^{\mathrm{c}}_{ab} - n_{ab})r_{xy}$ and $n_{ij}(1^{\mathrm{c}}_{ab} - n_{ab})$ stay non-negative for suitable initial values due to theorem 9. Hence, we get

$$\frac{\partial H[\mathbf{n},\mathbf{r}]}{\partial t} \leq 0 \tag{6.31}$$

with $\partial_t H[\mathbf{n},\mathbf{r}] = 0$ if and only if

$$(1^{\mathrm{c}}_{ij} - n_{ij})n_{ab}(r_{xy} + 1^{\mathrm{p}}_{xy}) = n_{ij}(1^{\mathrm{c}}_{ab} - n_{ab})r_{xy}, \qquad \forall\langle \mathcal{A}^{xy}_{ij\to ab}\rangle > 0, \tag{6.32a}$$

$$(1^{\mathrm{c}}_{ij} - n_{ij})n_{ab} = n_{ij}(1^{\mathrm{c}}_{ab} - n_{ab}), \qquad \forall\langle \mathcal{C}_{ij\to ab}\rangle > 0. \tag{6.32b}$$

We define the quantities $\mathbf{n}^*$ and $\mathbf{r}^*$ as the vectors of coefficients satisfying the above relations. In addition, $\mathbf{h}^*_{\mathrm{n}}$ and $\mathbf{h}^*_{\mathrm{r}}$ are $\mathbf{h}_{\mathrm{n}}$ and $\mathbf{h}_{\mathrm{r}}$ evaluated for $\mathbf{n}^*$ and $\mathbf{r}^*$. Since $2\mathbf{h}^*_{\mathrm{n}}\cdot\mathbf{C} + \mathbf{h}^*_{\mathrm{r}}\cdot\mathbf{D} = 0$, we obtain $2\mathbf{h}^*_{\mathrm{n}}\cdot(\mathbf{n}-\mathbf{n}^*) + \mathbf{h}^*_{\mathrm{r}}\cdot(\mathbf{r}-\mathbf{r}^*) = 0$. The expansion of $H$ in a Taylor series around $H^* = H[\mathbf{n}^*,\mathbf{r}^*]$ yields

$$H - H^* = (\mathbf{n}-\mathbf{n}^*)^2 \cdot \left.\frac{\partial \mathbf{h}_{\mathrm{n}}}{\partial \mathbf{n}}\right|_{\mathbf{n}=\mathbf{n}^*} + \frac{1}{2}(\mathbf{r}-\mathbf{r}^*)^2 \cdot \left.\frac{\partial \mathbf{h}_{\mathrm{r}}}{\partial \mathbf{r}}\right|_{\mathbf{r}=\mathbf{r}^*} + \ldots, \tag{6.33}$$

which is always non-negative for the monotonically increasing functions in $\mathbf{h}_{\mathrm{n}}$ and $\mathbf{h}_{\mathrm{r}}$ (6.29). Hence, $H - H^* \geq 0$ holds.

### 6.4.4　*Equilibrium Solution*

The equilibrium distribution functions to the BBP equations (6.1) for non-drifting particles are the well-known Fermi-Dirac and the Bose-Einstein distribution functions:

$$f_{\mathrm{FD}} = \left[\exp\left(\frac{E - E_{\mathrm{F}}}{k_{\mathrm{B}}T_{\mathrm{L}}}\right) + 1\right]^{-1}, \qquad g_{\mathrm{BE}} = \left[\exp\left(\frac{\hbar\omega_0}{k_{\mathrm{B}}T_{\mathrm{L}}}\right) - 1\right]^{-1} \tag{6.34}$$

with the temperature $T_{\mathrm{L}}$ and the Fermi energy $E_{\mathrm{F}}$. The following theorem provides relations for the equilibrium solution of the multigroup model and expressions for the coefficients $\mathbf{n}$ and $\mathbf{r}$ at equilibrium for non-drifting particle distributions.

**Theorem 12**　*The definition of the equilibrium state* $\mathbf{C} = \mathbf{0}$ *and* $\mathbf{D} = \mathbf{0}$ *is equivalent to*

$$(1^{\mathrm{c}}_{ij} - n_{ij})n_{ab}(r_{xy} + 1^{\mathrm{p}}_{xy}) = n_{ij}(1^{\mathrm{c}}_{ab} - n_{ab})r_{xy}, \qquad \forall\langle \mathcal{A}^{xy}_{ij\to ab}\rangle > 0, \tag{6.35a}$$

$$(1^{\mathrm{c}}_{ij} - n_{ij})n_{ab} = n_{ij}(1^{\mathrm{c}}_{ab} - n_{ab}), \qquad \forall\langle \mathcal{C}_{ij\to ab}\rangle > 0 \tag{6.35b}$$

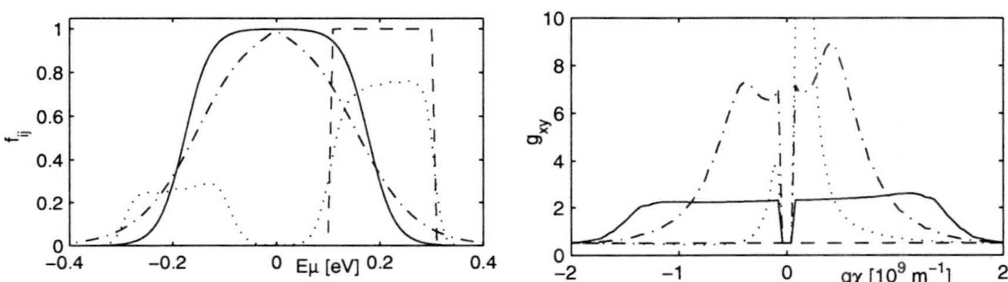

Fig. 6.1 Cuts of the electron and the phonon distribution functions $f_{ij}$ and $g_{xy}$ along the main direction: $(- - -)$ t=0 ps, $(\cdots)$ t=0.05 ps, $(- \cdot -)$ t=2 ps, $(\text{———})$ t=50 ps

$\forall (i,j) \in \mathcal{N} \times \mathcal{M}$ *and* $\forall (x,y) \in \mathcal{R} \times \mathcal{S}$. *These conditions are equivalent to*

$$\mathbf{n} = \mathbf{1}^{\mathrm{c}} \left[ \exp\left( \frac{\mathbf{E}^{\mathrm{c}} - E_{\mathrm{F}}}{k_{\mathrm{B}} T_{\mathrm{L}}} \right) + 1 \right]^{-1}, \quad \mathbf{r} = \mathbf{1}^{\mathrm{P}} \left[ \exp\left( \frac{\hbar\omega_0}{k_{\mathrm{B}} T_{\mathrm{L}}} \right) - 1 \right]^{-1}, \quad (6.36)$$

*for non-drifting particle distributions* $n_{i1} = n_{i2} = \ldots = n_{iM}$, $\forall i \in \mathcal{N}$ *and* $r_{x1} = r_{x2} = \ldots = r_{xS}$, $\forall x \in \mathcal{R}$.

*Proof:* With the help of proposition 4, one can easily verify that the relation (6.35) implies $\mathbf{C} = \mathbf{D} = \mathbf{0}$. On the other hand, $\mathbf{C} = \mathbf{D} = \mathbf{0}$ and, consequently, $\mathbf{h}_{\mathrm{n}} \cdot \mathbf{C} + \mathbf{h}_{\mathrm{r}} \cdot \mathbf{D} = 0$ are only fulfilled when (6.35) holds according to theorem 11. This completes the first part of the proof. For non-drifting particle distributions, the carrier density $\langle n^{\mathrm{c}} \rangle$ and the total energy density $\langle E^{\mathrm{tot}} \rangle$ are the only conserved macroscopic quantities in the coupled carrier-phonon system. Hence, $\mathbf{n}^{\mathrm{c}} \oplus \mathbf{0}$ and $2\mathbf{E}^{\mathrm{c}} \oplus \mathbf{E}^{\mathrm{P}}$ are collisional invariants. Since $\partial_t (2\mathbf{h}_{\mathrm{n}}^* \cdot \mathbf{n} + \mathbf{h}_{\mathrm{r}}^* \cdot \mathbf{r}) = 0$ as shown in the proof of theorem 11, $\mathbf{h}_{\mathrm{n}}^* \oplus \mathbf{h}_{\mathrm{r}}^*$ is a collisional invariant, too. Hence, it can be expressed in terms of $\mathbf{n}^{\mathrm{c}}$, $\mathbf{E}^{\mathrm{c}}$ and $\mathbf{E}^{\mathrm{P}}$ via

$$\ln n_{ij} - \ln(1_{ij}^{\mathrm{c}} - n_{ij}) = c_1 + c_2\, E_i, \quad \ln r_{xy} - \ln(r_{xy} + 1_{xy}^{\mathrm{P}}) = c_3\, \hbar\omega_0. \quad (6.37)$$

Setting $c_1 = -E_F/kbt$, $c_2 = 1/k_{\mathrm{B}} T_{\mathrm{L}}$ and $c_3 = 1/k_{\mathrm{B}} T_{\mathrm{L}}$ leads to the expressions given in theorem 12. The comparison to the equilibrium solutions of the BBP equations reveals that the equilibrium solutions of our multigroup equations are discretized versions of the Fermi-Dirac and the Bose-Einstein distributions.

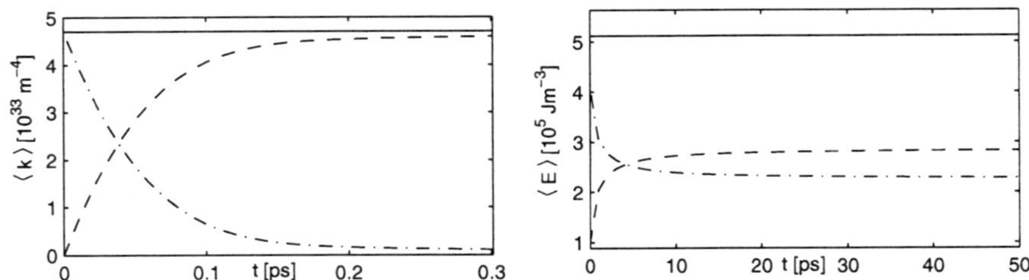

Fig. 6.2  Momentum $\langle k \rangle$ and energy density $\langle E \rangle$ versus time $t$: $(-\cdot-)$ electrons, $(---)$ phonons, $(\text{———})$ total.

## 6.5  Numerical Results

In this section, we demonstrate the applicability of our numerical scheme by studying the relaxation of hot electrons coupled with optical phonons. We use the non-parabolic energy band for the electron dispersion law with the effective mass $m^* = 0.1\, m_0$ and the non-parabolicity factor $\alpha^*=0.5$ eV$^{-1}$. The optical phonon energy $\hbar\omega_0$ equals 0.01 eV. For the scattering mechanism, we apply the screened polar optical interaction with the coupling constant $\mathcal{K} = 3 \times 10^{-14}$ Jms$^{-1}$ and $q_{\mathrm{D}} = 10^8$ m$^{-1}$.

The parameters in our transport equations are set to $N = 40$, $M = 40$, $R = 50$, $S = 40$, $E_{\mathrm{max}} = 0.4$ eV and $q_{\mathrm{max}} = 2 \times 10^9$ m$^{-1}$. The initial data for the coefficients $r_{ij}$ and $r_{xy}$ are obtained by forming moments of the distribution functions $f_0(E, \mu) = \Theta(E - E_1)\Theta(E_2 - E)\Theta(\mu)$ and $g_0(q, \chi) = 0.5$ with $E_1=0.1$ eV, $E_2=0.3$ eV and $\Theta$ denoting the Heaviside function. We use these unphysical initial values for the $r_{ij}$ and $r_{xy}$ for demonstrating the appearance of strong degeneracy effects and hot phonon phenomena together with highly anisotropic particle distribution in a test case.

In Fig. 6.1, we display cuts of the particles distribution functions along the main direction at different times. The functions $f_{ij}$ and $g_{xy}$ are approximated by $f_{ij} = n_{ij}/1_{ij}^{\mathrm{c}}$ and $g_{xy} = r_{xy}/1_{xy}^{\mathrm{P}}$. We observe the relaxation of $f_{ij}$ and $g_{xy}$ to stationary-state distributions, which are similar to the shifted Fermi-Dirac and the Bose-Einstein distributions.

Figure 6.2 illustrates the temporal evolutions of the momenta and the energy densities of electrons and phonons. Here, we find that the momentum and energy is transfered from the electrons to the phonons. Moreover, we observe the energy relaxation acting on a much longer time scale than the momentum relaxation. Finally, we point out that, besides the total energy density, the total momentum is conserved in our test case within the numerical accuracy. Hence, our model equations approximate this feature of the BBP equations in a satisfying way.

Now, we turn to more realistic simulations. Therefore, we present numerical results for the stationary state distributions of the coupled electron-longitudinal optical phonon system in GaAs for a small applied electric field at the temperature $T = 77$ K. This low temperature and the high electron density $\langle n^c \rangle = 5 \times 10^{17}$ cm$^{-3}$ are chosen for observing both degeneracy effects and hot phonon phenomena in one simulation. For the electronic band structure, we use the non-parabolic approximation (3.1). The force term, which allows us to regard an external electric field $\mathbf{E}$ in our simulation, is treated as described in chapter 3.

Two types of scattering mechanisms are included in our transport model: screened polar optical electron-phonon scattering and scattering by ionized impurities. All the material parameter used are given in Tab. 5.1 and they are taken from [Lundstrom (2000)]. The scattering function $c_1$ for screened polar optical electron-phonon interaction reads

$$c_1(\mathbf{q}) = \frac{\pi e^2 \omega_0}{V \varepsilon_0} \left( \frac{1}{\kappa_\infty} - \frac{1}{\kappa_0} \right) \frac{q^2}{\left( q^2 + q_D^2 \right)^2}, \tag{6.38}$$

where the screening parameter $q_D$ equals the inverse of the Debye length

$$q_D = \sqrt{\frac{e^2 \langle n^c \rangle}{\kappa_0 \varepsilon_0 k_B T_L}}. \tag{6.39}$$

For the considered case, we obtain $q_D = 3.251 \times 10^8$ m$^{-1}$. Ionized impurity scattering is described according to the Brooks-Herring model [Brooks and Herring (1951)]. Hence, the transition rate $c_2$ is given by

$$c_2(\mathbf{k}, \mathbf{k}') = \frac{2\pi}{\hbar} \frac{N_i e^4}{V \kappa_0^2 \varepsilon_0^2} \frac{1}{\left( |\mathbf{k}' - \mathbf{k}|^2 + q_D^2 \right)^2} \tag{6.40}$$

with the impurity density $N_i$, which is chosen as $N_i = \langle n^c \rangle$.

Additionally, a phonon-phonon relaxation term is included in the phonon equations of our multigroup transport model, which is based on a relaxation time approximation. Details on this interaction term are found in chapter 3. The phonon relaxation time $\tau_L$ is obtained from a spectroscop-

Table 6.1 Macroscopic quantities for transport simulations in GaAs.

| Simulation | $v_{\mathrm{D}}$ [$10^4$ ms$^{-1}$] | $\varepsilon^{\mathrm{c}}$ [meV] | $\langle n^{\mathrm{p}} \rangle$ [$10^{17}$ cm$^{-3}$] |
|---|---|---|---|
| A | 9.81 | 19.3 | 1.899 |
| B | 11.02 | 31.3 | 1.899 |
| C | 10.56 | 43.8 | 17.70 |

ically fitted two-channel-decay formula [Menendez and Cardona (1984)]:

$$\tau_{\mathrm{L}} = \tau_0 \left\{ 1 + \left[ \exp\left( \frac{0.65\hbar\omega_0}{k_{\mathrm{B}}T_{\mathrm{L}}} \right) - 1 \right]^{-1} \right. \tag{6.41}$$
$$\left. + \left[ \exp\left( \frac{0.35\hbar\omega_0}{k_{\mathrm{B}}T_{\mathrm{L}}} \right) - 1 \right]^{-1} \right\}^{-1} .$$

Setting the zero-temperature LO-phonon relaxation time $\tau_0 = 16$ ps [Damen *et al.* (1970)] leads to $\tau_{\mathrm{L}} = 13.16$ ps.

For our simulation, we have chosen the following model parameter: $N = 360$, $M = 24$, $n_{\mathrm{mul}} = 60$, $R = 180$, $S = 24$ and $q_{\max} = 1.365 \times 10^9$ m$^{-1}$. The applied electric field strength is set to $|\mathbf{E}| = 1$ kVcm$^{-1}$ and the initial date are discretized Fermi-Dirac and Bose-Einstein distributions according to (6.36). The temporal integration of our multigroup equation is performed with the help of a simple Runge-Kutta scheme; the stationary state is assumed to be reached after 50 ps.

By applying our multigroup model, we have performed the following transport calculations. In the simulation A, we use the low density approximation in the collision terms by neglecting the Pauli factor and keeping the phonon coefficients at their equilibrium values. The simulation B includes degeneracy effects into an equilibrium phonon calculation. Finally, the simulation C contains both, the degeneracy of the electron gas and non-equilibrium optical phonons.

In Tab. 6.1, we summarize the steady state values of the main transport quantities for the simulations A, B and C. The drift velocity of electrons is obtained by $v_{\mathrm{D}} = 4\pi\, \mathbf{v}^{\mathrm{c}} \cdot \mathbf{n} / \langle n^{\mathrm{c}} \rangle$ with

$$\mathbf{v}^{\mathrm{c}} = \left( \ldots, \frac{\mu_j}{m^*} \frac{\sqrt{2m^* E_i (1 + \alpha^* E_i)}}{1 + 2\alpha^* E_i}, \ldots \right), \tag{6.42}$$

the average electron energy $\varepsilon^{\mathrm{c}} = \langle E^{\mathrm{c}} \rangle / \langle n^{\mathrm{c}} \rangle$ and the phonon density $\langle n^{\mathrm{p}} \rangle$ are evaluated with the help of (6.14).

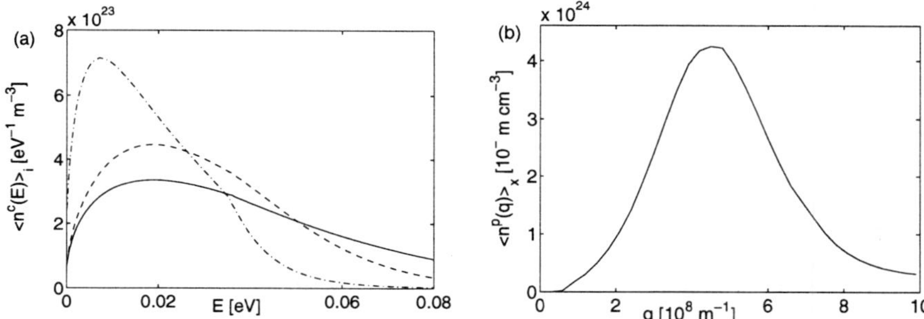

Fig. 6.3 The steady state energy distribution function $\langle n^c(E)\rangle_i$ of electrons versus energy $E$ (a) and the steady state momentum distribution function $\langle n^p(q)\rangle_x$ of optical phonons versus momentum (b) in GaAs. (-··-) simulation A, (- - -) simulation B, (——) simulation C.

While the drift velocities of electrons are almost the same for the simulations A, B and C, we observe significant changes in the average electron energies for the considered cases. As a consequence of the limited maximum occupancy of each allowed state when including the Pauli factor, electrons are pushed in the tail of the distribution, which explains the difference in the average energies for the cases A and B. The behavior of $\varepsilon^c$ for the simulation C compared to that in B can be clarified as follows. Electrons lose the energy gained in the electric field by emitting optical phonons, which leads to a higher phonon density (cf. Tab. 6.1). The increased phonon density shifts the ratio of absorption and emission rates towards that of the absorption, which increases the probability for the reabsoption of hot phonons by electrons. This behavior results in a higher electron energy for the simulation C compared to that of simulation B.

Figures 6.3a and 6.4a display the steady state energy distribution function $\langle n^c(E)\rangle_i$ and the steady state distribution function $f_{ij}$ of electrons versus the energy $E$ as well as the component of the wave vector $k_z$ in field direction for the simulations A, B and C. Moreover, Figs. 6.3b and 6.4b illustrate the steady state momentum distribution function $\langle n^p(q)\rangle_x$ and the steady state distribution function $g_{xy}$ of the optical phonons versus the component of the phonon wave vector $q_z$ in field direction for the simulation

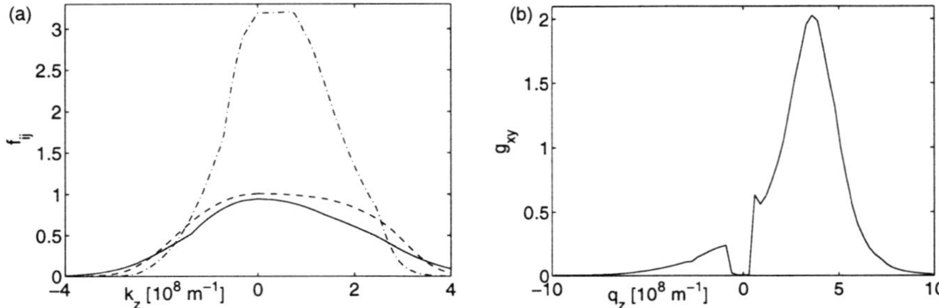

Fig. 6.4  Steady electron distribution function $f_{ij}$ (a) and steady state optical phonon distribution function $g_{xy}$ (b) versus the components of wave vectors in field direction. (-··-) simulation A, (- - -) simulation B, (—) simulation C.

C. These quantities are evaluated from the coefficients **n** and **r** via

$$\langle n^{\mathrm{c}}(E)\rangle_i = 4\pi \sum_{j=1}^{M} \frac{n_{ij}}{\Delta E}, \qquad\qquad f_{ij} = \frac{n_{ij}}{1_{ij}^{\mathrm{c}}}, \qquad\qquad (6.43\mathrm{a})$$

$$\langle n^{\mathrm{P}}(q)\rangle_x = 2\pi \sum_{y=1}^{S} \frac{r_{xy}}{\Delta q}, \qquad\qquad g_{xy} = \frac{r_{xy}}{1_{xy}^{\mathrm{P}}}. \qquad\qquad (6.43\mathrm{b})$$

In Figs. 6.3 and 6.4, we observe all the features of the electron and phonon distributions mentioned above. Both, the including of degeneracy effects and the regarding of hot phonons leads to a heating of electrons, as shown in Fig. 6.3a by the energy distribution function. Here, it is interesting to note that the kink present in the curve for simulation A at the optical phonon energy $\hbar\omega_0$ is completely washed out for the cases B and C. This kink is the result of a strong emission of phonons from the streaming tail of the distribution, which creates a strong push of electrons by phonon emission back into the low-energy part of the distribution. The Pauli principle compensates this process by prohibiting scattering into the highly populated states at low energies.

The main insight, we obtain from Fig. 6.4a, is the fact that the classical treatment of the electron gas by neglecting the Pauli factor is completely inapplicable in the considered case. Only the simulations B and C provide realistic results for the electron distribution function by being smaller than one. Moreover, Fig. 6.4b illustrates the importance of including hot phonons in the transport calculations. The phonon distribution function shows dramatic deviations of the value of the Bose-Einstein distribution, which is $4.87 \times 10^{-3}$ for the considered case. Hence, a realistic description of the carrier transport for this physical situation can only be achieved by regarding non-equilibrium phonon effects.

Chapter 7

# The Two-dimensional Electron Gas

## 7.1 Introduction

With the use of modern epitaxial growth techniques (MBE, MOCVD, etc.), the alloy composition can vary on an atomic scale, and very sophisticated layer structures consisting of several barriers and wells can be constructed. Figure 7.1 gives an example for a carrier in such a structure, a so-called quantum well. Electrons in quantum wells are termed two-dimensional electron gas (2DEG) because of the confined motion in the well. The consideration of 2DEGs is very important, since such a quantum confinement is observed in modern heterostructure devices such as the GaN-based HFET as well as in the conventional silicon MOSFET [Sze (1998); Sze (2002)]. For more information on the carrier transport in heterostructures, we refer to [Ando *et al.* (1982); Ridley (1991)]. In the following, we summarize the most important features of the transport in a two-dimensional electron gas. We note that all these considerations are aimed at the description of a 2D system at a $Al_x Ga_{1-x} N/GaN$ heterojunction. Hence, only these properties of 2DEGs that concern such structures will be mentioned.

## 7.2 General Theory of Transport in Confined Systems

### 7.2.1 *Dispersion Laws*

We assume that the electron motion is confined in the z-direction but it is free in the x-y plane. Such an electron motion can be analyzed by the

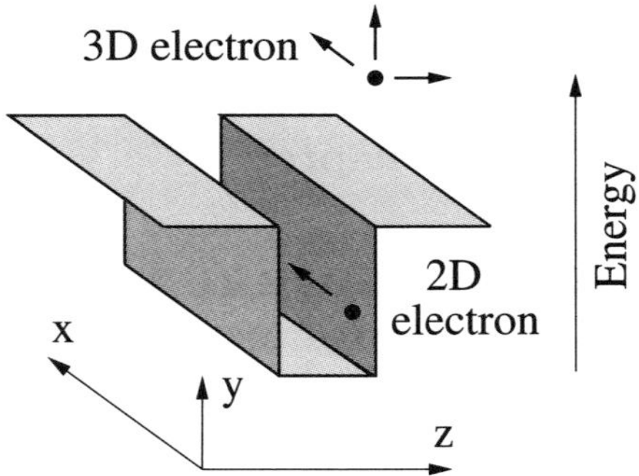


Fig. 7.1　Electrons confined in a quantum well.

three-dimensional effective mass Schrödinger equation,

$$\left[-\frac{\hbar^2}{2m^*}\Delta + E_c(z)\right]\psi(\mathbf{r}) = E\psi(\mathbf{r}), \tag{7.1}$$

where $E_c(z)$ is the energy minimum of the conduction band. The first step to solve (7.1) is to employ the separation of variables for reducing the dimension of the problem. Since the carriers are free to move in the x-y plane, it is very natural to try plane wave solutions in the x-y direction. Thus, we use the ansatz

$$\psi_\nu(\mathbf{k}_\parallel, \mathbf{r}) = \frac{1}{\sqrt{A}}\,\varphi_\nu(z)e^{i\mathbf{k}_\parallel \cdot \mathbf{r}_\parallel} \tag{7.2}$$

with the considered area $A$ and the components of the wave vector $\mathbf{k}_\parallel$ and of the position vector $\mathbf{r}_\parallel$ in the x-y plane. Moreover, $\varphi_\nu(z)$ should be normalized so that

$$\int_{-W/2}^{W/2} dz\,|\varphi_\nu(z)|^2 = 1, \tag{7.3}$$

where $W$ is the extension in z-direction. Inserting (7.2) into (7.1) leads to an equation for $\varphi_\nu(z)$,

$$\left[-\frac{\hbar^2}{2m^*}\frac{\partial^2}{\partial z^2} + E_c(z)\right]\varphi_\nu(z) = \varepsilon_\nu\varphi_\nu(z), \tag{7.4}$$

which is identical in form to the simple one-dimensional wave equation. The energy $\varepsilon_\nu$ is associated with the confinement of electrons in z-direction.

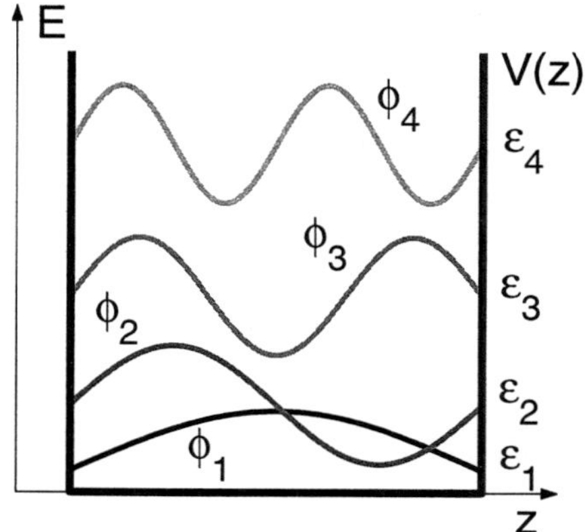



Fig. 7.2   Eigenenergies $\varepsilon_\nu$ and eigenfunctions $\varphi_\nu$ for the infinitely deep square quantum well.

Together with the kinetic energy of motion in the x-y plane, we find for the total energy

$$E_\nu(\mathbf{k}_\parallel) = \varepsilon_\nu + \frac{\hbar^2 k_\parallel^2}{2m^*}. \tag{7.5}$$

with $k_\parallel = |\mathbf{k}_\parallel|$. Hence, the energy dispersion law $E(\mathbf{k})$ for the three-dimensional case splits up into energy subbands $E_\nu(\mathbf{k}_\parallel)$ with index $\nu$, which consist of the discrete energies $\varepsilon_\nu$ of the bottoms of the subbands and the continuous part $\hbar^2 k_\parallel^2/2m^*$, which allows the electrons to increase their kinetic energies by remaining in the subband.

For determining a set of eigenfunctions $\varphi_\nu(z)$ and eigenvalues $\varepsilon_\nu$ by solving (7.4), the potential energy $E_c(z)$ specified. As a simple example, the square and infinitely deep quantum well of the width $W$ is considered. Its potential $V(z)$ is given by

$$E_c(z) = V(z) = \begin{cases} 0, & 0 \leq z \leq W, \\ \infty, & \text{elsewhere.} \end{cases} \tag{7.6}$$

Inserting (7.6) into the Schrödinger equations (7.4) with the boundary conditions $\varphi_i(z = 0) = \varphi_i(z = W) = 0$ leads to the infinite number of solutions

$$\varphi_\nu(z) = \sqrt{\frac{2}{W}} \sin k_\nu z \tag{7.7}$$

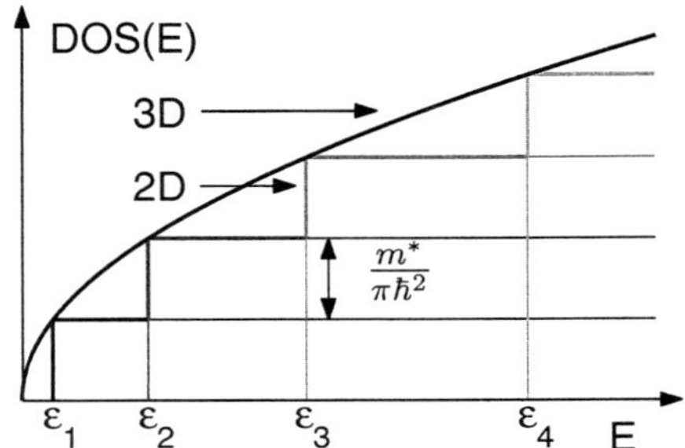


Fig. 7.3   Density of states versus energy for an infinite quantum well. The equivalent 3D density of states is included for comparison.

with the normalization factor $\sqrt{2/W}$ and $k_\nu$ restricted to the discrete values ($\nu \in \mathbb{N}$)

$$k_\nu = \frac{\nu\pi}{W}. \tag{7.8}$$

The corresponding energy due to the confinement in the z-direction is also restricted to

$$\varepsilon_\nu = \frac{\hbar^2 k_\nu^2}{2m^*} = \frac{\hbar^2}{2m^*}\left(\frac{\nu\pi}{W}\right)^2. \tag{7.9}$$

The eigenenergies $\varepsilon_\nu$ and eigenfunctions $\varphi_\nu$ for the infinitely deep square quantum well are displayed in Fig. 7.2. Moreover, the more realistic results for a confining potential at a $Al_xGa_{1-x}N/GaN$ heterojunction are shown in Fig. 7.10.

It is instructive to consider the density of states (DOS) of a 2DEG system, which is defined as

$$\mathcal{Z}(E) = 2\sum_{\mathbf{k}_\parallel,\nu} \delta[E - E_\nu(\mathbf{k}_\parallel)]. \tag{7.10}$$

Converting the sum over the continuous wave vector $\mathbf{k}_\parallel$ to an integral and taking advantage of (7.5) leads to

$$\begin{aligned}
\mathcal{Z}(E) &= 2\frac{1}{4\pi^2}\sum_\nu \int_0^\infty dk_\parallel\, k_\parallel \int_0^{2\pi} d\vartheta\, \delta\left[E - \varepsilon_\nu - \frac{\hbar^2 k_\parallel^2}{2m^*}\right] \\
&= \sum_\nu \frac{m^*}{\pi\hbar^2}\Theta(E - \varepsilon_\nu),
\end{aligned} \tag{7.11}$$

where $\Theta$ is the unit step function. The density of states is thus a staircase in which the density of the states for individual subbands is constant in the energy characteristic of a two-dimensional system. Figure 7.3 shows the density of states in a square, infinitely deep quantum well. As shown by comparison to the parabolic 3D density of states, the 2D density of states is a piecewise-continuous approximations to the 3D case.

### 7.2.2 *Scattering Mechanisms*

Phonons at heterojunctions are assumed to be three-dimensional, since they are not affected by the confining potential and the mechanical parameters of the involved materials do not differ significantly. Hence, the phonon wave vector $\mathbf{q}$ is given in the form

$$\mathbf{q} = \mathbf{q}_{\parallel} + q_z \mathbf{e}_z \tag{7.12}$$

with the component $\mathbf{q}_{\parallel}$ in the x-y plane and the component $q_z$ in normal direction to the interface with the unit vector $\mathbf{e}_z$. The transition rate for electron-phonon scattering continues to be given by Fermi's golden rule (2.22), but the matrix elements to the perturbation potential $H'$,

$$\langle \mathbf{k}'_{\parallel}, \mu | H' | \mathbf{k}_{\parallel}, \nu \rangle = \int_V d^3 r \, \psi_\mu^*(\mathbf{k}'_{\parallel}, \mathbf{r}) H' \psi_\nu(\mathbf{k}_{\parallel}, \mathbf{r}), \tag{7.13}$$

must be evaluated with the wave function given by (7.2). For $H'$ of the form (2.25), we find

$$\langle \mathbf{k}'_{\parallel}, \mu | H' | \mathbf{k}_{\parallel}, \nu \rangle \tag{7.14}$$

$$= \frac{1}{WA} \sum_{\mathbf{q}_{\parallel}} \sum_{q_z} \int_A d^2 r_{\parallel} \int_{-W/2}^{W/2} dz \, \varphi_\mu^*(z) \varphi_\nu(z) e^{i q_z z} U(\mathbf{q}) e^{i(\mathbf{k}_{\parallel} - \mathbf{k}'_{\parallel} + \mathbf{q}_{\parallel}) \cdot \mathbf{r}_{\parallel}}$$

$$= \sum_{\mathbf{q}_{\parallel}} \frac{1}{2\pi} \int_{-W/2}^{W/2} dq_z \int_{-\infty}^{\infty} dz \, \varphi_\mu^*(z) \varphi_\nu(z) e^{i q_z z} U(\mathbf{q}) \delta_{\mathbf{k}'_{\parallel}, \mathbf{k}_{\parallel} + \mathbf{q}_{\parallel}}.$$

Defining the form factor

$$I_{\mu\nu}(q_z) = \int_{-W/2}^{W/2} dz \, \varphi_\mu^*(z) \varphi_\nu(z) e^{i q_z z}, \tag{7.15}$$

and inserting the last result into Fermi's golden rule for a typical electron-phonon interaction potential (cf. (2.84)) leads to the transition rate for the two-dimensional electron gas

$$S^{\nu \to \mu}(\mathbf{k}_{\|}, \mathbf{k}'_{\|}) = \sum_{\mathbf{q}_{\|}} \frac{1}{2\pi} \int_{-W/2}^{W/2} dq_z \, |I_{\mu\nu}(q_z)|^2 \, s(\mathbf{q}) \left[ g(\mathbf{q}) + \frac{1}{2} \mp \frac{1}{2} \right]$$

$$\times \, \delta_{\mathbf{k}'_{\|}, \mathbf{k}_{\|} \pm \mathbf{q}_{\|}} \, \delta[E_\mu(\mathbf{k}'_{\|}) - E_\nu(\mathbf{k}_{\|}) \mp \hbar\omega(\mathbf{q})], \qquad (7.16)$$

transferring electrons from the state $\mathbf{k}_{\|}$ in the subband $\nu$ to $\mathbf{k}'_{\|}$ in the subband $\mu$. The scattering function $s(\mathbf{q})$ is the same as for the 3D case, since phonons are supposed to be three-dimensional. The wave vector component of an electron in x-y plane $\mathbf{k}_{\|}$ is changed by $\mathbf{q}_{\|}$ after phonon emission or absorption, but such a conservation rule does not hold in the z-direction. Instead, the form factor appears corresponding to the uncertainty principle. The cases $\mu = \nu$ are called intra-subband scattering, since initial and final energy subbands are the same. On the other hand, $\mu \neq \nu$ corresponds to an inter-subband transition. This split up of one scattering mechanism into intra-subband and inter-subband scattering represents one of the main differences between transport of a two-dimensional and a three-dimensional electron gas.

Corresponding to (2.31), we define the scattering rate $1/\tau^{\nu \to \mu}(k_{\|})$ for scattering in the two-dimensional electron gas by integrating the transition rate $S^{\nu \to \mu}(\mathbf{k}_{\|}, \mathbf{k}'_{\|})$ (7.16) with respect to the final states $\mathbf{k}'_{\|}, \mu$. Hence, we set

$$\frac{1}{\tau^\nu(k_{\|})} = \frac{A}{4\pi^2} \sum_\mu \int d^2 k'_{\|} \, S^{\nu \to \mu}(\mathbf{k}_{\|}, \mathbf{k}'_{\|}). \qquad (7.17)$$

### 7.2.2.1   *Acoustic Deformation Potential Scattering*

Assuming elastic acoustic deformation potential scattering (cf. (2.36)), the scattering function $s_{\mathrm{ADP}}$ in the equipartition approximation is given by

$$s_{\mathrm{ADP}}(\mathbf{q}) = \frac{2\pi D_{\mathrm{A}}^2 k_{\mathrm{B}} T_{\mathrm{L}}}{\rho v_{\mathrm{s}}^2 A}. \qquad (7.18)$$

This leads to the transition rate for 2DEGs due to acoustic deformation potential scattering

$$S_{\mathrm{ADP}}^{\nu \to \mu}(\mathbf{k}_{\|}, \mathbf{k}'_{\|}) = \frac{D_{\mathrm{A}}^2 k_{\mathrm{B}} T_{\mathrm{L}}}{\hbar \rho v_{\mathrm{s}}^2 A} \delta[E_\mu(\mathbf{k}'_{\|}) - E_\nu(\mathbf{k}_{\|})] \int_{-\infty}^{\infty} dq_z \, |I_{\mu\nu}(q_z)|^2. \quad (7.19)$$

Inserting (7.15) into the remaining integral yields

$$\int_{-\infty}^{\infty} dq_z \, |I_{\mu\nu}(q_z)| \tag{7.20}$$

$$= \int_{-W/2}^{W/2} dz \, \varphi_\mu^*(z)\varphi_\nu(z) \int_{-W/2}^{W/2} dz' \, \varphi_\mu(z')\varphi_\nu^*(z') \int_{-\infty}^{\infty} dq_z \, e^{iq_z(z-z')}$$

$$= 2\pi \int_{-W/2}^{W/2} dz \, \varphi_\mu^*(z)\varphi_\nu(z) \int_{-W/2}^{W/2} dz' \, \varphi_\mu(z')\varphi_\nu^*(z')\delta(z-z')$$

$$= 2\pi \int_{-W/2}^{W/2} dz \, |\varphi_\mu(z)|^2 |\varphi_\nu(z)|^2 .$$

Next, we define

$$\frac{1}{W_{\mu\nu}} = \int_{-W/2}^{W/2} dz \, |\varphi_\mu(z)|^2 |\varphi_\nu(z)|^2 , \tag{7.21}$$

where the quantity $W_{\mu\nu}$ describes the effective extent of the interaction in z-direction. The combination of (7.19) and (7.21) yields the transition rate

$$S_{\mathrm{ADP}}^{\nu\to\mu}(\mathbf{k}_\parallel, \mathbf{k}_\parallel') = \frac{2\pi D_{\mathrm{A}}^2 k_{\mathrm{B}} T_{\mathrm{L}}}{\hbar \rho v_{\mathrm{s}}^2 A} \frac{1}{W_{\mu\nu}} \delta[E_\mu(\mathbf{k}_\parallel') - E_\nu(\mathbf{k}_\parallel)] \tag{7.22}$$

and the scattering rate

$$\frac{1}{\tau_{\mathrm{ADP}}^\nu(E)} = \frac{\pi D_{\mathrm{A}}^2 k_{\mathrm{B}} T_{\mathrm{L}}}{\hbar \rho v_{\mathrm{s}}^2 A} \sum_\mu \frac{1}{W_{\mu\nu}} \frac{m^*}{\pi \hbar^2} \Theta(E - \varepsilon_\mu). \tag{7.23}$$

For the eigenfunctions $\varphi_\nu(z)$ (7.7) of the infinitely deep square well, the quantity $W_{\mu\nu}$ is given by

$$\frac{1}{W_{\mu\nu}} = \begin{cases} \dfrac{3}{2W}, & \nu = \mu, \\[2mm] \dfrac{1}{W}, & \nu \neq \mu. \end{cases} \tag{7.24}$$

This result allows us to evaluate the scattering rate for acoustic deformation potential scattering of a 2DEG in GaN with the material parameter of Tab. 7.1, as it is illustrated in Fig. 7.4 for the lowest energy subband in comparison to that of 3D electrons. Here, we assume the width of the quantum well to equal $W = 10$ nm and a lattice temperature of $T_{\mathrm{L}} = 300$ K.

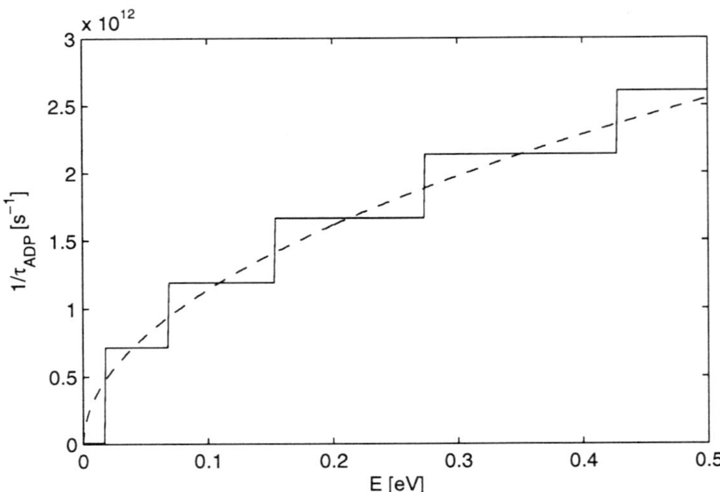

Fig. 7.4 Acoustic deformation potential scattering rate in a square quantum well. (———): lowest subband of the 2D electron gas, $(---)$: 3D electron gas.

### 7.2.2.2 *Piezoelectric Scattering*

Piezoelectric scattering is related to the scattering function

$$s_{\mathrm{PZ}}(\mathbf{q}) = \frac{2\pi k_{\mathrm{B}} T_{\mathrm{L}}}{\hbar \rho v_{\mathrm{s}}^2 A} \left( \frac{e e_{\mathrm{PZ}}}{\kappa_0 \varepsilon_0} \right)^2 \frac{1}{q^2} \tag{7.25}$$

in the elastic equipartition approximation as shown in (2.58). The combination of (7.25) and (7.16) yields the transition rate for two-dimensional piezoelectric scattering

$$S_{\mathrm{PZ}}^{\nu \to \mu}(\mathbf{k}_\|, \mathbf{k}'_\|) = \sum_{\mathbf{q}_\|} \int_{-\infty}^{\infty} dq_z \, |I_{\mu\nu}(q_z)|^2 \frac{k_{\mathrm{B}} T_{\mathrm{L}}}{\hbar \rho v_{\mathrm{s}}^2 A} \left( \frac{e e_{\mathrm{PZ}}}{\kappa_0 \varepsilon_0} \right)^2$$

$$\times \frac{1}{q_\|^2 + q_z^2} \delta_{\mathbf{k}'_\|, \mathbf{k}_\| \pm \mathbf{q}_\|} \delta[E_\mu(\mathbf{k}'_\|) - E_\nu(\mathbf{k}_\|)]. \tag{7.26}$$

A usual simplification of this expression is based on the insight that the form factor $I_{\mu\nu}$ only differs significantly from zero for small $q_z$. This allows us to approximate

$$\int_{-\infty}^{\infty} dq_z \, \frac{|I_{\mu\nu}(q_z)|^2}{q_\|^2 + q_z^2} \approx \frac{1}{q_\|^2} \int_{-\infty}^{\infty} dq_z \, |I_{\mu\nu}(q_z)|^2. \tag{7.27}$$

Hence, we find for the simplified transition rate for two-dimensional piezo-electric scattering

$$S_{\mathrm{PZ}}^{\nu\to\mu}(\mathbf{k}_{\|},\mathbf{k}_{\|}') = \sum_{\mathbf{q}_{\|}} \frac{2\pi k_{\mathrm{B}} T_{\mathrm{L}}}{\hbar\rho v_{\mathrm{s}}^2 A}\left(\frac{ee_{\mathrm{PZ}}}{\kappa_0\varepsilon_0}\right)^2$$

$$\times \frac{1}{W_{\mu\nu} q_{\|}^2}\delta_{\mathbf{k}_{\|}',\mathbf{k}_{\|}\pm\mathbf{q}_{\|}}\delta[E_\mu(\mathbf{k}_{\|}') - E_\nu(\mathbf{k}_{\|})] \qquad (7.28)$$

$$= \frac{2\pi k_{\mathrm{B}} T_{\mathrm{L}}}{\hbar\rho v_{\mathrm{s}}^2 A}\left(\frac{ee_{\mathrm{PZ}}}{\kappa_0\varepsilon_0}\right)^2 \frac{1}{W_{\mu\nu} |\mathbf{k}_{\|}' - \mathbf{k}_{\|}|^2}\,\delta[E_\mu(\mathbf{k}_{\|}') - E_\nu(\mathbf{k}_{\|})].$$

### 7.2.2.3 *Polar Optical Phonon Scattering*

Next, we will discuss the polar optical scattering at heterojunctions. Here, the scattering function $S_{\mathrm{POP}}(\mathbf{q})$ reads

$$S_{\mathrm{POP}}(\mathbf{q}) = \frac{\pi e^2 \omega_0}{\varepsilon_0 A}\left(\frac{1}{\kappa_\infty} - \frac{1}{\kappa_0}\right)\frac{1}{q^2}. \qquad (7.29)$$

Inserting this expression into Fermi's golden rule (7.16) ends in

$$S_{\mathrm{POP}}^{\nu\to\mu}(\mathbf{k}_{\|},\mathbf{k}_{\|}') = \sum_{\mathbf{q}_{\|}} \frac{e^2\omega_0}{2\varepsilon_0 A}\left(\frac{1}{\kappa_\infty} - \frac{1}{\kappa_0}\right)\int_{-\infty}^{\infty} dq_z\,\frac{|I_{\mu\nu}(q_z)|^2}{q_{\|}^2 + q_z^2}$$

$$\times \left[g(\mathbf{q}) + \frac{1}{2} \mp \frac{1}{2}\right]\delta_{\mathbf{k}_{\|}',\mathbf{k}_{\|}\pm\mathbf{q}_{\|}}\delta[E_\mu(\mathbf{k}_{\|}') - E_\nu(\mathbf{k}_{\|}) \mp \hbar\omega_0]. \qquad (7.30)$$

Here, the main difficulty is the evaluation of the integral

$$J_{\mu\nu}(q_{\|}) = \int_{-\infty}^{\infty} dq_z\,\frac{|I_{\mu\nu}(q_z)|^2}{q_{\|}^2 + q_z^2}, \qquad (7.31)$$

since the approximations used for the piezoelectric scattering are not valid for this case. Instead, we proceed as suggested in [Price (1981)]. For $q_z \ll q_{\|}$, we estimate as it was done for piezoelectric scattering,

$$J_{\mu\nu}(q_{\|}) = \frac{1}{q_{\|}^2}\int_{-\infty}^{\infty} dq_z\,|I_{\mu\nu}(q_z)|^2 = \frac{1}{W_{\mu\nu} q_{\|}^2}. \qquad (7.32)$$

For $q_z \gg q_{\|}$ and $\mu = \nu$, the integral can by approximated by

$$J_{\nu\nu}(q_{\|}) = |I_{\nu\nu}(0)|^2\int_{-\infty}^{\infty} dq_z\,\frac{1}{q_{\|}^2 + q_z^2} = |I_{\nu\nu}(0)|^2\frac{\pi}{q_{\|}} = \frac{\pi}{q_{\|}} \qquad (7.33)$$

because of the normalization of the $\varphi_\nu(z)$. Finally, we consider the case $q_z \gg q_{\parallel}$ and $\mu \neq \nu$. With the help of

$$\int_{-\infty}^{\infty} dq_z \, \frac{e^{iq_z|z-z'|}}{q_{\parallel}^2 + q_z^2} = 2\pi i \operatorname{Res}_{q_z \to iq_{\parallel}} \left( \frac{e^{iq_z|z-z'|}}{q_{\parallel}^2 + q_z^2} \right) \tag{7.34}$$

$$= 2\pi i \lim_{q_z \to iq_{\parallel}} (q_z - iq_{\parallel}) \left( \frac{e^{iq_z|z-z'|}}{(q_z - iq_{\parallel})(q_z + iq_{\parallel})} \right) = \frac{\pi}{q_{\parallel}} e^{-q_{\parallel}|z-z'|},$$

we obtain

$$J_{\mu\nu}(q_{\parallel}) = \int_{-W/2}^{W/2} dz \int_{-W/2}^{-W/2} dz' \frac{\pi}{q_{\parallel}} e^{-q_{\parallel}|z-z'|} \varphi_\nu^*(z')\varphi_\mu^*(z')\varphi_\nu(z)\varphi_\mu(z). \tag{7.35}$$

Expanding the exponential function in a Taylor series up to the second term leads to

$$J_{\mu\nu}(q_{\parallel}) = \int_{-W/2}^{-W/2} dz \int_{-W/2}^{-W/2} dz' \frac{\pi}{q_{\parallel}} (1 - q_{\parallel}|z - z'|)\varphi_\nu^*(z')\varphi_\mu^*(z')\varphi_\nu(z)\varphi_\mu(z)$$

$$= \frac{\pi}{q_{\parallel}} (\alpha_{\mu\nu} - \beta_{\mu\nu} q_{\parallel}) \tag{7.36}$$

with

$$\begin{pmatrix} \alpha_{\mu\nu} \\ \beta_{\mu\nu} \end{pmatrix} = \int_{-W/2}^{W/2} dz \int_{-W/2}^{W/2} dz' \begin{pmatrix} 1 \\ |z - z'| \end{pmatrix} \varphi_\nu^*(z')\varphi_\mu^*(z')\varphi_\nu(z)\varphi_\mu(z). \tag{7.37}$$

Combining (7.32), (7.33) and (7.36) yields an interpolation formula for $J_{\mu\nu}(q_{\parallel})$:

$$J_{\mu\nu}(q_{\parallel}) = \begin{cases} \dfrac{\pi}{q_{\parallel} + \pi W_{\mu\nu} q_{\parallel}^2}, & \mu = \nu, \\[2ex] \dfrac{\pi(\alpha_{\mu\nu} - \beta_{\mu\nu} q_{\parallel})}{q_{\parallel} + \pi W_{\mu\nu}(\alpha_{\mu\nu} - \beta_{\mu\nu} q_{\parallel})q_{\parallel}^2}, & \mu \neq \nu. \end{cases} \tag{7.38}$$

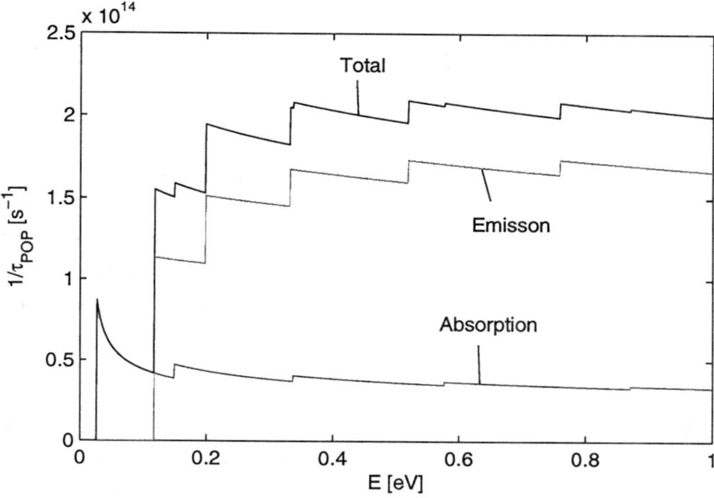



Fig. 7.5  Polar optical scattering rate in a GaN square quantum well for the first subband of the 2DEG.

This implies that the transition rate for polar optical scattering in 2DEGs may be written as

$$
S_{\mathrm{POP}}^{\nu\to\mu}(\mathbf{k}_\|,\mathbf{k}_\|') = \sum_{\mathbf{q}_\|} \frac{e^2\omega_0}{2\varepsilon_0 A}\left(\frac{1}{\kappa_\infty}-\frac{1}{\kappa_0}\right) \tag{7.39}
$$

$$
\times\left[\delta_{\mu\nu}\frac{\pi}{q_\|+\pi W_{\mu\nu}q_\|^2}+(1-\delta_{\mu\nu})\frac{\pi(\alpha_{\mu\nu}-\beta_{\mu\nu}q_\|)}{q_\|+\pi W_{\mu\nu}(\alpha_{\mu\nu}-\beta_{\mu\nu}q_\|)q_\|^2}\right]
$$

$$
\times\left[g(\mathbf{q})+\frac{1}{2}\mp\frac{1}{2}\right]\delta_{\mathbf{k}_\|',\mathbf{k}_\|\pm\mathbf{q}_\|}\delta[E_\mu(\mathbf{k}_\|')-E_\nu(\mathbf{k}_\|)\mp\hbar\omega_0].
$$

The corresponding scattering rate is given by

$$
\frac{1}{\tau_{\mathrm{POP}}^\nu(k_\|)}=\frac{e^2\omega_0}{2\varepsilon_0 A}\left(\frac{1}{\kappa_\infty}-\frac{1}{\kappa_0}\right)\mathcal{Z}(E_{\mu\nu}^\pm)\frac{1}{2}\int_0^{2\pi}d\vartheta \tag{7.40}
$$

$$
\times\left[g(\pm\Delta k_{\mu\nu}^\pm(\vartheta))+\frac{1}{2}\mp\frac{1}{2}\right]\left[\delta_{\mu\nu}\frac{\pi}{\Delta k_{\mu\nu}^\pm(\vartheta)+\pi W_{\mu\nu}\Delta k_{\mu\nu}^\pm(\vartheta)^2}\right.
$$

$$
\left.+(1-\delta_{\mu\nu})\frac{\pi(\alpha_{\mu\nu}-\beta_{\mu\nu}\Delta k_{\mu\nu}^\pm(\vartheta))}{\Delta k_{\mu\nu}^\pm(\vartheta)+\pi W_{\mu\nu}(\alpha_{\mu\nu}-\beta_{\mu\nu}\Delta k_{\mu\nu}^\pm(\vartheta))\Delta k_{\mu\nu}^\pm(\vartheta)^2}\right],
$$

where

$$E^{\pm}_{\mu\nu} = E_\nu(k_{\parallel}) \pm \hbar\omega_0 - \varepsilon_\mu, \tag{7.41a}$$

$$k^{\pm}_{\mu\nu} = \frac{1}{\hbar}\sqrt{2m^* E^{\pm}_{\mu\nu}}, \tag{7.41b}$$

$$\Delta k^{\pm}_{\mu\nu}(\vartheta) = \left[(k_{\parallel} - k^{\pm}_{\mu\nu}\cos\vartheta)^2 + (k^{\pm}_{\mu\nu}\sin\vartheta)^2\right]^{\frac{1}{2}}. \tag{7.41c}$$

Figure 7.5 shows the polar optical scattering rate for electrons confined in a GaN infinitely deep square quantum well of the width $W = 8$ nm. An equilibrium phonon distribution at the temperature $T = 600$ K is assumed.

### 7.2.2.4  *Screening Effects*

So far, we have discussed the electronic properties of quantum confined systems in terms of a one-electron picture. However, in transport and other nonequilibrium phenomena, one has to consider the dynamic response of the system of electrons to the presence of an external perturbation. Their induced charge gives rise to a polarization field in addition to the external field and the net effect is described in terms of the frequency and wave vector dependent dielectric function $\epsilon(\mathbf{q}, \omega)$ of the material.

Based on the random phase approximation, one can deduce that the effective perturbation potential $V^{\mathrm{tot}}_{\mu\nu}(\mathbf{q})$ residing on the 2DEG associated with the external potential $V^{\mathrm{ext}}_{\mu\nu}(\mathbf{q})$ for transitions between the subbands $\mu$ and $\nu$ may be written as

$$V^{\mathrm{tot}}_{\mu\nu}(\mathbf{q}) = \sum_{\mu'\nu'} \varepsilon^{-1}_{\mu'\nu',\mu\nu}(\mathbf{q},\omega) V^{\mathrm{ext}}_{\mu'\nu'}(\mathbf{q}) \tag{7.42}$$

as it is performed in [Ferry and Goodnick (1997)]. The four-dimensional dielectric matrix $\varepsilon_{\mu'\nu',\mu\nu}$ is defined by

$$\varepsilon_{\mu'\nu',\mu\nu}(\mathbf{q},\omega) = \delta_{\mu\mu'}\delta_{\nu\nu'} - \frac{e^2}{2\kappa_0\varepsilon_0 q} F^{\mu\nu}_{\mu'\nu'}(\mathbf{q}) L_{\mu'\nu'}(\mathbf{q},\omega) \tag{7.43}$$

with

$$F^{\mu\nu}_{\mu'\nu'}(\mathbf{q}) = \int_{-W/2}^{W/2} dz \int_{-W/2}^{W/2} dz'\, \varphi_\mu(z)\varphi_\nu^*(z)\varphi_{\mu'}^*(z')\varphi_{\nu'}(z') e^{-q_z(z-z')}$$

$$L_{\mu\nu}(\mathbf{q},\omega) = 2\sum_{\mathbf{k}_{\parallel}} \frac{f^\mu(\mathbf{k}_{\parallel}) - f^\mu(\mathbf{k}_{\parallel} + \mathbf{q}_{\parallel})}{E_\nu(\mathbf{k}_{\parallel} + \mathbf{q}_{\parallel}) - E_\mu(\mathbf{k}_{\parallel}) - \hbar\omega - i\alpha\hbar}. \tag{7.44}$$

Here, $f^\nu$ is the distribution function for the subband $\nu$ and $\alpha$ is a small convergence parameter. The long-wavelength limit, $\mathbf{q} \to 0$, is reached when the wavelength $2\pi/q$ is large compared to the spatial extent of the subband envelope functions. Thus, in the limit of a purely 2D system of zero width, the long-wavelength limit is valid for all $\mathbf{q}$. Since the eigenfunctions $\varphi_\nu(z)$ are orthonormal functions with respect to the subband indices, we find

$$\lim_{\mathbf{q}\to 0} F^{\mu\nu}_{\mu'\nu'}(\mathbf{q}) = \delta_{\mu\mu'}\delta_{\nu\nu'}. \tag{7.45}$$

This implies that intersubband transitions are completely unscreened in the long wavelength limit

$$V^{\text{tot}}_{\mu\nu}(\mathbf{q}) = V^{\text{ext}}_{\mu\nu}(\mathbf{q}), \quad \mu \neq \nu, \tag{7.46}$$

whereas intrasubband transitions reduce to

$$V^{\text{ext}}_{\nu\nu}(\mathbf{q}) = V^{\text{tot}}_{\nu\nu}(\mathbf{q}) + \sum_{\nu'} \frac{e^2}{2\kappa_0\varepsilon_0 q_\parallel} L_{\nu'\nu'}(\mathbf{q}_\parallel,\omega) V^{\text{tot}}_{\nu'\nu'}(\mathbf{q}). \tag{7.47}$$

Neglecting the intersubband polarizabilities and the correction terms due to intrasubband polarizabilities of the other subbands, the multidimensional dielectric matrix reduces to a scalar relation as in the bulk case:

$$V^{\text{tot}}_{\nu\nu}(\mathbf{q}) = \frac{V^{\text{ext}}_{\nu\nu}(\mathbf{q})}{\epsilon(\mathbf{q}_\parallel,\omega)} \tag{7.48}$$

where

$$\epsilon(\mathbf{q}_\parallel,\omega) = 1 + \sum_{\nu} \frac{e^2}{2\kappa_0\varepsilon_0 q_\parallel} L_{\nu\nu}(\mathbf{q}_\parallel,\omega). \tag{7.49}$$

For static perturbation potentials (i.e. $\omega = 0$), we find that

$$\lim_{\mathbf{q}\to 0} L_{\nu\nu}(\mathbf{q}_\parallel,0) = \frac{m^*}{\pi\hbar^2} f_\nu(0) \tag{7.50}$$

with the occupancy of the bottom of the $\nu$-th subband $f_\nu(0)$. Combining (7.49) and (7.50) leads to the final expression for the dielectric function

$$\epsilon(\mathbf{q}_\parallel) = 1 + \frac{q_s}{q_\parallel} \tag{7.51}$$

used in the further calculations with the screening parameter

$$q_s = \sum_{\nu} \frac{m^* e^2}{2\pi\kappa_0\varepsilon_0\hbar^2} f^\nu(0). \tag{7.52}$$

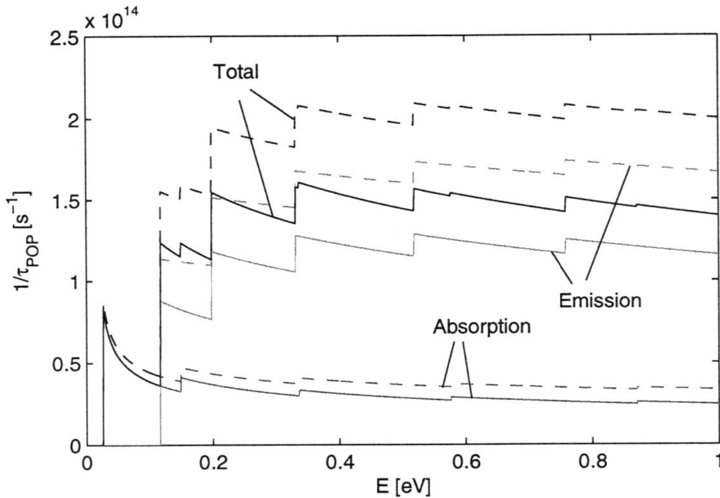


Fig. 7.6    Screened polar optical scattering rate in a GaN square quantum well for the first subband of the 2DEG. (——): with screening, (– – –): without screening.

Hence, the intrasubband transition rates of scattering mechanisms associated with electric fields must be divided by $\epsilon(\mathbf{q}_{\parallel})^2$ for taking into account screening effects.

In Fig. 7.6 we illustrate the differences between including and neglecting the screening in the scattering rates of polar optical scattering for electrons confined in GaN in an infinitely deep square quantum well of the width $W = 8$ nm. The screening parameter is set to $q_s = 10^8$ m$^{-1}$ and an equilibrium phonon distribution at the temperature $T = 600$ K is assumed.

### 7.2.3    BBP Equations for 2D Systems

Electrons of a confined system are described by the distribution function $\tilde{f}^\nu(\mathbf{r}_{\parallel}, z, \mathbf{k}_{\parallel}, t)$, which is the probability density to find an electron in the subband $\nu$ with the wave vector component $\mathbf{k}_{\parallel}$ parallel to the heterojunction interface at the position $\mathbf{r}$ at time $t$. The position vector $\mathbf{r}$ is split up into its components $\mathbf{r}_{\parallel}$ and $z$, parallel and normal to the interface, respectively. According to the considerations of Sec. 7.2.1, all the information on the $z$-dependence of the electron distribution is contained in the envelope wave functions $\varphi_\nu(z)$. Hence, $\tilde{f}^\nu(\mathbf{r}_{\|}, z, \mathbf{k}_{\|}, t)$ can be written in the form

$$\tilde{f}^\nu(\mathbf{r}_{\|}, z, \mathbf{k}_{\|}, t) = f^\nu(\mathbf{r}_{\|}, \mathbf{k}_{\|}, t)|\varphi_\nu(z)|^2. \tag{7.53}$$

The integration of this expression with respect to $z$,

$$\int_{-W/2}^{W/2} dz\, \tilde{f}^\nu(\mathbf{r}_{\|}, z, \mathbf{k}_{\|}, t)$$

$$= f^\nu(\mathbf{r}_{\|}, \mathbf{k}_{\|}, t) \int_{-W/2}^{W/2} dz\, |\varphi_\nu(z)|^2 = f^\nu(\mathbf{r}_{\|}, \mathbf{k}_{\|}, t), \tag{7.54}$$

reveals that the distribution function $f^\nu(\mathbf{r}_{\|}, \mathbf{k}_{\|}, t)$ is the probability density for finding a carrier of the subband $\nu$ in the state $\mathbf{k}_{\|}$ at the position $\mathbf{r}_{\|}$ at time $t$ for any $z$. This two-dimensional distribution function for electrons will be considered in the following.

Similarly, we can construct a two-dimensional distribution function $g$ for the three-dimensional phonons at a heterojunction. We define $\tilde{g}(\mathbf{r}_{\|}, z, \mathbf{q}_{\|}, q_z, t)$ as the distribution function of the phonons depending on position $\mathbf{r}_{\|}$, $z$, wave vector $\mathbf{q}_{\|}$, $q_z$ and time $t$. The function $\tilde{g}$ can be split up into a Bose-Einstein distribution $\hat{g}_{\mathrm{BE}}$, which describe the lattice background, and a portion $\hat{g}_{\mathrm{e-p}}$ due to electron-phonon interaction. Since phonons do not "feel" the heterojunction, $\hat{g}_{\mathrm{BE}}$ does not depend on $z$, while $\hat{g}_{\mathrm{e-p}}$ exhibits the same z-dependence as the electrons,

$$\tilde{g}(\mathbf{r}_{\|}, z, \mathbf{q}_{\|}, q_z, t) = \hat{g}_{\mathrm{BE}}(\mathbf{r}_{\|}, \mathbf{q}_{\|}, q_z, t) + \hat{g}_{\mathrm{e-p}}(\mathbf{r}_{\|}, \mathbf{q}_{\|}, q_z, t)|\varphi_\nu(z)|^2. \tag{7.55}$$

Integrating $\tilde{g}$ with respect to $z$ leads to the new distribution function $W\hat{g}$,

$$W\hat{g}(\mathbf{r}_{\|}, \mathbf{q}_{\|}, q_z, t) = \int_{-W/2}^{W/2} dz\, \tilde{g}(\mathbf{r}_{\|}, z, \mathbf{q}_{\|}, q_z, t) \tag{7.56}$$

$$= \int_{-W/2}^{W/2} dz\, \hat{g}_{\mathrm{BE}}(\mathbf{r}_{\|}, \mathbf{q}_{\|}, q_z, t) + \hat{g}_{\mathrm{e-p}}(\mathbf{r}_{\|}, \mathbf{q}_{\|}, q_z, t) \int_{-W/2}^{W/2} dz\, |\varphi_\nu(z)|^2$$

$$= W\hat{g}_{\mathrm{BE}}(\mathbf{r}_{\|}, \mathbf{q}_{\|}, q_z, t) + \hat{g}_{\mathrm{e-p}}(\mathbf{r}_{\|}, \mathbf{q}_{\|}, q_z, t).$$

This result is averaged with respect to $q_z$. Therefore, we define

$$g(\mathbf{r}_\|, \mathbf{q}_\|, t) = \frac{W}{2\pi} \int_{-\infty}^{\infty} dq_z\, \hat{g}(\mathbf{r}_\|, \mathbf{q}_\|, q_z t), \qquad (7.57a)$$

$$g_{\text{BE}}(\mathbf{r}_\|, \mathbf{q}_\|, t) = \frac{W}{2\pi} \int_{-\infty}^{\infty} dq_z\, \hat{g}_{\text{BE}}(\mathbf{r}_\|, \mathbf{q}_\|, q_z t), \qquad (7.57b)$$

$$g_{\text{e-p}}(\mathbf{r}_\|, \mathbf{q}_\|, t) = \frac{1}{2\pi} \int_{-\infty}^{\infty} dq_z\, \hat{g}_{\text{e-p}}(\mathbf{r}_\|, \mathbf{q}_\|, q_z t) \qquad (7.57c)$$

and find

$$g(\mathbf{r}_\|, \mathbf{q}_\|, t) = \frac{1}{2\pi} \int_{-\infty}^{\infty} dq_z\, W \hat{g}(\mathbf{r}_\|, \mathbf{q}_\|, q_z, t) \qquad (7.58)$$

$$= \frac{1}{2\pi} \int_{-\infty}^{\infty} dq_z\, W \hat{g}_{\text{BE}}(\mathbf{r}_\|, \mathbf{q}_\|, q_z, t) + \frac{1}{2\pi} \int_{-\infty}^{\infty} dq_z\, \hat{g}_{\text{e-p}}(\mathbf{r}_\|, \mathbf{q}_\|, q_z, t)$$

$$= g_{\text{BE}}(\mathbf{r}_\|, \mathbf{q}_\|, t) + g_{\text{e-p}}(\mathbf{r}_\|, \mathbf{q}_\|, t).$$

This final two-dimensional phonon distribution function $g$, which equals the probability density to find phonons at position $\mathbf{r}_\|$ and state $\mathbf{q}_\|$ at time $t$ for arbitrary $z$ and $q_z$ is the one, we use in the following. It should be noted that evaluating transition rates with the help of $g$ instead of $\hat{g}$ is only approximatively correct. While the integration with respect to $z$ is exact in a way that one does not loose information on the phonon distribution, the integration with respect to $q_z$ constitutes an averaging procedure. This means that replacing $\hat{g}$ by $g$ is only valid, when $\hat{g}$ varies slowly with $q_z$. In this chapter, we treat the dynamics of optical phonons at a heterojunction by means of this approximation. Since their distribution function is basically set up by a background Bose-Einstein distribution, which does not depend on $\mathbf{q}$ and therefore on $q_z$ in the Einstein approximation of the phonon energy, this approximation should be valid in the considered case.

The Bloch-Boltzmann-Peierls equations for the carrier-phonon transport in a confined system are constructed in analogy to the three-dimensional case. They read for the two-dimensional electron distribution function $f^\nu$ of the subband $\nu$ and for the two-dimensional phonon distribution function $g$

$$\left[ \frac{\partial}{\partial t} + \mathbf{v}_\|^\nu(\mathbf{k}_\|) \cdot \nabla_{\mathbf{r}_\|} - \frac{e}{\hbar} \mathbf{E}_\|(\mathbf{r}_\|) \cdot \nabla_{\mathbf{k}_\|} \right] f^\nu(\mathbf{r}_\|, \mathbf{k}_\|, t) = \mathcal{C}[f^\nu, g], \qquad (7.59a)$$

$$\left[ \frac{\partial}{\partial t} + \mathbf{u}_\|(\mathbf{q}_\|) \cdot \nabla_{\mathbf{r}_\|} \right] g(\mathbf{r}_\|, \mathbf{q}_\|, t) = \mathcal{D}[g, f^\nu]. \qquad (7.59b)$$

Here, $\mathbf{E}_\parallel$ is the electric field strength parallel to the heterojunction interface and the symbols $\mathbf{v}_\parallel^\nu(\mathbf{k}_\parallel)$ and $\mathbf{u}_\parallel(\mathbf{q}_\parallel)$ which label the electron and the phonon group velocities, respectively, are defined by

$$\mathbf{v}_\parallel(\mathbf{k}_\parallel) = \frac{1}{\hbar}\frac{\partial E_\nu(\mathbf{k}_\parallel)}{\partial \mathbf{k}_\parallel}, \tag{7.60a}$$

$$\mathbf{u}_\parallel(\mathbf{q}_\parallel) = \frac{\partial \omega(\mathbf{q}_\parallel)}{\partial \mathbf{q}_\parallel}. \tag{7.60b}$$

The electron collision term for electron-phonon interaction containing the scattering function $s(\mathbf{q}_\parallel)$ is given by

$$\mathcal{C}[f^\nu, g] = \sum_\mu \sum_{\mathbf{q}_\parallel, \mathbf{k}_\parallel'} s(\mathbf{q}_\parallel) \tag{7.61}$$

$$\times \Big\{ f^\mu(\mathbf{k}_\parallel')[1 - f^\nu(\mathbf{k}_\parallel)]g(\mathbf{q}_\parallel)\delta_{\mathbf{k}_\parallel', \mathbf{k}_\parallel - \mathbf{q}_\parallel}\delta[E^\mu(\mathbf{k}_\parallel') - E^\nu(\mathbf{k}_\parallel) + \hbar\omega_{\mathrm{LO}}]$$

$$+ f^\mu(\mathbf{k}_\parallel')[1 - f^\nu(\mathbf{k}_\parallel)][g(\mathbf{q}_\parallel) + 1]\delta_{\mathbf{k}_\parallel', \mathbf{k}_\parallel + \mathbf{q}_\parallel}\delta[E^\mu(\mathbf{k}_\parallel') - E^\nu(\mathbf{k}_\parallel) - \hbar\omega_{\mathrm{LO}}]$$

$$- f^\nu(\mathbf{k}_\parallel)[1 - f^\mu(\mathbf{k}_\parallel')]g(\mathbf{q}_\parallel)\delta_{\mathbf{k}_\parallel', \mathbf{k}_\parallel + \mathbf{q}_\parallel}\delta[E^\mu(\mathbf{k}_\parallel') - E^\nu(\mathbf{k}_\parallel) - \hbar\omega_{\mathrm{LO}}]$$

$$- f^\nu(\mathbf{k}_\parallel)[1 - f^\mu(\mathbf{k}_\parallel')][g(\mathbf{q}_\parallel) + 1]\delta_{\mathbf{k}_\parallel', \mathbf{k}_\parallel - \mathbf{q}_\parallel}\delta[E^\mu(\mathbf{k}_\parallel') - E^\nu(\mathbf{k}_\parallel) + \hbar\omega_{\mathrm{LO}}]\Big\}.$$

The corresponding phonon collision term reads

$$\mathcal{D}[g, f^\nu] = 2\sum_\mu \sum_{\mathbf{k}_\parallel, \mathbf{k}_\parallel'} s(\mathbf{q}_\parallel)f^\nu(\mathbf{k}_\parallel)[1 - f^\mu(\mathbf{k}_\parallel')] \tag{7.62}$$

$$\times \Big\{ [g(\mathbf{q}_\parallel) + 1]\delta_{\mathbf{k}_\parallel', \mathbf{k}_\parallel - \mathbf{q}_\parallel}\delta[E^\mu(\mathbf{k}_\parallel') - E^\nu(\mathbf{k}_\parallel) + \hbar\omega_{\mathrm{LO}}]$$

$$- g(\mathbf{q}_\parallel)\delta_{\mathbf{k}_\parallel', \mathbf{k}_\parallel + \mathbf{q}_\parallel}\delta[E^\mu(\mathbf{k}_\parallel') - E^\nu(\mathbf{k}_\parallel) - \hbar\omega_{\mathrm{LO}}]\Big\}.$$

The macroscopic quantities are closely related to the electron and phonon distribution functions. For instance, the density $\langle n^\nu(\mathbf{r}_\parallel, t)\rangle$, the velocity $\langle v^\nu(\mathbf{r}_\parallel, t)\rangle$ and the energy $\langle E^\nu(\mathbf{r}_\parallel, t)\rangle$ of electrons belonging to the subband $\nu$ and the phonon density $\langle n^{\mathrm{p}}(\mathbf{r}_\parallel, t)\rangle$ can be evaluated via

$$\langle n^\nu(\mathbf{r}_\parallel, t)\rangle = \frac{1}{2\pi^2}\int d^2k\, f^\nu(\mathbf{r}_\parallel, \mathbf{k}_\parallel, t), \tag{7.63a}$$

$$\langle \mathbf{v}_\parallel^\nu(\mathbf{r}_\parallel, t)\rangle = \frac{1}{2\pi^2\langle n^\nu(\mathbf{r}_\parallel, t)\rangle}\int d^2k\, \mathbf{v}_\parallel^\nu f^\nu(\mathbf{r}_\parallel, \mathbf{k}_\parallel, t), \tag{7.63b}$$

$$\langle E^\nu(\mathbf{r}_\parallel, t)\rangle = \frac{1}{2\pi^2\langle n^\nu(\mathbf{r}_\parallel, t)\rangle}\int d^2k\, E_\nu(\mathbf{k}_\parallel)f^\nu(\mathbf{r}_\parallel, \mathbf{k}_\parallel, t), \tag{7.63c}$$

$$\langle n^{\mathrm{p}}(\mathbf{r}_\parallel, t)\rangle = \frac{1}{4\pi^2}\int d^2q\, g(\mathbf{r}_\parallel, \mathbf{q}_\parallel, t). \tag{7.63d}$$

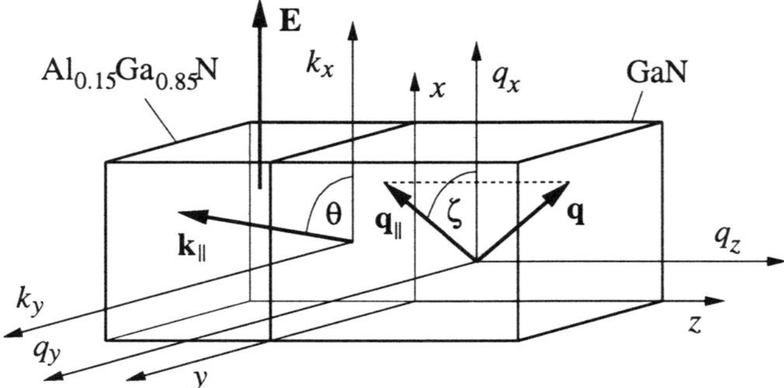


Fig. 7.7   Schematic illustration of the AlGaN/GaN heterojunction including the chosen coordinate systems for space and wave vectors of electrons and phonons.

## 7.3   Multigroup Equations to the 2D-BBP Equations

In this section, we introduce multigroup equations for investigating the transport of electrons and phonons at a heterojunction for the spatially homogeneous case, which means that $f^\nu$ and $g$ do not depend on $\mathbf{r}_\parallel$. To begin with, we state the approximations used for describing the dispersion laws for the confined electrons and the 3D phonons. The electron energy $E^\nu(\mathbf{k}_\parallel)$ in the $\nu$th energy subband and the electron wave vector $\mathbf{k}_\parallel$ are related by the spherical parabolic energy momentum rule (7.5). This implies that the modulus of the electron wave vector in the subband $\nu$ is determined by

$$k_\parallel^\nu(E) = \frac{1}{\hbar}\sqrt{2m^*(E - \varepsilon_\nu)}\,\Theta(E - \varepsilon_\nu) \tag{7.64}$$

for the given energy $E$. Here, $\Theta$ denotes the Heaviside step function. The energies of longitudinal optical (LO) phonons $\hbar\omega_{LO}$ are related to the phonon wave vector $\mathbf{q}$ via the usual Einstein approximations (cf. Sec. 2.3)

$$\hbar\omega_{LO}(\mathbf{q}) = \hbar\omega_{LO}. \tag{7.65}$$

The electron system is described by a set of one-particle distributions functions $f^\nu(\mathbf{k}_\parallel, t)$, which gives the probability to find an electron in the infinitesimal volume $d^2 k_\parallel$ around $\mathbf{k}_\parallel$ at time $t$ in the the $\nu$th subband. The evolution equation for $f^\nu(\mathbf{k}_\parallel, t)$ is the 2D BTE (7.59a). By expressing $\mathbf{k}_\parallel$

in polar coordinates

$$\mathbf{k}_{\parallel}^{\nu} = \left( k_{\parallel}^{\nu}(E)\cos\theta,\, k_{\parallel}^{\nu}(E)\sin\theta \right) \tag{7.66}$$

with the polar angle $\theta$ between $\mathbf{k}_{\parallel}$ and $\mathbf{E}_{\parallel}$ (cf. Fig. 7.7) and introducing the new unknown function

$$F^{\nu}(E,\theta,t) = \mathcal{Z}_{\mathrm{el}}^{\nu}(E) f^{\nu}(E,\theta,t) \tag{7.67}$$

with the electron density of states

$$\mathcal{Z}_{\mathrm{el}}^{\nu}(E) = \frac{m^*}{4\pi^2\hbar^2}\,\Theta(E-\varepsilon_{\nu}), \tag{7.68}$$

we obtain the energy dependent formulation of the 2D BTE. It reads for $\mathbf{E}_{\parallel} = (\mathcal{E},0)$:

$$\frac{\partial F^{\nu}}{\partial t} + \frac{\partial}{\partial E}\left[ -e\mathcal{E}\frac{\hbar k_{\parallel}^{\nu}}{m^*}\cos\theta F^{\nu} \right] + \frac{\partial}{\partial\theta}\left[ \frac{e\mathcal{E}}{\hbar k_{\parallel}^{\nu}}\sin\theta F^{\nu} \right] = \mathcal{C}_1[F^{\nu}] + \mathcal{C}_2[F^{\nu}]. \tag{7.69}$$

In this equation, only two scattering mechanisms are included for simplicity. An extension to more precise descriptions of the physics of the particle transport can easily be performed. The collision term $\mathcal{C}_1[F^{\nu}]$

$$\mathcal{C}_1[F^{\nu}] = \sum_{\mu}\int_0^{\infty} dE' \int_0^{2\pi} d\theta' \tag{7.70}$$

$$\times \Big\{ c_1(\mathbf{k}_{\parallel}^{\nu} - \mathbf{k}_{\parallel}^{\mu\,\prime})F^{\mu}(\mathbf{k}_{\parallel}^{\mu\,\prime})[\mathcal{Z}_{\mathrm{el}}^{\nu}(E) - F^{\nu}(\mathbf{k}_{\parallel}^{\nu})]g(\mathbf{k}_{\parallel}^{\nu} - \mathbf{k}_{\parallel}^{\mu\,\prime})\,\delta(E' - E + \hbar\omega_{\mathrm{LO}})$$

$$+ c_1(\mathbf{k}_{\parallel}^{\mu\,\prime} - \mathbf{k}_{\parallel}^{\nu})F^{\mu}(\mathbf{k}_{\parallel}^{\mu\,\prime})[\mathcal{Z}_{\mathrm{el}}^{\nu}(E) - F^{\nu}(\mathbf{k}_{\parallel}^{\nu})][g(\mathbf{k}_{\parallel}^{\mu\,\prime} - \mathbf{k}_{\parallel}^{\nu}) + 1]\,\delta(E' - E - \hbar\omega_{\mathrm{LO}})$$

$$- c_1(\mathbf{k}_{\parallel}^{\mu\,\prime} - \mathbf{k}_{\parallel}^{\nu})F^{\nu}(\mathbf{k}_{\parallel}^{\nu})[\mathcal{Z}_{\mathrm{el}}^{\mu}(E') - F^{\mu}(\mathbf{k}_{\parallel}^{\mu\,\prime})]g(\mathbf{k}_{\parallel}^{\mu\,\prime} - \mathbf{k}_{\parallel}^{\nu})\,\delta(E' - E - \hbar\omega_{\mathrm{LO}})$$

$$- c_1(\mathbf{k}_{\parallel}^{\nu} - \mathbf{k}_{\parallel}^{\mu\,\prime})F^{\nu}(\mathbf{k}_{\parallel}^{\nu})[\mathcal{Z}_{\mathrm{el}}^{\mu}(E') - F^{\mu}(\mathbf{k}_{\parallel}^{\mu\,\prime})][g(\mathbf{k}_{\parallel}^{\nu} - \mathbf{k}_{\parallel}^{\mu\,\prime}) + 1]\,\delta(E' - E + \hbar\omega_{\mathrm{LO}}) \Big\}$$

with the scattering function $c_1$ depending on the interaction mechanism (e.g., polar optical scattering) and $\mathbf{k}_{\parallel}^{\nu} = \mathbf{k}_{\parallel}^{\nu}(E,\theta)$, $\mathbf{k}_{\parallel}^{\mu\,\prime} = \mathbf{k}_{\parallel}^{\mu}(E',\theta')$ couples the electron-phonon system. On the other hand, $\mathcal{C}_2[F^{\nu}]$ regards to an elastic scattering mechanism (e.g. acoustic deformation potential and piezoelectric scattering) with the scattering function $c_2$ and it is given by

$$C_2[F^\nu] = \sum_\mu \int_0^\infty dE' \int_0^{2\pi} d\theta' \left\{ c_2(\mathbf{k}_\|^{\mu\,\prime}, \mathbf{k}_\|^\nu)\, F^\mu(\mathbf{k}_\|^{\mu\,\prime})[\mathcal{Z}_{\mathrm{el}}^\nu(E) - F^\nu(\mathbf{k}_\|^\nu)] \right.$$

$$\left. - c_2(\mathbf{k}_\|^\nu, \mathbf{k}_\|^{\mu\,\prime})\, F^\nu(\mathbf{k}_\|^\nu)[\mathcal{Z}_{\mathrm{el}}^\mu(E') - F^\mu(\mathbf{k}_\|^{\mu\,\prime})] \right\} \delta(E' - E), \qquad (7.71)$$

where $\mathbf{k}_\|^\nu = \mathbf{k}_\|^\nu(E, \theta)$ and $\mathbf{k}_\|^{\mu\,\prime} = \mathbf{k}_\|^\mu(E', \theta')$.

Similarly as for electrons, we represent the wave vector $\mathbf{q}$ of LO phonons in cylindrical coordinates

$$\mathbf{q} = (q_\| \cos\zeta, q_\| \sin\zeta, q_z), \qquad (7.72)$$

where $q_\|$ is the modulus of $\mathbf{q}_\|$, i.e. the projection of $\mathbf{q}$ onto the (x-y) plane, and $\zeta$ is the polar angle between $\mathbf{q}_\|$ and $\mathbf{E}_\|$ as illustrated in Fig. 7.7. Here, we introduce the 2D LO phonon distribution $g(\mathbf{q}_\|, t)$, which is obtained from the 3D distribution function by averaging it with respect to $q_z$. The dynamics of LO phonons is governed by a BTE similar to that of electrons. Taking advantage of (7.72) and the definition

$$G(q, \zeta, t) = \mathcal{Z}_{\mathrm{LO}}(q_\|) g(q, \zeta, t) \qquad (7.73)$$

with the LO phonon density of states

$$\mathcal{Z}_{\mathrm{LO}}(q_\|) = \frac{q_\|}{4\pi^2} \qquad (7.74)$$

leads to

$$\frac{\partial G}{\partial t} = \mathcal{D}_1[G] + \mathcal{D}_2[G]. \qquad (7.75)$$

The phonon-phonon interaction term $\mathcal{D}_2$ is represented in the relaxation time approximation

$$\mathcal{D}_2[G] = -\frac{G(\mathbf{q}_\|) - G_{\mathrm{BE}}(\mathbf{q}_\|)}{\tau_{\mathrm{LO}}} \qquad (7.76)$$

with the Bose-Einstein distribution $G_{\mathrm{BE}} = \mathcal{Z}_{\mathrm{LO}}\, g_{\mathrm{BE}}$ and the relaxation time $\tau_{\mathrm{LO}}$. Electron-phonon interaction is described by the collision term

$$\mathcal{D}_1[G] = \sum_{\nu,\mu} \int_0^\infty dE \int_0^{2\pi} d\theta \qquad (7.77)$$

$$\times \left\{ c_1(\mathbf{k}_\|^\nu, \mathbf{k}_\|^\nu - \mathbf{q}_\|) F^\nu(\mathbf{k}_\|^\nu)[1 - f^\mu(\mathbf{k}_\|^\nu - \mathbf{q}_\|)][G(\mathbf{q}_\|) + \mathcal{Z}_{\mathrm{LO}}(q_\|)]\, \delta(E' - E + \hbar\omega_{\mathrm{LO}}) \right.$$

$$\left. - c_1(\mathbf{k}_\|^\nu, \mathbf{k}_\|^\mu + \mathbf{q}_\|) F^\nu(\mathbf{k}_\|^\nu)[1 - f^\mu(\mathbf{k}_\|^\nu + \mathbf{q}_\|)][G(\mathbf{q}_\|)]\, \delta(E' - E - \hbar\omega_{\mathrm{LO}}) \right\}.$$

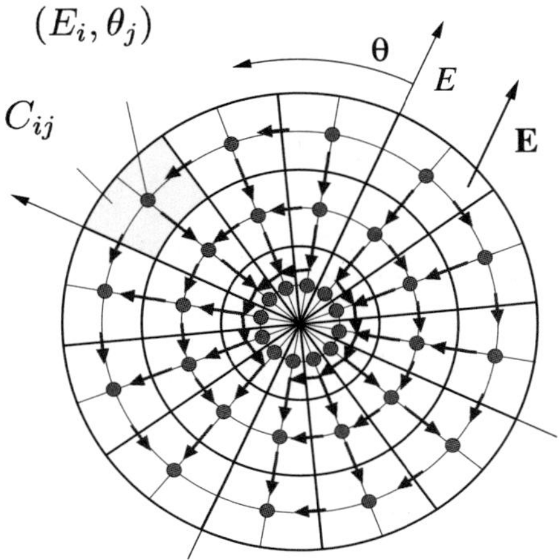


Fig. 7.8   Partition of the $\mathbf{k}_\parallel$ space and possible flows of the 2DEG according to the electric field.

In order to formulate multigroup equations to the 2D-BBP equations (7.69) and (7.73), we introduce the discretization of the independent variables $E$, $\theta$, $q_\parallel$ and $\zeta$ according to

$$E_{i+1/2} = \varepsilon_1 + i\Delta E, \; i = 0, 1, \ldots, N, \; \Delta E = \frac{\hbar\omega_{\mathrm{LO}}}{n_{\mathrm{mul}}}, \; n_{\mathrm{mul}} \in \mathbb{N}, \quad (7.78\mathrm{a})$$

$$E_i = \left[ i - \frac{1}{2} \right] \Delta E, \; i = 1, 2, \ldots, N, \quad (7.78\mathrm{b})$$

$$\theta_{j+1/2} = j\Delta\theta, \; j = 0, 1, \ldots, M, \Delta\theta = \frac{2\pi}{M}, \quad (7.78\mathrm{c})$$

$$\theta_j = \left[ j - \frac{1}{2} \right] \Delta\theta, \; j = 0, 1, \ldots, M, \quad (7.78\mathrm{d})$$

$$q_{x+1/2} = j\Delta q, \; x = 0, 1, \ldots, R, \Delta q = \frac{q_{\mathrm{max}}}{R}, \quad (7.78\mathrm{e})$$

$$\zeta_{y+1/2} = y\Delta\zeta, \; y = 0, 1, \ldots, S, \Delta\zeta = \frac{2\pi}{S} \quad (7.78\mathrm{f})$$

as it is illustrated in Fig. 7.8. Here, $E_{\mathrm{max}} = N\Delta E$ and $q_{\mathrm{max}}$ must be chosen in a way that $f^\nu(E_{\mathrm{max}})$ is negligible for all $\nu$, $\theta$ and $t$ and that $g(q_{\mathrm{max}})$ is undisturbed by the electron-phonon interaction for all $\zeta$ and $t$.

The distribution functions of electrons and phonons are approximated as the finite sums

$$\mathcal{Z}_{\mathrm{el}}^{\nu}(E)f^{\nu}(E,\theta,t) = \sum_{i=1}^{N}\sum_{j=1}^{M} n_{ij}^{\nu}(t)\,\delta(E-E_i)\delta(\theta-\theta_j), \tag{7.79a}$$

$$\mathcal{Z}_{\mathrm{el}}^{\nu}(E)[1-f^{\nu}(E,\theta,t)] = \sum_{i=1}^{N}\sum_{j=1}^{M}[1_{ij}^{\nu}-n_{ij}^{\nu}(t)]\,\lambda_i^{\mathrm{E}}(E)\delta(\theta-\theta_j), \tag{7.79b}$$

$$\mathcal{Z}_{\mathrm{ph}}(q_{\|})g(q_{\|},\zeta,t) = \sum_{x=1}^{R}\sum_{y=1}^{S} r_{xy}(t)\,\lambda_x^{\mathrm{q}}(q_{\|})\lambda_y^{\zeta}(\zeta), \tag{7.79c}$$

$$\mathcal{Z}_{\mathrm{ph}}(q_{\|})[g(q_{\|},\zeta,t)+1] = \sum_{x=1}^{R}\sum_{y=1}^{S}[1_{xy}^{\mathrm{p}}-r_{xy}(t)]\,\lambda_x^{\mathrm{q}}(q_{\|})\lambda_y^{\zeta}(\zeta) \tag{7.79d}$$

with $N \times M$ coefficients $N_{ij}^{\nu}$ for each electron subband and $R \times S$ phonon coefficients $r_{xy}$. We remark that $n_{ij}^{\nu} = 0$ for $E_i < \varepsilon_{\nu}$ for all $\theta$ and $t$. In ansatz (7.79), we use the abbreviations

$$\lambda_i^{\mathrm{E}}(E) = \begin{cases} \Delta E^{-1}, & \text{if } E \in [E_{i-1/2}, E_{i+1/2}], \\ 0, & \text{if } E \notin [E_{i-1/2}, E_{i+1/2}], \end{cases} \tag{7.80a}$$

$$\lambda_x^{\mathrm{q}}(q_{\|}) = \begin{cases} \Delta q^{-1}, & \text{if } q_{\|} \in [q_{x-1/2}, q_{x+1/2}], \\ 0, & \text{if } q_{\|} \notin [q_{x-1/2}, q_{x+1/2}], \end{cases} \tag{7.80b}$$

$$\lambda_y^{\zeta}(\zeta) = \begin{cases} \Delta\zeta^{-1}, & \text{if } \zeta \in [\zeta_{y-1/2}, \zeta_{y+1/2}], \\ 0, & \text{if } \zeta \notin [\zeta_{y-1/2}, \zeta_{y+1/2}] \end{cases} \tag{7.80c}$$

and

$$1_{ij}^{\nu} = \int_{E_{i-1/2}}^{E_{i+1/2}} dE \int_{\theta_{j-1/2}}^{\theta_{j+1/2}} d\theta\, \mathcal{Z}_{\mathrm{el}}^{\nu}(E) \tag{7.81a}$$

$$= \frac{m^*}{4\pi^2\hbar^2}[\Theta(E_{i+1/2}-\varepsilon_{\nu})E_{i+1/2} - \Theta(E_{i-1/2}-\varepsilon_{\nu})E_{i-1/2}]\Delta\theta,$$

$$1_{xy}^{\mathrm{p}} = \int_{q_{x-1/2}}^{q_{x+1/2}} dq_{\|} \int_{\zeta_{y-1/2}}^{\zeta_{y+1/2}} d\zeta\, \mathcal{Z}_{\mathrm{ph}}(q_{\|}) = \frac{1}{8\pi^2}[q_{x+1/2}^2 - q_{x-1/2}^2]\Delta\zeta. \tag{7.81b}$$

Forming moments of (8.15) reveals that the macroscopic quantities electron density $\langle n^{\nu}\rangle$, drift velocity $\langle v^{\nu}\rangle$ in direction of the electric field and mean energy $\langle E^{\nu}\rangle$ in the $\nu$th subband as well as the phonon density $\langle n^{\mathrm{p}}\rangle$

are simply given by

$$\langle n^\nu(t)\rangle = 2\sum_{i=1}^{N}\sum_{j=1}^{M} n_{ij}^\nu(t), \tag{7.82a}$$

$$\langle v^\nu(t)\rangle = \frac{2\hbar}{m^*\langle n^\nu(t)\rangle}\sum_{i=1}^{N}\sum_{j=1}^{M} k_\parallel^\nu(E_i)\cos\theta_j\, n_{ij}^\nu(t), \tag{7.82b}$$

$$\langle E^\nu(t)\rangle = \frac{2}{\langle n^\nu(t)\rangle}\sum_{i=1}^{N}\sum_{j=1}^{M} E_i\, n_{ij}^\nu(t), \tag{7.82c}$$

$$\langle n^{\mathrm{P}}(t)\rangle = \sum_{x=1}^{R}\sum_{y=1}^{S} r_{xy}(t). \tag{7.82d}$$

The evolution equations for the coefficients $n_{ij}^\nu$ and $r_{xy}$ are obtained by inserting (7.79) into the 2D-BBP equations (7.69) and (7.73) and integrating the result over the cells $C_{ij} = [E_{i-1/2}, E_{i+1/2}] \times [\theta_{j-1/2}, \theta_{j+1/2}]$ and $D_{xy} = [q_{x-1/2}, q_{x+1/2}] \times [\zeta_{y-1/2}, \zeta_{y+1/2}]$, respectively. This procedure yields, when applying an upwind scheme (cf. Fig. 7.8) with linear approximations for the fluxes at the boundaries of $C_{ij}$ controlled by a MinMod slope limiter [LeVeque (1992)],

$$\frac{\partial n_{ij}^\nu}{\partial t} + a_1^\nu(E_{i+1/2},\theta_j)[n_{ij}^{\mathrm{E},\nu}]^+ - a_1^\nu(E_{i-1/2},\theta_j)[n_{ij}^{\mathrm{E},\nu}]^+ \tag{7.83}$$

$$+a_2^\nu(E_i,\theta_{j+1/2})[n_{ij}^{\theta,\nu}]^+ - a_2^\nu(E_i,\theta_{i-1/2})[n_{ij}^{\theta,\nu}]^- = \mathcal{C}_1[n_{ij}^\nu] + \mathcal{C}_2[n_{ij}^\nu],$$

$$\frac{\partial r_{xy}}{\partial t} = 2\mathcal{D}_1[r_{xy}] + \mathcal{D}_2[r_{xy}]$$

with

$$a_1^\nu(E,\theta) = -\frac{e\mathcal{E}}{m^*}\hbar k_\parallel^\nu(E)\cos\theta, \tag{7.84a}$$

$$a_2^\nu(E,\theta) = \frac{e\mathcal{E}}{\hbar k_\parallel^\nu(E)}\sin\theta. \tag{7.84b}$$

Moreover, we have defined

$$[n_{ij}^{\mathrm{E},\nu}]^{+} = \begin{cases} n_{ij}^{\nu} + \dfrac{s_{ij}^{\mathrm{E},\nu}}{2}, & \text{if } a_1^{\nu}(E_i,\theta_j) > 0, \\[2ex] n_{i+1,j}^{\nu} - \dfrac{s_{i+1,j}^{\mathrm{E},\nu}}{2}, & \text{if } a_1^{\nu}(E_i,\theta_j) < 0, \end{cases} \tag{7.85a}$$

$$[n_{ij}^{\mathrm{E},\nu}]^{-} = \begin{cases} n_{ij}^{\nu} - \dfrac{s_{ij}^{\mathrm{E},\nu}}{2}, & \text{if } a_1^{\nu}(E_i,\theta_j) < 0, \\[2ex] n_{i-1,j}^{\nu} + \dfrac{s_{i-1,j}^{\mathrm{E},\nu}}{2}, & \text{if } a_1^{\nu}(E_i,\theta_j) > 0 \end{cases} \tag{7.85b}$$

and

$$[n_{ij}^{\theta,\nu}]^{+} = \begin{cases} n_{ij}^{\nu} + \dfrac{s_{ij}^{\theta,\nu}}{2}, & \text{if } a_2^{\nu}(E_i,\theta_j) > 0, \\[2ex] n_{i,j+1}^{\nu} - \dfrac{s_{i,j+1}^{\theta,\nu}}{2}, & \text{if } a_2^{\nu}(E_i,\theta_j) < 0, \end{cases} \tag{7.86a}$$

$$[n_{ij}^{\mathrm{E},\nu}]^{-} = \begin{cases} n_{ij}^{\nu} - \dfrac{s_{ij}^{\theta,\nu}}{2}, & \text{if } a_2^{\nu}(E_i,\theta_j) < 0, \\[2ex] n_{i,j-1}^{\nu} + \dfrac{s_{i,j-1}^{\theta,\nu}}{2}, & \text{if } a_2^{\nu}(E_i,\theta_j) > 0 \end{cases} \tag{7.86b}$$

with the slopes

$$s_{ij}^{\mathrm{E},\nu} = \mathrm{MM}(n_{i+1,j}^{\nu} - n_{ij}^{\nu}, n_{ij}^{\nu} - n_{i-1,j}^{\nu}), \tag{7.87a}$$

$$s_{ij}^{\theta,\nu} = \mathrm{MM}(n_{i,j+1}^{\nu} - n_{ij}^{\nu}, n_{ij}^{\nu} - n_{i,j-1}^{\nu}) \tag{7.87b}$$

and the MinMod limiter

$$\mathrm{MM}(a,b) = \begin{cases} 0, & \text{if } ab < 0, \\ \max(a,b), & \text{if } a < 0, \\ \min(a,b), & \text{if } a > 0. \end{cases} \tag{7.88}$$

The electron collision terms $C_1[n_{ij}^{\nu}]$ and $C_2[n_{ij}^{\nu}]$ read

$$\mathcal{C}_1[n_{ij}^{\nu}] = \sum_{\mu} \sum_{x=1}^{R} \sum_{y=1}^{S} \sum_{a=1}^{N} \sum_{b=1}^{M} \tag{7.89a}$$

$$\left\{ \langle S_{ab\to ij,xy}^{\mu\to\nu}\rangle^{+} n_{ab}^{\mu}[1_{ij}^{\nu} - n_{ij}^{\nu}] \, r_{xy} + \langle S_{ab\to ij,xy}^{\mu\to\nu}\rangle^{-} n_{ab}^{\mu}[1_{ij}^{\nu} - n_{ij}^{\nu}][r_{xy} + 1_{xy}^{\mathrm{P}}] \right.$$

$$\left. - \langle S_{ij\to ab,xy}^{\nu\to\mu}\rangle^{+} n_{ij}^{\nu}[1_{ab}^{\mu} - n_{ab}^{\mu}] r_{xy} - \langle S_{ij\to ab,xy}^{\nu\to\mu}\rangle^{-} n_{ij}^{\nu}[1_{ab}^{\mu} - n_{ab}^{\mu}][r_{xy} + 1_{xy}^{\mathrm{P}}] \right\},$$

$$\mathcal{C}_2[n_{ij}^{\nu}] = \sum_{\mu} \sum_{a=1}^{N} \sum_{b=1}^{M} \tag{7.89b}$$

$$\left\{ \langle C_{ab\to ij}^{\mu\to\nu}\rangle n_{ab}^{\mu}[1_{ij}^{\nu} - n_{ij}^{\nu}] - \langle C_{ij\to ab}^{\nu\to\mu}\rangle n_{ij}^{\nu}[1_{ab}^{\mu} - n_{ab}^{\mu}] \right\}$$

with the collision coefficients

$$\langle S^{\nu\to\mu}_{ij\to ab,xy}\rangle^{+} = \int_{C_{ij}} dE\, d\theta \int_{C_{ab}} dE'\, d\theta' \frac{c_1(\mathbf{k}^{\mu}_{\parallel}{}' - \mathbf{k}^{\nu}_{\parallel})}{\mathcal{Z}_{\mathrm{LO}}(|\mathbf{k}^{\mu}_{\parallel}{}' - \mathbf{k}^{\nu}_{\parallel}|)} \tag{7.90a}$$
$$\times\, \delta(E - E_i)\lambda^{\mathrm{E}}_a(E')\delta(\theta - \theta_j)\delta(\theta' - \theta_b)\lambda^{\mathrm{q}}_x(|\mathbf{k}^{\mu}_{\parallel}{}' - \mathbf{k}^{\nu}_{\parallel}|)$$
$$\times\, \lambda^{\zeta}_y(\mathrm{acos}|\mathbf{k}^{\mu}_{\parallel}{}' - \mathbf{k}^{\nu}_{\parallel}|^{-1}\mathbf{e}_x \cdot (\mathbf{k}^{\mu}_{\parallel}{}' - \mathbf{k}^{\nu}_{\parallel})]\,\delta(E' - E - \hbar\omega_{\mathrm{LO}}),$$

$$\langle S^{\nu\to\mu}_{ij\to ab,xy}\rangle^{-} = \int_{C_{ij}} dE\, d\theta \int_{C_{ab}} dE'\, d\theta' \frac{c_1(\mathbf{k}^{\nu}_{\parallel} - \mathbf{k}^{\mu}_{\parallel}{}')}{\mathcal{Z}_{\mathrm{LO}}(|\mathbf{k}^{\nu}_{\parallel} - \mathbf{k}^{\mu}_{\parallel}{}'|)} \tag{7.90b}$$
$$\times\, \delta(E - E_i)\lambda^{\mathrm{E}}_a(E')\delta(\theta - \theta_j)\delta(\theta' - \theta_b)\lambda^{\mathrm{q}}_x(|\mathbf{k}^{\nu}_{\parallel} - \mathbf{k}^{\mu}_{\parallel}{}'|)$$
$$\times\, \lambda^{\zeta}_y(\mathrm{acos}|\mathbf{k}^{\nu}_{\parallel} - \mathbf{k}^{\mu}_{\parallel}{}'|^{-1}\mathbf{e}_x \cdot (\mathbf{k}^{\nu}_{\parallel} - \mathbf{k}^{\mu}_{\parallel}{}')]\,\delta(E' - E + \hbar\omega_{\mathrm{LO}}),$$

$$\langle C^{\mu\to\nu}_{ab\to ij}\rangle = \int_{C_{ij}} dE\, d\theta \int_{C_{ab}} dE'\, d\theta'\, c_2(\mathbf{k}^{\nu}_{\parallel}, \mathbf{k}^{\mu}_{\parallel}{}') \tag{7.90c}$$
$$\times\, \delta(E - E_i)\lambda^{\mathrm{E}}_a(E')\delta(\theta - \theta_j)\delta(\theta' - \theta_b)\delta(E' - E).$$

In addition, the phonon collision terms $D_1[r_{xy}]$ and $D_2[r_{xy}]$ are given by

$$\mathcal{D}_1[r_{xy}] = \sum_{\nu,\mu}\sum_{i=1}^{N}\sum_{j=1}^{M}\sum_{a=1}^{N}\sum_{b=1}^{M} n^{\nu}_{ij}[1^{\mu}_{ab} - n^{\mu}_{ab}] \tag{7.91}$$
$$\times\left\{\langle S^{\nu\to\mu}_{ij\to ab,xy}\rangle^{-}[r_{xy} + 1^{\mathrm{P}}_{xy}] - \langle S^{\nu\to\mu}_{ij\to ab,xy}\rangle^{+}r_{xy}\right\},$$

$$\mathcal{D}_2[r_{xy}] = \frac{1}{\tau_{\mathrm{LO}}}[1^{\mathrm{P}}_{xy}g_0 - r_{xy}] \tag{7.92}$$

with

$$g_0 = \left[\exp\frac{\hbar\omega_{\mathrm{LO}}}{k_{\mathrm{B}}T_{\mathrm{L}}} - 1\right]^{-1}. \tag{7.93}$$

For computing the screening parameter for polar optical and piezoelectric scattering, it is necessary to determine the electron distribution functions at the bottoms of the subbands (cf. (7.52)). These quantities $f^{\nu}(0)$ are approximated via

$$f^{\nu}(0) = \sum_{j=1}^{M} \frac{n^{\nu}_{I^{\nu}_{\min},j}}{1^{\nu}_{I^{\nu}_{\min},j}}, \tag{7.94}$$

where $I^{\nu}_{\min}$ is the energy index so that $\varepsilon_{\nu} \in [E_{I^{\nu}_{\min}-1/2}, E_{I^{\nu}_{\min}+1/2}]$.

## 7.4 Transport in $Al_x Ga_{1-x} N/GaN$

In this section, we present the physical model, which the simulations of the particle transport at a $Al_x Ga_{1-x} N/GaN$ heterojunction are based on. Here, we closely follow the considerations given in [Ramonas *et al.* (2003)], which enables us to validate the results obtained by our deterministic multigroup approach by comparing them to the Monte Carlo results given there. All the calculations are performed for an electron density $n_{el} = 10^{12}$ cm$^{-2}$ and a lattice temperature $T_L = 300$ K.

We consider the transport of a 2DEG formed at a wurtzite, Ga-face AlGaN/GaN heterojunction consisting of a 25 nm $Al_{0.15} Ga_{0.85} N$ undoped layer and a thick undoped GaN layer. The c-axis is taken to be perpendicular to the AlGaN/GaN interface resulting in one transverse effective mass $m^*$ and one set of energy subbands with the energy eigenvalues $\varepsilon_\nu$ and the normalized envelope wave functions $\varphi_\nu$. In contrast to the confined electrons, we treat the phonons as three-dimensional particles, which is reasonable for the small changes in the mechanical parameters at the AlGaN/GaN heterojunction. In Fig. 7.7, we display the geometry used in our considerations. The z-axis is chosen normal to the AlGaN/GaN interface at $z = 0$. Hence, electrons are confined in z-direction and they move semiclassically free in the (x,y) plane parallel to the heterojunction. This homogeneous transport is driven by an electric field **E** in x-direction.

We include scattering mechanisms for the electrons caused by acoustic and longitudinal optical phonons in our transport model. Scattering by ionized impurities is neglected since we deal with an undoped heterojunction, where the 2DEG is induced only by spontaneous and piezoelectric polarization charges. In the considered range of the electric field strength, electron scattering into upper valleys is supposed to be negligible. We employ a one-valley multi-subband spherical parabolic model band structure. Electron real space transfer and sharing effects are neglected in our calculations.

The electrons interact with acoustic phonons through deformation potential and electrostatic polarization associated with atom vibrations. In wurtzite structures, the deformation potential in the central valley is a diagonal second rank tensor. The value of $D_{zz}$ is in general different from $D_{xx} = D_{yy}$. However, experimental data are not available for these quantities. Therefore, we assume equal diagonal elements and treat the deformation potential tensor as a scalar quantity [Kolnik *et al.* (1995)]. This implies that the scattering function $c_{ADP}$ for acoustic deformation potential

Table 7.1  Material parameter for the 2DEG simulation.

| Quantity | Symbol | Unit | GaN | Al$_{0.15}$Ga$_{0.85}$N |
|---|---|---|---|---|
| Electron effective mass[a] | $m^*$ | | 0.22 $m_e$ | |
| Static dielectric constant[a] | $\kappa_0$ | | 8.9 | |
| HF dielectric constant[b] | $\kappa_\infty$ | | 5.23 | |
| Lattice constants[c] | $a$ | nm | 0.3189 | 0.3177 |
| | $c$ | nm | 5.185 | |
| Acoustic deform. potential[b] | $D_A$ | eV | 8.3 | |
| Long. sound velocity[b] | $v_l$ | ms$^{-1}$ | 6560 | |
| Trans. sound velocity[b] | $v_t$ | ms$^{-1}$ | 2680 | |
| Mass density[b] | $\rho$ | kgm$^{-3}$ | 6150 | |
| Longitudinal optical phonon energy[b] | $\hbar\omega_{LO}$ | meV | 91.2 | |
| Transverse optical phonon energy[d] | $\hbar\omega_{TO}$ | meV | 69.5 | |
| Piezoelectric constants[c] | $e_{31}$ | Cm$^{-2}$ | -0.49 | -0.506 |
| | $e_{33}$ | Cm$^{-2}$ | 0.73 | 0.839 |
| | $e_{15}$ | Cm$^{-2}$ | -0.3 | |
| Spontaneous polarization [c] | $P_{sp}$ | Cm$^{-2}$ | -0.029 | -0.0368 |
| Elastic constants[c] | $c_{13}$ | GPa | 103.0 | 103.75 |
| | $c_{33}$ | GPa | 405.0 | 400.2 |
| Averaged elastic constants[e] | $c_{LA}$ | GPa | 265 | |
| | $c_{TA}$ | GPa | 44.2 | |
| Phonon relaxation time[f] | $\tau_{LO}$ | ps | 1 | |
| Interface barrier[f] | $\Delta E_c$ | eV | 0.6 | |

[a] Ref. [Gaska *et al.* (1998)], [b] Ref. [O'Leary *et al.* (1998)],
[c] Ref. [Ambacher *et al.* (1999)], [d] Ref. [Bulutay *et al.* (2000)],
[e] Ref. [Ridley *et al.* (2000)], [f] Ref. [Ramonas *et al.* (2003)]

scattering (cf. (7.22), (7.90c)) reads in the elastic approximation as

$$c_{ADP}(\mathbf{k}_\parallel^\nu, \mathbf{k}_\parallel^{\mu\,\prime}) = \frac{2\pi D_A^2 k_B T_L}{\hbar \rho v_s^2} \frac{1}{W_{\mu\nu}}. \tag{7.95}$$

The strength of the piezoelectric scattering is determined by the dimensionless electromechanical coupling coefficient $K^2$, which is related to the piezoelectric constant $e_{PZ}$ in (7.28) via [Lundstrom (2000)]

$$\frac{K^2}{1-K^2} = \frac{e_{PZ}^2}{\kappa_0 \varepsilon_0 \rho v_s^2} \tag{7.96}$$

with $v_s = (v_l v_t^2)^{1/3}$. The quantity $K^2$ contains contributions of the longitudinal (LA) and the transverse (TA) acoustic phonons. The effective piezoelectric constants for LA and TA modes are

$$e_{\mathrm{LA}} = (2e_{15} + e_{31}) \sin^2 \theta \cos \theta + e_{33} \cos^3 \theta, \tag{7.97a}$$

$$e_{\mathrm{LA}} = (e_{33} - e_{15} - e_{31}) \cos^2 \theta \sin \theta + e_{15} \sin^3 \theta, \tag{7.97b}$$

where $\theta$ is the angle between the direction of propagation and the c-axis [Ridley *et al.* (2000)]. Following [Hutson (1981)] by performing the spherical averages, one obtains

$$K^2 = \frac{\langle e_{\mathrm{LA}}^2 \rangle}{\kappa_0 \varepsilon_0 c_{\mathrm{LA}}} + \frac{\langle e_{\mathrm{TA}}^2 \rangle}{\kappa_0 \varepsilon_0 c_{\mathrm{TA}}}, \tag{7.98}$$

where $c_{\mathrm{LA}}$ and $c_{\mathrm{TA}}$ are angular averages of the elastic constants describing the propagation of LA and TA waves. The averaged piezoelectric constants are given by

$$\langle e_{\mathrm{LA}}^2 \rangle = \frac{1}{105} [8(2e_{15} + e_{31})^2 + 12(2e_{15} + e_{31})e_{33} + 15e_{33}^2], \tag{7.99a}$$

$$\langle e_{\mathrm{TA}}^2 \rangle = \frac{1}{105} [6(e_{33} - e_{15} - e_{31})^2 + 16(e_{33} - e_{15} - e_{31})e_{15} + 48e_{15}^2]. \tag{7.99b}$$

This implies that the scattering function $c_{\mathrm{PZ}}$ (cf. (7.28), (7.90c)) for screened piezoelectric scattering reads

$$c_{\mathrm{PZ}}(\mathbf{k}_\parallel^\nu, \mathbf{k}_\parallel^{\mu\,\prime}) = \frac{2\pi k_{\mathrm{B}} T_{\mathrm{L}}}{\hbar \rho v_{\mathrm{s}}^2} \left( \frac{e e_{\mathrm{PZ}}}{\kappa_0 \varepsilon_0} \right)^2 \frac{1}{W_{\mu\nu}} \tag{7.100}$$

$$\times \left[ \frac{\delta_{\mu\nu}}{(|\mathbf{k}_\parallel^{\mu\,\prime} - \mathbf{k}_\parallel^\nu| + q_{\mathrm{s}})^2} + \frac{1 - \delta_{\mu\nu}}{|\mathbf{k}_\parallel^{\mu\,\prime} - \mathbf{k}_\parallel^\nu|^2} \right]$$

with the screening parameter

$$q_{\mathrm{s}} = \sum_\nu \frac{m^* e^2}{2\pi \kappa_0 \varepsilon_0 \hbar^2} f^\nu(0). \tag{7.101}$$

Electron optical phonon coupling in wurtzite crystals is different from the well-known cubic case. The electrons interact with both longitudinal optical (LO) and transverse optical (TO) modes rather than with a single LO mode as in cubic lattices. However, it has been shown that the scattering rate for TO scattering is more than two orders of magnitude smaller than that for LO scattering. Moreover, the LO scattering rate in the cubic approximation is valid regardless of the chosen point in the Brillouin zone [Bulutay *et al.* (2000)]. Hence, we use the cubic approximation and the formulation of Price for the transition rates, which leads to the scattering function $c_{\mathrm{POP}}$ (cf. (7.39), (7.90a)) for screened polar optical scattering:

$$c_{\mathrm{POP}}(q_\parallel) = \frac{e^2 \omega_0}{2\varepsilon_0} \left( \frac{1}{\kappa_\infty} - \frac{1}{\kappa_0} \right) \tag{7.102}$$

$$\times \left[ \frac{\pi \delta_{\mu\nu}\, q_\parallel^2}{(q_\parallel + \pi W_{\mu\nu} q_\parallel^2)(q_\parallel + q_s)^2} + \frac{\pi(1 - \delta_{\mu\nu})(\alpha_{\mu\nu} - \beta_{\mu\nu} q_\parallel)}{q_\parallel + \pi W_{\mu\nu}(\alpha_{\mu\nu} - \beta_{\mu\nu} q_\parallel)q_\parallel^2} \right].$$

The simulations shown in this section are obtained by taking into account the four lowest energy subbands. As concerns the parameters used in our numerical scheme, we set $N = 60$, $M = 24$, $n_{\mathrm{mul}} = 10$. This implies that $E_{\max} = 0.55$ eV. Additionally, we use $R = 50$ and $S = 24$ with the maximum wave vector $q_{\max} = 3.6 \times 10^9$ m$^{-1}$. Initial date for the coefficients $n_{ij}^\nu$ and $r_{xy}$ are obtained by integrating the Fermi-Dirac distribution and the Bose-Einstein distribution over the cells $C_{ij}$ and $D_{xy}$, respectively. The stationary state is assumed be be reached after 10 ps after the onset of the electric field.

### 7.4.1 *Self-consistent Solution for Confining Potential*

We solve the coupled system of the effective mass Schrödinger and the Poisson equations self-consistently for computing the confining potential and the energy subband structures needed for our transport equations. Therefore, we proceed as follows [Ferry and Goodnick (1997)]. For a 2DEG confined in a quantum well at a heterojunction, the electrons move in the potential $V(z)$ given by

$$V(z) = -e\Psi(z) + \Delta E_{\mathrm{c}}\Theta(z - z_0). \tag{7.103}$$

Here, $e$ is the elementary charge, $\Psi(z)$ denotes the electrostatic potential depending on the position $z$ normal to the AlGaN/GaN interface at $z_0 = 0$ nm (cf. Fig. 7.10), the symbols $\Delta E_{\mathrm{c}}$ and $\Theta$ label the interface barrier and the Heaviside step function. The electrostatic potential is related to the charge distributions by the Poisson equation

$$\frac{d}{dz}\left[ \kappa_0(z) \frac{d}{dz} \right] \Psi(z) = -\frac{e}{\varepsilon_0}[N_{\mathrm{f}}(z) - n(z)], \tag{7.104}$$

where $\kappa_0$ is the static dielectric constant, $\varepsilon_0$ is the permittivity of free space, $N_{\mathrm{f}}(z)$ is the space dependent fixed charge given by the structure of the heterojunction and $n(z)$ is the charge distribution of the quantum

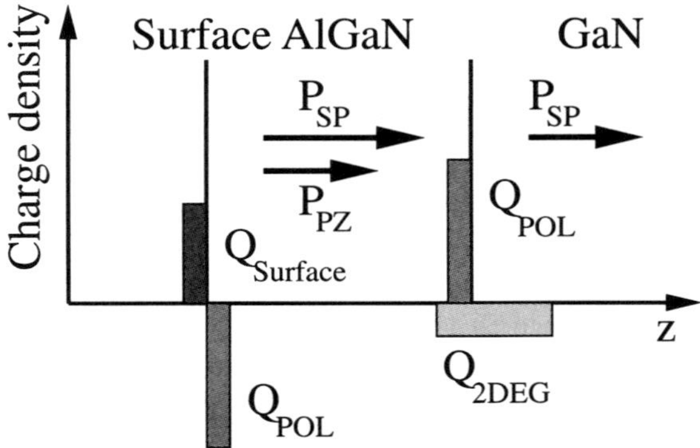


Fig. 7.9    Charge distribution and polarization at a $Al_xGa_{1-x}N/GaN$ heterojunction.

confined electrons. This quantity can be evaluated via

$$n(z) = \sum_i \frac{m^*(z)k_B T_L}{\pi\hbar^2} \ln\left[1 + \exp\left(\frac{E_F - \varepsilon_i}{k_B T_L}\right)\right] |\varphi_i(z)|^2 \qquad (7.105)$$

with the lattice temperature $T_L$ and the Fermi energy $E_F$ for parabolic energy subbands (7.5) and Fermi statistics taking into account degeneracy effects of the electron gas. The charge of the two-dimensional confined electrons depends on the subband envelope functions and the eigenenergies. These quantities are the solutions of the one-dimensional effective mass Schrödinger equation

$$\left\{-\frac{\hbar^2}{2}\frac{d}{dz}\left[\frac{1}{m^*(z)}\frac{d}{dz}\right] + V(z)\right\}\varphi_i(z) = \varepsilon_i\varphi_i(z). \qquad (7.106)$$

We apply an iteration procedure to obtain solutions to (7.104) and (7.106). To begin with, we solve the Poisson equation (7.104) with a trial 2DEG charge distribution $n(z)$. This leads to a confining potential $V(z)$ via (7.103), which is used in the Schrödinger equation (7.106) to determine eigenenergies and wave functions. The 2DEG density is corrected with the help of these quantities according to (7.105). All these calculations are performed by using standard numerical methods. Subsequent iterations lead to the final self-consistent solution for $V(z)$, $\varepsilon_i$ and $\varphi_i$ with the required accuracy.

Our calculations are performed for the lattice temperature $T_L{=}300$ K. All required constants are found in Tab. 7.1. As shown in [Ambacher *et al.*

(1999)], spontaneous and piezoelectric polarization play an important role in the quantum-confinement of electrons at AlGaN/GaN interfaces. Hence, we take into account the polarization charge $\sigma$ for the fixed charges $N_\mathrm{f}$ in (7.104), while the surface charge is neglected. The spatial distribution of the considered charge is illustrated in Fig. 7.9. In the considered configuration, $\sigma$ is positive and free electrons tend to compensate for it. Following [Ramonas *et al.* (2003)], we use a value for the interface band offset $\Delta E_\mathrm{c}$ twice as high as found in the literature [Ambacher *et al.* (1999)]. This prevents electrons entering the AlGaN layer considerably and we can neglect electron sharing effects. This assumption is justified at not to high electric fields.

The polarization charge due to piezoelectric and spontaneous polarization at the $\mathrm{Al}_x\mathrm{Ga}_{1-x}\mathrm{N}/\mathrm{GaN}$ interface is obtained by applying continuum mechanics for solid states [Weißmantel and Hamann (1995)]. We consider a thin stressed $\mathrm{Al}_x\mathrm{Ga}_{1-x}\mathrm{N}$ layer, which is grown on the thick relaxed GaN layer at the position $z_0$. Here, the material parameters of GaN alter unsteadily to these of $\mathrm{Al}_x\mathrm{Ga}_{1-x}\mathrm{N}$ with the aluminum concentration $x$. Concerning the portion of the total polarization charge due to spontaneous polarization $P_\mathrm{sp}(z)$ depending on position $z$, we simply assume that this quantity is given by

$$P_\mathrm{sp}(z) = \begin{cases} P_\mathrm{sp}(0), & \text{if } z < z_0, \\ P_\mathrm{sp}(0.15), & \text{if } z > z_0. \end{cases} \tag{7.107}$$

Crystal lattices of wurtzite-type are described by the two lattice constants $a$ and $c$ [Weißmantel and Hamann (1995)]. In the considered case, these constants equal their values of bulk GaN in the whole heterostructure leading to a stressed $\mathrm{Al}_x\mathrm{Ga}_{1-x}\mathrm{N}$ layer. This implies that the strain vector $\mathbf{e}$ in the main axis system reads in the usual notation

$$\mathbf{e} = (\Delta a(z), \Delta a(z), \Delta c(z), 0, 0, 0) \tag{7.108}$$

with

$$\Delta a(z) = \begin{cases} 0, & \text{if } z < z_0, \\ \dfrac{a(0) - a(0.15)}{a(0.15)}, & \text{if } z > z_0 \end{cases} \tag{7.109}$$

and a corresponding expression for $\Delta c$. The atomic displacements $\Delta a$ and $\Delta c$ are not independent of each other, since they are related by Hook's law:

$$\sigma = \mathsf{C} \cdot \mathbf{e}. \tag{7.110}$$

For a vanishing vector of external stresses $\sigma = 0$ and with the help of the tensor of elastic coefficients for wurtzite-type lattices

$$
\mathsf{C} = \begin{pmatrix}
C_{11} & C_{12} & C_{13} & 0 & 0 & 0 \\
C_{12} & C_{11} & C_{13} & 0 & 0 & 0 \\
C_{13} & C_{13} & C_{33} & 0 & 0 & 0 \\
0 & 0 & 0 & C_{44} & 0 & 0 \\
0 & 0 & 0 & 0 & C_{44} & 0 \\
0 & 0 & 0 & 0 & 0 & C_{66}
\end{pmatrix},
\tag{7.111}
$$

we easily find that

$$
\Delta c = -2\frac{C_{13}}{C_{33}}\Delta a.
\tag{7.112}
$$

The vector of piezoelectric polarization $\mathbf{P}_{\mathrm{pz}}$ is related to the strain vector via

$$
\mathbf{P}_{\mathrm{pz}} = \mathbf{e}\cdot\mathbf{e}
\tag{7.113}
$$

with the tensor of piezoelectric coefficients for wurtzite-type lattices

$$
\mathsf{e} = \begin{pmatrix}
0 & 0 & 0 & 0 & e_{15} & 0 \\
0 & 0 & 0 & e_{15} & 0 & 0 \\
e_{31} & e_{31} & e_{33} & 0 & 0 & 0
\end{pmatrix}.
\tag{7.114}
$$

Hence, the component $P_{\mathrm{PZ}}$ of the piezoelectric polarization in z-direction as a function of the position reads

$$
P_{\mathrm{pz}} = 2\Delta a(z)\left[e_{31}(z) - e_{33}(z)\frac{C_{13}(z)}{C_{33}(z)}\right].
\tag{7.115}
$$

Assuming discontinuous behavior of the emerging material parameter and adding $P_{\mathrm{sp}}$ and $P_{\mathrm{pz}}$ to obtain the total polarization $P(z)$ leads to

$$
P(z) = P_{\mathrm{sp}}(0) + \Theta(z - z_0)[P_{\mathrm{sp}}(0.15) - P_{\mathrm{sp}}(0) + P_{\mathrm{pz}}(0.15)]
\tag{7.116}
$$

with the Heaviside step function $\Theta$. As a final result, we find the polarization charge $Q_{\mathrm{pol}} = \partial_z P$ associated to $P$ to equal

$$
Q_{\mathrm{pol}}(z) = \left\{ P_{\mathrm{sp}}(0.15) - P_{\mathrm{sp}}(0) + 2\Delta a(z)\left[e_{31}(z) - e_{33}(z)\frac{C_{13}(z)}{C_{33}(z)}\right]\right\}\delta(z - z_0).
\tag{7.117}
$$

Figure 7.10 displays the resulting self-consistent solution for the confining potential $V(z)$ and the first four eigenenergies and envelope wave

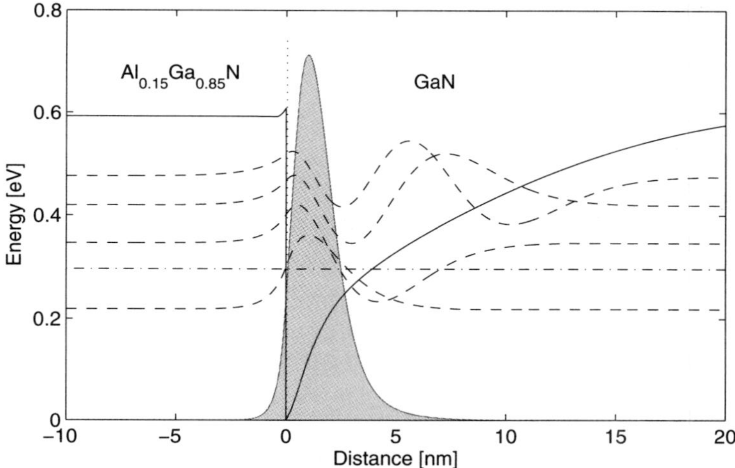


Fig. 7.10   The potential profile (solid line), the first four wave functions (dashed lines) and the profile of the 2DEG density (filled) versus position. Wave functions and electron density are plotted in arbitrary units, the zero of each wave function is the corresponding eigenenergy. The dashed-dotted line refers to the Fermi energy.

functions, which are used in our transport simulations. Moreover, this figure shows the self-consistent solution for the 2DEG density. The scattering probabilities in our transport model are calculated for the equilibrium self-consistent wave functions. For saving computational time, no field-induced modulations on the $\varepsilon_i$ and $\varphi_i$ are taken into account. This assumption is valid for the range of electric field strengths under consideration.

### 7.4.2   *Transport Properties*

In Fig. 7.11, we show the results for the electron velocity-field characteristics at a Al$_x$Ga$_{1-x}$N/GaN heterojunction at 300 K simulated by means of our multigroup equations. Three cases are considered: including the degeneracy of the 2DEG but neglecting hot phonon effects, neglecting the degeneracy of the electron gas and taking into account hot phonon effects, and including both, the degeneracy of the 2DEG and hot phonons into the transport model. The simulations show that the electron gas degeneracy as well as the hot phonon effects influence the electron drift velocity in the investigated range of the electric field and that both effects must be taken into account for good agreement with experimental data [Matulionis

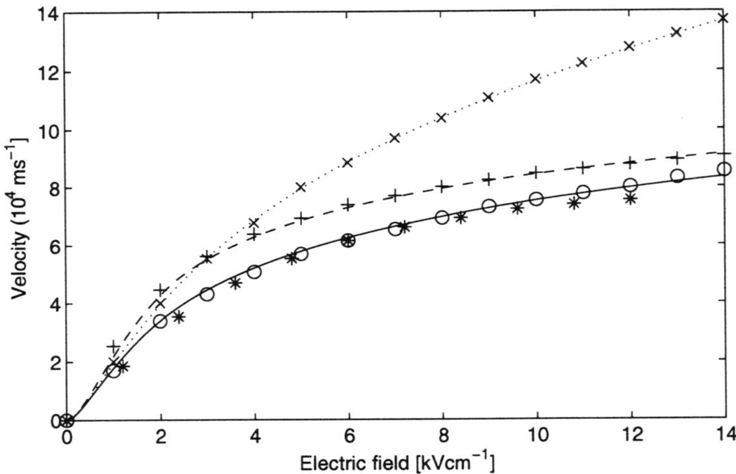

Fig. 7.11   The electron velocity-field characteristics for the $Al_xGa_{1-x}N$-GaN heterojunction at 300 K. The lines refer to results from the multigroup method, symbols to MC calculations. $(-\cdot-, \times)$: degenerated gas, equilibrium phonons; $(---, +)$: nondegenerated gas, hot phonons; $(\longrightarrow, \bigcirc)$: degenerated gas, hot phonons; $*$: experiment.

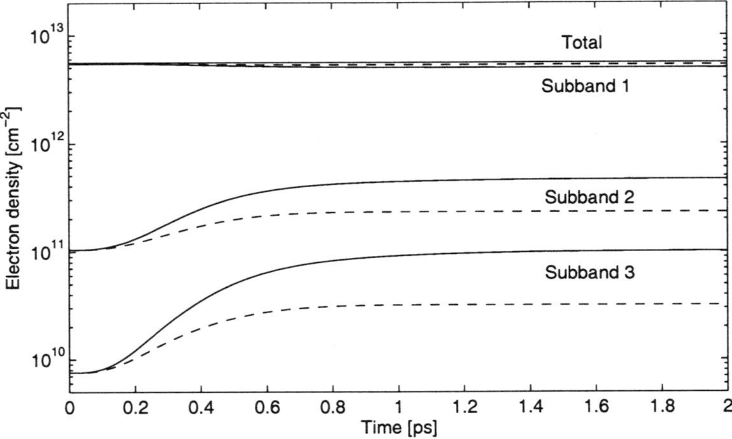

Fig. 7.12   Temporal evolution of the electron densities in the three lowest subbands of a 2DEG formed at an AlGaN/GaN heterojunction after the onset of the electric field with $\mathcal{E} = 10$ kVcm$^{-1}$. Solid lines: hot phonons; dashed lines: equilibrium phonons.

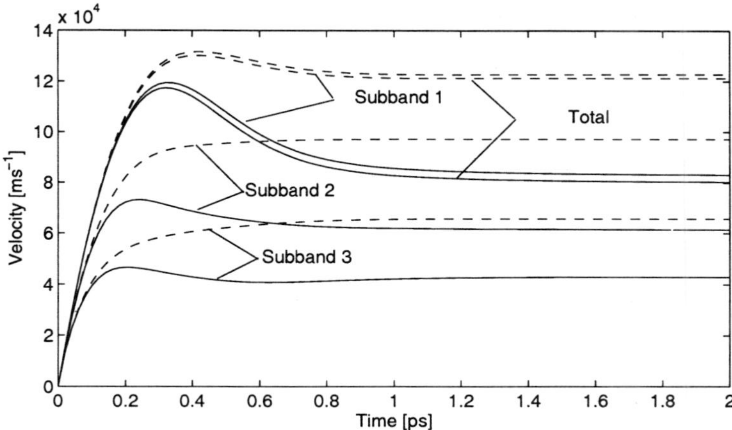

Fig. 7.13  Temporal evolution of the drift velocities in the three lowest subbands of a 2DEG formed at an AlGaN/GaN heterojunction after the onset of the electric field with $\mathcal{E} = 10$ kVcm$^{-1}$. Solid lines: hot phonons; dashed lines: equilibrium phonons.

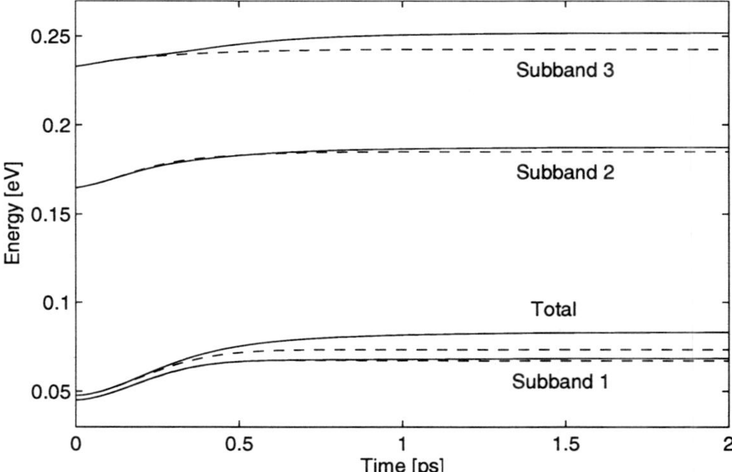

Fig. 7.14  Temporal evolution of the mean energies in the three lowest subbands of a 2DEG formed at an AlGaN/GaN heterojunction after the onset of the electric field with $\mathcal{E} = 10$ kVcm$^{-1}$. Solid lines: hot phonons; dashed lines: equilibrium phonons.

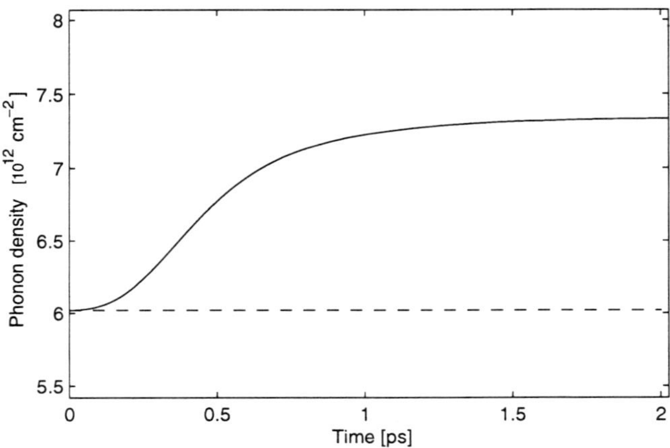

Fig. 7.15   Temporal evolution of the phonon densities at an AlGaN/GaN heterojunction after the onset of the electric field with $\mathcal{E} = 10 \text{ kVcm}^{-1}$. Solid lines: hot phonons; dashed lines: equilibrium phonons.

*et al.* (2002)a]. Moreover, we observe excellent agreement of our results with those of MC calculations [Ramonas *et al.* (2003)].

When degeneracy is neglected, the drift velocity exceeds that obtained by the full model. This can be explained in terms of the angular dependence of the final electron states after phonon emission. Small angle scattering by phonon emission processes is the dominant scattering mechanism for electrons with wave vectors in direction of the drift velocity, if the degeneracy is neglected. When the Pauli principle is applied, the small angle scattering rate reduces dramatically, since the corresponding final states are occupied. Consequently, the probability for large angle scattering is enhanced, and electrons are forced to scatter to final states in opposite direction of the drift velocity, which implies a strong negative contribution to the drift velocity. Since the mean electron energy increases as the electric field increases, electron gas degeneracy effects decrease with the rising electric field. Hence, the drift velocity curves calculated with and without degeneracy tend to merge at high fields.

When the hot phonon effects are not taken into account, the calculated drift velocity exceeds considerably the experimental data and the simulation results for the complete model. Hence, hot phonon effects can be regarded as quite important in nitride heterostructures. Due to the enhanced phonon occupation number (cf. Fig. 7.17), hot phonons support a stronger scattering of electrons. The scattering caused by the optical phonon absorption increases because of the phonon reabsorption; and the scattering caused by phonon emission increases due to the stimulated emission. This explains the essential reduction in the drift velocity, when hot phonons are taken into account. The stronger the electric field, the more pronounced are the hot phonon effects.

Finally, we note a deviation of the calculated drift velocity from the experimental data in Fig. 7.11 for fields higher than 10 kVcm$^{-1}$. It is known that electron sharing influences the experimental results on the drift velocity at stronger electric fields [Matulionis *et al.* (2002)b]. The sharing of electrons by the Al$_x$Ga$_{1-x}$N and the GaN layer has not been taken into account in this simulation, which leads to the higher calculated drift velocity in comparison to the experimentally determined one for high fields.

In Figs. 7.12, 7.13, 7.14 and 7.15, we depict the temporal evolution of the density, the velocity and the mean energy of electrons confined at a Al$_x$Ga$_{1-x}$N/GaN heterojunction as well as the LO phonon density in response to the onset of an electric field pulse of the strength $\mathcal{E} = 10$ kVcm$^{-1}$. Solid lines refer to calculations taking into account hot phonons; the dashed curves are obtained by assuming equilibrium phonon. In both simulation, degeneracy of the 2DEG is regarded. For times $t < 0.2$ ps, the results of the two considered cases almost agree, while the increasing phonon density (cf. Fig. 7.15) leads to significant differences in the macroscopic quantities for later times in correspondence with the stationary state values displayed in Fig. 7.11.

The most interesting result of these simulations is the temporal behavior of the drift velocity. We observe a velocity overshoot as it is expected for the relatively high electric field. In the case of the equilibrium phonon calculation, this velocity overshoot is caused by the ballistic transport of electrons right after the onset of the electric field, when the distribution function is shifted by the electric field, but hardly altered by scattering events. This ballistic motion contributes to the velocity overshoot in the hot phonon simulation as well, however, the dominant reason for the decrease of the velocity with increasing time is the enhanced phonon scattering caused by the non-equilibrium phonons. Their effect is strong enough to cause a velocity overshoot, not only in the lowest subband as it is the case in the equilibrium phonon simulation, but also for the higher one. Moreover, we note that the maximum velocity achieved is much lower in the case of taking into account non-equilibrium phonons. Hence, simulations aimed at designing GaN-based heterostructure semiconductor devices, which take

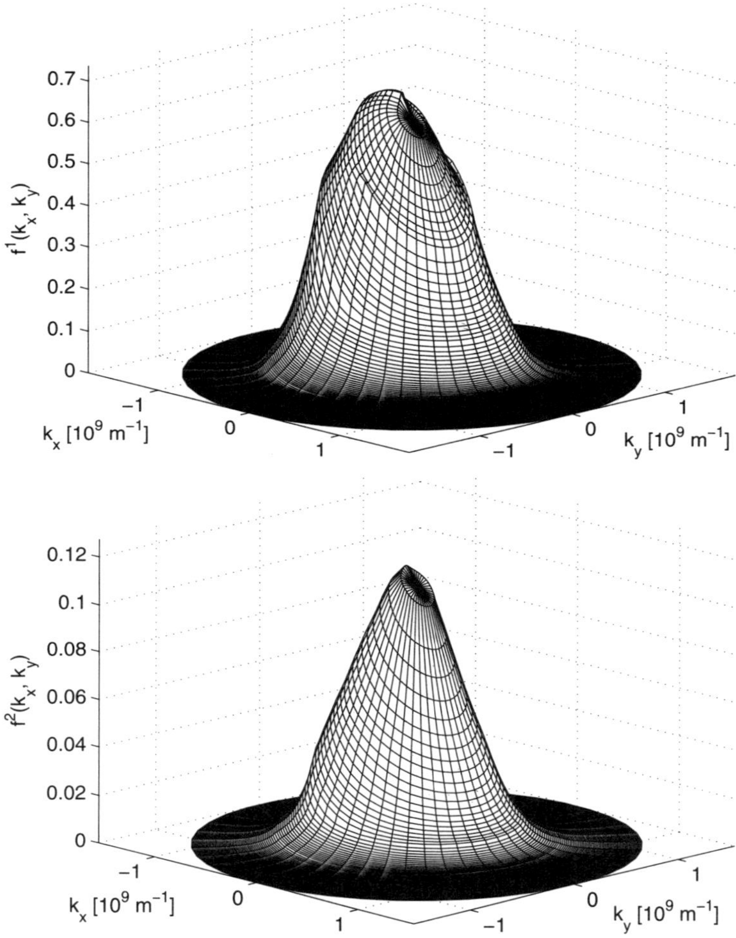

Fig. 7.16 Stationary-state electron distribution functions $f^1$, $f^2$ in the two lowest energy subbands versus $\mathbf{k}_\parallel$ at an AlGaN/GaN heterojunction under the influence of an electric field $\mathcal{E} = 10\ \mathrm{kVcm^{-1}}$.

advantage of velocity overshoots for reducing switching times, must include hot phonon effects for not overestimating the achievable performance of such devices.

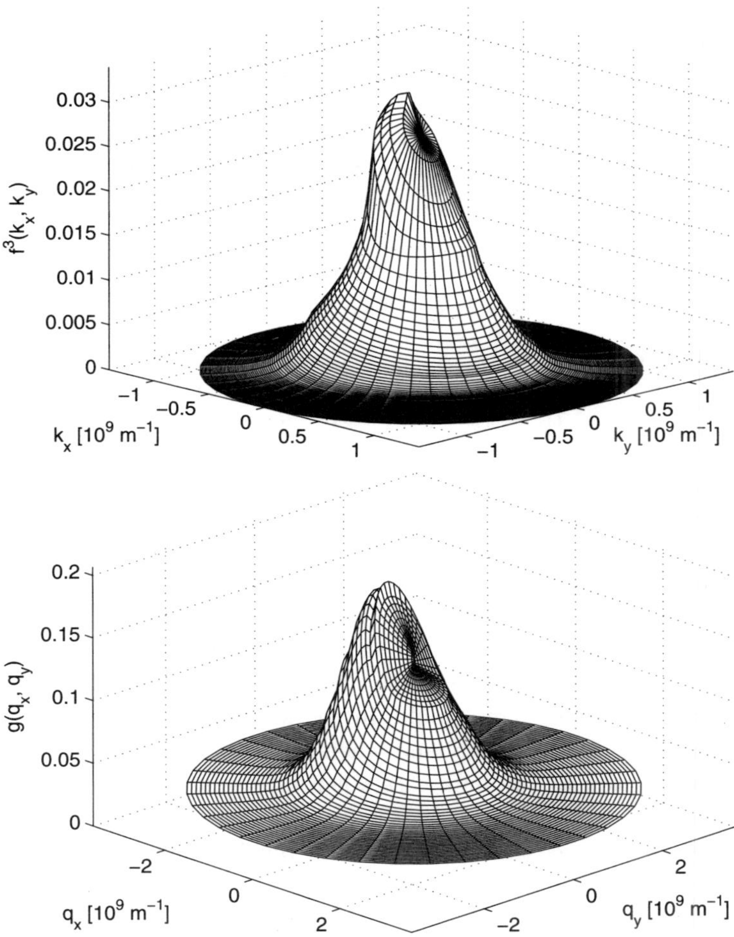

Fig. 7.17  Stationary-state electron distribution function $f^3$ in the third energy subband versus $\mathbf{k}_\|$ and the stationary-state LO phonon distribution function $g$ versus $\mathbf{q}_\|$ at an AlGaN/GaN heterojunction under the influence of an electric field $\mathcal{E} = 10 \text{ kVcm}^{-1}$.

### 7.4.3  *Distribution Functions*

One of the advantages of handling the Bloch-Boltzmann-Peierls equations with the help of deterministic solution methods is the availability of the particle distribution functions in noise-free resolution. Moreover, the consideration of two-dimensional transport problems allows the illustration of the whole information on the distribution function without an averaging procedure. Hence, we think that the following figures are quite illustrative.

Figures 7.16 and 7.17 depict the stationary-state electron distribution functions for the three lowest energy bands and the associated longitudinal optical phonon distribution function at the $Al_xGa_{1-x}N/GaN$ heterojunction. In this simulation, the electric field is set to $\mathcal{E} = 10$ kVcm$^{-1}$. The distribution functions $f_{ij}^\nu$ and $g_{xy}$ in the points

$$\mathbf{k}_{ij}^\nu = k_\parallel^\nu(E_i)(\cos\theta_j, \sin\theta_j), \tag{7.118a}$$

$$\mathbf{q}_{xy} = q_x(\cos\zeta_y, \sin\zeta_y) \tag{7.118b}$$

are approximated via

$$f_{ij}^\nu = \frac{n_{ij}^\nu}{1_{ij}^\nu}\Theta(E_i - \varepsilon_\nu), \quad g_{xy} = \frac{r_{xy}^\nu}{1_{xy}^{\mathrm{p}}}. \tag{7.119}$$

In Fig. 7.16, we observe that the electron distribution functions can be seen as shifted Fermi-Dirac distributions with some abrupt changes in their slopes with rising energies. This finer structure is related to the onset of possible intersubband scattering as it is the case, for instance, in InP for energies that allow intervalley scattering (cf. chapter 4). In contrast to bulk polar semiconductors with their large differences in the masses of $\Gamma$-valley and $L$-valley electrons, the density of states is the same for all subbands for 2DEGs. This results in the less pronounced decline in the distribution function of a subband when starting to overlap with another one in comparison to the bulk case when intervalley scattering begins to take place.

In Fig. 7.17, we see the reason for the strong influence of the hot phonon effect on the electron drift velocity. The phonon distribution function is significantly enhanced in comparison to its equilibrium Bose-Einstein value given in the undisturbed regions for large $\mathbf{q}_\parallel$. Another interesting feature of $g$ is its behavior for very small $\mathbf{q}_\parallel$. In bulk semiconductors, there exists a value $q_{\min} > 0$ so that scattering by phonons with $|\mathbf{q}| < q_{\min}$ is prohibited because of the conservation of momentum and energy. This feature is for instance visible in Fig. 6.1. In confined systems, $q_{\min}$ tends to zero, since the constant optical phonon energy is converted into potential energy of electrons by intersubband scattering processes. Thus, the phonon distribution function does not exhibit an undisturbed region centered at $\mathbf{q}_\parallel = 0$ in the case of two-dimensional transport, but only a small inversion as it is observable in Fig. 7.17.

Chapter 8

# The Multigroup-WENO Solver for Semiconductor Device Simulation

## 8.1 Introduction

With the increasing miniaturization of modern semiconductor devices, the application of kinetic Boltzmann-type equations to simulate the charge transport becomes mandatory, since the hydrodynamic models used for large devices lose their validities [Markowich *et al.* (1990)].

In this chapter, we propose a deterministic multigroup-WENO solver for the non-stationary two-dimensional Boltzmann-Poisson system for semiconductor devices. The multigroup approach, which is applied to determine the dependence of the electron distribution function on the electron wave vector, has been used with great success for investigating the particle transport in bulk semiconductors as illustrated in the previous chapters.

On the other hand, modern semiconductors devices are featured by changes of their composition on a short length scale. Hence, suitable numerical methods for dealing with the spatial dependence of the distribution function must be applied to cope, e.g., with abrupt changes in the doping concentration. Consequently, we combine our multigroup transport equations with a weighted essentially non-oscillatory (WENO) code [Jiang and Shu (1996)] for approximating the spatial derivatives in the diffusion term of the BTEs. We present an approach based on a multivalley-model for approximating the band structure of the considered semiconductor. In addition, we construct the transport equations in a way that they allow us to correctly describe the anisotropy of scattering mechanisms. In addition, the Poisson equations is coupled with the BTEs to determine the electric field strength in the device self-consistently.

## 8.2   The Boltzmann-Poisson System

In this section, we summarize the basic equations that constitute the Boltzmann-Poisson system for semiconductor devices. Concerning the approximations which we apply to describe the conduction bands of a semiconductor, we use the non-parabolic, spherical dispersion law

$$E_\nu(\mathbf{k}) = \Delta_{0\nu} + \frac{1}{2\alpha_\nu}\left[\sqrt{1 + 2\alpha_\nu \frac{\hbar^2 k^2}{m_\nu^*}} - 1\right]. \tag{8.1}$$

It relates the energy $E_\nu$ in the valley $\nu$ and the electron wave vector $\mathbf{k}$ with $k = |\mathbf{k}|$ [Nag (1980)], where $\Delta_{0\nu}$ denotes the gap between the minimum of the valley $\nu$ and the zero energy level. The symbols $m_\nu^*$ and $\alpha_\nu$ refer to the effective mass and the non-parabolicity factor of the considered valley, respectively.

In semiconductors, the electronic states are occupied according to the electron distribution functions $f_\nu(t, \mathbf{r}, \mathbf{k})$, which are the probability densities to find an electron at time $t$ at the position $\mathbf{r}$ in the valley $\nu$ with the wave vector $\mathbf{k}$. The evolutions of these functions are governed by the Boltzmann transport equations [Lundstrom (2000)]:

$$\frac{\partial f_\nu(t, \mathbf{r}, \mathbf{k})}{\partial t} + \mathbf{v}_\nu \cdot \nabla_\mathbf{r} f_\nu(t, \mathbf{r}, \mathbf{k}) - \frac{e}{\hbar}\, \mathbf{E} \cdot \nabla_\mathbf{k} f_\nu(t, \mathbf{r}, \mathbf{k}) = \mathcal{C}[f_\nu, f_\mu]. \tag{8.2}$$

Here, the group velocity is defined as

$$\mathbf{v}_\nu(\mathbf{k}) = \frac{1}{\hbar}\nabla_\mathbf{k} E_\nu(\mathbf{k}) = \frac{\hbar \mathbf{k}}{m_\nu^*}\{1 + 2\alpha_\nu[E_\nu(\mathbf{k}) - \Delta_{0\nu}]\} \tag{8.3}$$

in correspondence with (8.1).

The electrostatic potential $V$ and the related electric field strength $\mathbf{E}$ are found as the solution of the Poisson equation

$$\nabla \cdot [\varepsilon_{\rm st}(\mathbf{r})\varepsilon_0 \nabla V(t, \mathbf{r})] = -e[N_{\rm D}(\mathbf{r}) - n(t, \mathbf{r})], \quad \mathbf{E}(t, \mathbf{r}) = -\nabla V(t, \mathbf{r}), \tag{8.4}$$

where $e$ and $\varepsilon_{\rm st}\varepsilon_0$ label the positive elementary charge and the dielectric constant of the semiconductor, respectively. The donor density $N_{\rm D}$ is fixed in time, while the time-dependent electron density $n$ is found by forming moments of the electron distribution functions:

$$n(t, \mathbf{r}) = \sum_\nu \frac{Z_\nu}{4\pi^3} \int_{\mathbb{R}^3} d\mathbf{k} f_\nu(t, \mathbf{r}, \mathbf{k}). \tag{8.5}$$

The collision terms in the BTEs (8.2) for the scattering mechanisms $\xi$ read in the low density approximation

$$\mathcal{C}^\xi[f_\nu, f_\mu] = \int_{\mathbb{R}^3} d\mathbf{k}' \left[ S^\xi_{\mu\nu}(\mathbf{k}', \mathbf{k}) f_\mu(\mathbf{k}') - S^\xi_{\nu\mu}(\mathbf{k}, \mathbf{k}') f_\nu(\mathbf{k}) \right]. \tag{8.6}$$

With the aim of transforming the BTE (8.2) in a conservative form, we introduce the following change of variables. We express the electron wave vector $\mathbf{k}_\nu(E, \mu, \varphi)$ in the valley $\nu$ in spherical coordinates as a function of the energy $E$, measured from the energy reference, the cosine of the polar angle $\mu$ and the azimuth angle $\varphi$:

$$\mathbf{k}_\nu(E, \mu, \varphi) = k_\nu(E)(\mu, \sqrt{1-\mu^2}\cos\varphi, \sqrt{1-\mu^2}\sin\varphi), \tag{8.7a}$$

$$k_\nu(E) = \frac{\sqrt{2m_\nu^*}}{\hbar}\sqrt{(E - \Delta_{0\nu})[1 + \alpha_\nu(E - \Delta_{0\nu})]}\,\Theta(E - \Delta_{0\nu}) \tag{8.7b}$$

with the Heaviside step function $\Theta$. Furthermore, we take into account that $f_\nu$ only depends on the variables $t$, $x$, $y$, $E$, $\mu$ and $\varphi$ for spatially two-dimensional problems with inhomogeneities in the x- and y-direction. Based on the change of variables (8.7) and the features of $f_\nu$, some algebra reveals that the BTE (8.2) can be transformed into

$$\frac{\partial F_\nu}{\partial t} + \frac{\partial}{\partial x}(g_1^\nu F_\nu) + \frac{\partial}{\partial y}(g_2^\nu F_\nu) + \frac{\partial}{\partial E}(g_4^\nu F_\nu) \tag{8.8}$$

$$+ \frac{\partial}{\partial \mu}(g_4^\nu F_\nu) + \frac{\partial}{\partial \varphi}(g_5^\nu F_\nu) = \mathcal{C}[F_\nu, F_\mu].$$

This expression contains the new unknown function

$$F_\nu(t, x, y, E, \mu, \varphi) = \mathcal{Z}_\nu(E) f_\nu(t, x, y, E, \mu, \varphi) \tag{8.9}$$

with the Jacobian of the transformation (8.7),

$$\mathcal{Z}_\nu(E) = \frac{m_\nu^*}{\hbar^3}\sqrt{2m_\nu^*(E - \Delta_{0\nu})[1 + \alpha_\nu(E - \Delta_{0\nu})]} \tag{8.10}$$

$$\times [1 + 2\alpha_\nu(E - \Delta_{0\nu})]\Theta(E - \Delta_{0\nu}),$$

which equals the density of states except for a constant factor. The rewritten collision integral reads

$$\mathcal{C}^\xi[F_\nu, F_\mu] = \int_{\mathcal{B}} d\mathbf{S}' \left[ S^\xi_{\mu\nu}(\mathbf{k}'_\mu, \mathbf{k}_\nu)\mathcal{Z}_\nu(E)F'_\mu - S^\xi_{\nu\mu}(\mathbf{k}_\nu, \mathbf{k}'_\mu)\mathcal{Z}_\mu(E')F_\nu \right] \tag{8.11}$$

with $d\mathbf{S}' = dE'd\mu'd\varphi'$, $\mathcal{B} = [0, \infty) \times [-1, 1] \times [0, 2\pi]$, $\mathbf{k}_\nu = \mathbf{k}_\nu(E, \mu, \varphi)$, $\mathbf{k}'_\mu = \mathbf{k}_\mu(E', \mu', \varphi')$, $F_\nu = F_\nu(t, x, y, E, \mu, \varphi)$ and $F'_\mu = F_\mu(t, x, y, E', \mu', \varphi')$.

In addition, the functions $g_l^\nu$, $l = 1, 2, \ldots, 5$ are defined as

$$g_1^\nu(\cdot) = \frac{\hbar}{m_\nu^*} \frac{k^\nu(E)}{1 + 2\alpha_\nu(E - \Delta_{0\nu})} \mu, \tag{8.12a}$$

$$g_2^\nu(\cdot) = \frac{\hbar}{m_\nu^*} \frac{k^\nu(E)}{1 + 2\alpha_\nu(E - \Delta_{0\nu})} \sqrt{1 - \mu^2} \cos\varphi, \tag{8.12b}$$

$$g_3^\nu(\cdot) = -\frac{e}{m_\nu^*} \frac{k^\nu(E)}{1 + 2\alpha_\nu(E - \Delta_{0\nu})} \left[ E_x(t, x, y)\mu + E_y(t, x, y)\sqrt{1 - \mu^2} \cos\varphi \right], \tag{8.12c}$$

$$g_4^\nu(\cdot) = -\frac{e}{\hbar} \frac{\sqrt{1 - \mu^2}}{k_\nu(E)} \left[ E_x(t, x, y)\sqrt{1 - \mu^2} - E_y(t, x, y)\mu\cos\varphi \right], \tag{8.12d}$$

$$g_5^\nu(\cdot) = \frac{e}{\hbar} \frac{E_y(t, x, y)}{k_\nu(E)\sqrt{1 - \mu^2}} \sin\varphi. \tag{8.12e}$$

The functions $g_3^\nu$, $g_4^\nu$ and $g_5^\nu$ contain the x- and y-component of the electric field, determined from the 2D Poisson equation

$$E_x(t, x, y) = -\frac{\partial}{\partial x}V(t, x, y), \quad E_y(t, x, y) = -\frac{\partial}{\partial y}V(t, x, y), \tag{8.13a}$$

$$\frac{\partial}{\partial x}\left[ \varepsilon_{st}(x, y)\varepsilon_0 \frac{\partial}{\partial x}V(t, x, y) \right] + \frac{\partial}{\partial y}\left[ \varepsilon_{st}(x, y)\varepsilon_0 \frac{\partial}{\partial y}V(t, x, y) \right]$$
$$= -e[N_D(x, y) - n(t, x, y)]. \tag{8.13b}$$

## 8.3 The Multigroup-WENO Scheme

In this section, we introduce the numerical scheme used to solve the BP system (8.8) and (8.13b). To begin with, we treat the dependences of the unknown functions $F_\nu$ on the independent variables $E$, $\mu$ and $\varphi$ by means of the multigroup approach. Therefore, we discretize these variables according to

$$E_{i+\frac{1}{2}}^\nu = \Delta_{0\nu} + i\Delta E, \quad i = 0, 1, \ldots, N_\nu, \quad \Delta E = \frac{\hbar\omega}{n_{\text{mul}}}, \tag{8.14a}$$

$$\mu_{j+\frac{1}{2}} = j\Delta\mu - 1, \quad j = 0, 1, \ldots, M, \quad \Delta\mu = \frac{2}{M}, \tag{8.14b}$$

$$\varphi_{k+\frac{1}{2}} = k\Delta\varphi, \quad k = 0, 1, \ldots, R, \quad \Delta\varphi = \frac{\pi}{R} \tag{8.14c}$$

with $M = 2\,m_{\text{mul}}$, $R = 2\,r_{\text{mul}}$ and $n_{\text{mul}}, m_{\text{mul}}, r_{\text{mul}} \in \mathbb{N}$. In (8.14c), we take advantage of the mirror symmetry with respect to $\varphi$ in the considered two-dimensional case. The choice of $M$ and $R$ as even integers guarantees a well-defined wind direction, as it is needed below. The definition of $\Delta E$ allows us to evaluate the collision integral of the inelastic electron-phonon interaction without a smearing of the Dirac distribution related to the energy conservation. The integer numbers $N_\nu$ must be chosen in a way so that the maximum energies $E_{\max}^\nu = E_{N_\nu}^\nu$ guarantee that $F_\nu(t, x, y, E_{\max}^\nu, \mu, \varphi)$ are negligible for all $t$, $x$, $y$, $\mu$, $\varphi$ and $\nu$.

Next, we approximate the unknown functions $F_\nu$ as the finite sums

$$F_\nu(t, x, y, E, \mu, \varphi) \approx \sum_{i=1}^{N_\nu} \sum_{j=1}^{M} \sum_{k=1}^{R} n_{ijk}^\nu(t, x, y) \tag{8.15}$$
$$\times \, \delta(E - E_i^\nu)\delta(\mu - \mu_j)\delta(\varphi - \varphi_k)$$

containing $N_\nu \times M \times R$ unknown coefficients $n_{ijk}^\nu$ and the Dirac distributions $\delta$ with the poles $E_i^\nu = \Delta_{0\nu} + (i + 1/2)\Delta E$, $i = 1, 2, \ldots, N_\nu$, $\mu_j = (j + 1/2)\Delta\mu - 1$, $j = 1, 2, \ldots, M$ and $\varphi_k = (k + 1/2)\Delta\varphi$, $k = 1, 2, \ldots, R$. The ansatz (8.15) implies that the macroscopic quantity $\langle m(t, x, y)\rangle$ to the microscopic one, $m(E, \mu, \varphi)$, can be evaluated via

$$\langle m(t, x, y)\rangle = \sum_\nu \frac{Z_\nu}{2\pi^3} \sum_{i=1}^{N_\nu} \sum_{j=1}^{M} \sum_{k=1}^{R} m^\nu(E_i^\nu, \mu_j, \varphi_k) \, n_{ijk}^\nu(t, x, y), \tag{8.16}$$

as it can easily be verified by forming moments of (8.15).

The evolution equations of the coefficients $n_{ijk}^\nu$ are constructed by applying the method of weighted residuals [Lapidus and Pinder (1982)]. The ansatz (8.15) is inserted into the BTE (8.8) and the result is successively integrated over the cells $C_{ijk}^\nu = [E_{i-1/2}^\nu, E_{i+1/2}^\nu] \times [\mu_{j-1/2}, \mu_{j+1/2}] \times [\varphi_{k-1/2}, \varphi_{k+1/2}]$. In addition, we use an upwind scheme in the force term together with a linear approximation of the fluxes through the boundaries of the cells controlled by a MinMod slope limiter [LeVeque (1992)]. This procedure yields a closed set of $N_\nu \times M \times R$ partial differential equations for each of the valleys $\nu$. It reads

$$\frac{\partial n_{ijk}^\nu}{\partial t} + \frac{\partial}{\partial x}(g_{1,ijk}^\nu n_{ijk}^\nu) + \frac{\partial}{\partial y}(g_{2,ijk}^\nu n_{ijk}^\nu) \tag{8.17}$$
$$+ h_{3,i+\frac{1}{2},jk}^\nu - h_{3,i-\frac{1}{2},jk}^\nu + h_{4,i,j+\frac{1}{2},k}^\nu - h_{4,i,j-\frac{1}{2},k}^\nu + h_{5,ij,k+\frac{1}{2}}^\nu - h_{5,ij,k-\frac{1}{2}}^\nu$$
$$= \sum_{a=1}^{N_\nu} \sum_{b=1}^{M} \sum_{c=1}^{R} \left[ \langle S_{abc,ijk}^{\mu\nu}\rangle n_{abc}^\mu - \langle S_{ijk,abc}^{\nu\mu}\rangle n_{ijk}^\nu \right]$$

with $i = 1, 2, \ldots, N_\nu$, $j = 1, 2, \ldots, M$ and $k = 1, 2, \ldots, R$. The collision coefficients for the scattering mechanism $\xi$ are evaluated via

$$\langle S^{\xi,\nu\mu}_{ijk,abc}\rangle = \int_{C^\nu_{ijk}} d\mathbf{S} \int_{C^\mu_{abc}\cup\tilde{C}^\mu_{abc}} d\mathbf{S}'\, \mathcal{Z}_\mu(E')\, S^\xi_{\nu\mu}(\mathbf{k}_\nu,\mathbf{k}'_\mu) \qquad (8.18)$$

$$\times\, \delta(E - E^\nu_i)\delta(\mu - \mu_j)\delta(\varphi - \varphi_k)$$

with $\tilde{C}^\mu_{abc} = [E^\mu_{a-\frac{1}{2}}, E^\mu_{a+\frac{1}{2}}] \times [\mu_{b-\frac{1}{2}}, \mu_{b+\frac{1}{2}}] \times [2\pi - \varphi_{k+\frac{1}{2}}, 2\pi - \varphi_{k-\frac{1}{2}}]$. The numerical fluxes $h^\nu_{3,i\pm\frac{1}{2},jk}$ are given by

$$h^\nu_{3,i+\frac{1}{2},jk} = \begin{cases} g^\nu_{3,i+\frac{1}{2},jk}(n^\nu_{ijk} + s^\nu_{ijk}), & g^\nu_{3,ijk} > 0, \\ g^\nu_{3,i+\frac{1}{2},jk}(n^\nu_{i+1,jk} - s^\nu_{i+1,jk}), & g^\nu_{3,ijk} < 0, \end{cases} \qquad (8.19a)$$

$$h^\nu_{3,i-\frac{1}{2},jk} = \begin{cases} g^\nu_{3,i-\frac{1}{2},jk}(n^\nu_{ijk} - s^\nu_{ijk}), & g^\nu_{3,ijk} < 0, \\ g^\nu_{3,i-\frac{1}{2},jk}(n^\nu_{i-1,jk} + s^\nu_{i-1,jk}), & g^\nu_{3,ijk} > 0, \end{cases} \qquad (8.19b)$$

$$s^\nu_{ijk} = \frac{1}{2}\,\mathrm{MM}(n^\nu_{i+1,jk} - n^\nu_{ijk}, n^\nu_{ijk} - n^\nu_{i-1,jk}) \qquad (8.19c)$$

with the MinMod scheme

$$\mathrm{MM}(a,b) = \begin{cases} \mathrm{sgn}(a)\min(|a|,|b|), & ab > 0, \\ 0, & \text{otherwise.} \end{cases} \qquad (8.20)$$

Similar expressions are used to determine $h^\nu_{4,i,j\pm1/2,k}$ and $h^\nu_{5,ij,k\pm1/2}$.

For treating the spatial dependence of the coefficients $n^\nu_{ijk}$ in the multigroup equations (8.17), we apply a fifth-order WENO scheme to cope with the strong variations of the electron distributions in space due to, e.g., sharp doping profiles [Carrillo *et al.* (2003)b; Jiang and Shu (1996)]. Therefore, we discretize the spatial variables equidistantly,

$$x_n = n\Delta x, \qquad n = 0, 1, \ldots, P, \qquad \Delta x = \frac{L_x}{P}, \qquad (8.21a)$$

$$y_m = m\Delta y, \qquad m = 0, 1, \ldots, Q, \qquad \Delta y = \frac{L_y}{Q}, \qquad (8.21b)$$

where $L_x$ and $L_y$ are the dimensions of the device. Next, we approximate the spatial derivatives in (8.17), for instance, that with respect to $y$, as

$$\frac{\partial}{\partial y}[g^\nu_{2,ijk} n^\nu_{ijk}(t, x_n, y_m)] \approx \frac{1}{\Delta y}\left(\hat{h}_{m+\frac{1}{2}} - \hat{h}_{m-\frac{1}{2}}\right) \qquad (8.22)$$

for fixed indices $i$, $j$, $k$, $n$ and fixed time $t$. Therefore, these parameters are not written down in the numerical fluxes $\hat{h}_{m\pm1/2}$ to keep the description concise. We note that the wind direction, i.e., the sign of $g_{2,ijk}^{\nu}$, is well-defined because of the chosen discretization (8.14) and it is independent of $m$. Thus, for fixed $i$, $j$, $k$, the wind direction is fixed. Without the loss of generality, we can assume that $g_{2,ijk}^{\nu} > 0$. Otherwise, the procedure must be carried out mirror symmetrically with respect to the point $y_{m+1/2}$. We define

$$h_m = g_{2,ijk}^{\nu} n_{ijk}^{\nu}(t, x_n, y_m), \quad m = -3, -2, \ldots, Q+3 \tag{8.23}$$

for fixed $i$, $j$, $k$, $n$ and obtain the numerical flux by

$$\hat{h}_{m+\frac{1}{2}} = \omega_1 \hat{h}_{m+\frac{1}{2}}^{(1)} + \omega_2 \hat{h}_{m+\frac{1}{2}}^{(2)} + \omega_3 \hat{h}_{m+\frac{1}{2}}^{(3)}. \tag{8.24}$$

Here, $\hat{h}_{m+\frac{1}{2}}^{(1)}$, $\hat{h}_{m+\frac{1}{2}}^{(2)}$ and $\hat{h}_{m+\frac{1}{2}}^{(3)}$ are three fluxes on three different stencils:

$$\hat{h}_{m+\frac{1}{2}}^{(1)} = \frac{1}{3}h_{m-2} - \frac{7}{6}h_{m-1} + \frac{11}{6}h_m, \tag{8.25a}$$

$$\hat{h}_{m+\frac{1}{2}}^{(2)} = -\frac{1}{6}h_{m-1} + \frac{5}{6}h_m + \frac{1}{3}h_{m+1}, \tag{8.25b}$$

$$\hat{h}_{m+\frac{1}{2}}^{(3)} = \frac{1}{3}h_m + \frac{5}{6}h_{m+1} - \frac{1}{6}h_{m+2}. \tag{8.25c}$$

In addition, (8.24) contains the nonlinear weights

$$\omega_i = \frac{\tilde{\omega}_i}{\sum_{l=1}^{3} \tilde{\omega}_l}, \quad \tilde{\omega}_l = \frac{\gamma_l}{(\varepsilon + \beta_l)^2}, \quad i,l = 1,2,3 \tag{8.26}$$

with the linear weights $\gamma_1 = 1/10$, $\gamma_2 = 3/5$ and $\gamma_3 = 3/10$. The smoothness indicators $\beta_l$ equal

$$\beta_1 = \frac{13}{12}(h_{m-2} - 2h_{m-1} + h_m)^2 + \frac{1}{4}(h_{m-2} - 4h_{m-1} + 3h_m)^2, \tag{8.27a}$$

$$\beta_2 = \frac{13}{12}(h_{m-1} - 2h_m + h_{m+1})^2 + \frac{1}{4}(h_{m-1} - h_{m+1})^2, \tag{8.27b}$$

$$\beta_3 = \frac{13}{12}(h_m - 2h_{m+1} + h_{m+2})^2 + \frac{1}{4}(3h_m - 4h_{m+1} + h_{m+2})^2, \tag{8.27c}$$

where the parameter $\varepsilon = 10^{-6}$ is introduced for preventing the denominator to become zero.

In (8.23), fluxes are defined in points, the so-called ghost points, which do not belong to the considered device. They must be chosen to suitably model the demanded boundary conditions. For instance, we consider a boundary at $y = y_Q$. For simulating reflecting boundaries, we set

$$n_{ijk}^{\nu}(t, x_n, y_{Q+m}) = n_{ijk}^{\nu}(t, x_n, y_{Q+1-m}), \quad m = 1,2,3, \tag{8.28a}$$

which implies that $\hat{h}_{Q+1/2} = 0$. The Schottky contacts are modeled as totally absorbing contacts. Hence, we apply

$$n_{ijk}^{\nu}(t, x_n, y_{Q+m}) = \begin{cases} n_{ijk}^{\nu}(t, x_n, y_Q), & g_{ijk}^{\nu} > 0, \\ 0, & g_{ijk}^{\nu} < 0, \end{cases} \qquad m = 1, 2, 3 \qquad (8.28\text{b})$$

for determining the fluxes in the ghost-points at such contacts. Finally, the ohmic contacts are simulated via

$$n_{ijk}^{\nu}(t, x_n, y_{Q+m}) = \begin{cases} n_{ijk}^{\nu}(t, x_n, y_Q), & g_{ijk}^{\nu} > 0, \\ n_{ijk}^{M,\nu}(x_n, y_Q), & g_{ijk}^{\nu} < 0, \end{cases} \qquad m = 1, 2, 3. \qquad (8.28\text{c})$$

Here, the coefficient $n_{ijk}^{M,\nu}$ are obtained from Maxwell distributions, which are normalized to the donor density

$$n_{ijk}^{M,\nu}(x_n, y_m) = N_{\mathrm{D}}(x_n, y_m) \frac{\displaystyle\int_{D_{ijk}^{\nu}} d\mathbf{S}\, \mathcal{Z}_{\nu}(E) \exp\left(-\frac{E}{k_{\mathrm{B}} T_{\mathrm{L}}}\right)}{\displaystyle\sum_{\mu} \frac{Z_{\mu}}{4\pi^3} \int_{B} d\mathbf{S}\, \mathcal{Z}_{\mu}(E) \exp\left(-\frac{E}{k_{\mathrm{B}} T_{\mathrm{L}}}\right)}. \qquad (8.29)$$

In addition, we set $\hat{h}_{Q+1/2} = C(t, x_n) g_{ijk}^{\nu} n_{ijk}^{M,\nu}(x_n, y_Q)$ for $g_{ijk}^{\nu} < 0$, where the factor $C(t, x_n)$ is determined in a way that the charge neutrality at the contact is fulfilled.

The combination of the multigroup equations (8.17) with the WENO scheme (8.22) leads to a set of $N_{\nu} \times M \times R \times P \times Q$ ordinary differential equations in time for each of the considered valleys $\nu$. The time integration is performed by means of the TVD forward Euler scheme [Shu and Osher (1988)], where the time step $\Delta t$ is chosen so that the CFL condition for guaranteeing stability of the numerical procedure is fulfilled. Thus, we determine the time step via

$$\Delta t = \frac{C_{\mathrm{CFL}}}{\max\left(\frac{|g_1^{\nu}|}{\Delta x} + \frac{|g_2^{\nu}|}{\Delta y} + \frac{|g_3^{\nu}|}{\Delta E} + \frac{|g_4^{\nu}|}{\Delta \mu} + \frac{|g_5^{\nu}|}{\Delta \varphi}\right)} \qquad (8.30)$$

with $C_{\mathrm{CFL}} = 0.8$ and update it in the simulation to take into account the variation of the electric field (cf. (8.12)). The Poisson equation (8.13b) is solved with the help of a standard Successive Over-Relaxation (SOR) scheme [Jacoboni and Lugli (1989)] at each time step with the modified charge density obtained from the multigroup-WENO equations. At the initial time $t_0$, the electron distribution functions is set equal to the equilibrium Maxwellians (8.29), i.e., $n_{ijk}^{\nu}(t_0, x_n, y_m) = n_{ijk}^{M,\nu}(x_n, y_m)$.

# Chapter 9

# Simulation of Silicon Devices

## 9.1 Introduction

In this chapter we present numerical results obtained with the help of our multigroup WENO solver. All of the calculations are carried out for silicon at the temperature $T_{\mathrm{L}} = 300$ K. The conduction band of silicon is modeled by six equivalent energy valleys in the Kane approximation. Since we are modeling electrons moving in the same symmetry-type of valleys, the electron distribution function $f$ represents an average probability function among these valleys. The collision term used includes acoustic deformation potential scattering and optical intervalley scattering between equivalent valleys. Therefore, the transition rate for the scattering of electrons in silicon reads

$$
\begin{aligned}
S(\mathbf{k}, \mathbf{k}') =& K_0(\mathbf{k}, \mathbf{k}')\delta[E(\mathbf{k}') - E(\mathbf{k})] \\
&+ n_{\mathrm{q}} K(\mathbf{k}, \mathbf{k}')\delta[E(\mathbf{k}') - E(\mathbf{k}) - \hbar\omega_0] \\
&+ (n_{\mathrm{q}} + 1) K(\mathbf{k}, \mathbf{k}')\delta[E(\mathbf{k}') - E(\mathbf{k}) + \hbar\omega_0].
\end{aligned}
\tag{9.1}
$$

Table 9.1  Material parameters used for the simulation of silicon devices

| Quantity | Symbol | Unit | Value |
|---|---|---|---|
| Mass density | $\rho$ | $\mathrm{kgm}^{-3}$ | 2330 |
| Effective mass ratio | $m^*/m_0$ | | 0.32 |
| Non-parabolicity factor | $\alpha^*$ | $(\mathrm{eV})^{-1}$ | 0.5 |
| Relative dielectric constant Si | $\varepsilon_{\mathrm{Si}}$ | | 11.7 |
| Relative dielectric constant SiO$_2$ | $\varepsilon_{\mathrm{SiO_2}}$ | | 3.9 |
| Optical phonon energy | $\hbar\omega_0$ | eV | 0.063 |
| Intervalley coupling constant | $D_{\mathrm{IV}}$ | $10^{10}\ \mathrm{eVm}^{-1}$ | 11.4 |
| Acoustic deformation potential | $D_{\mathrm{A}}$ | eV | 9 |
| Sound velocity | $v_{\mathrm{s}}$ | $\mathrm{ms}^{-1}$ | 9040 |
| Lattice temperature | $T_{\mathrm{L}}$ | K | 300 |

155

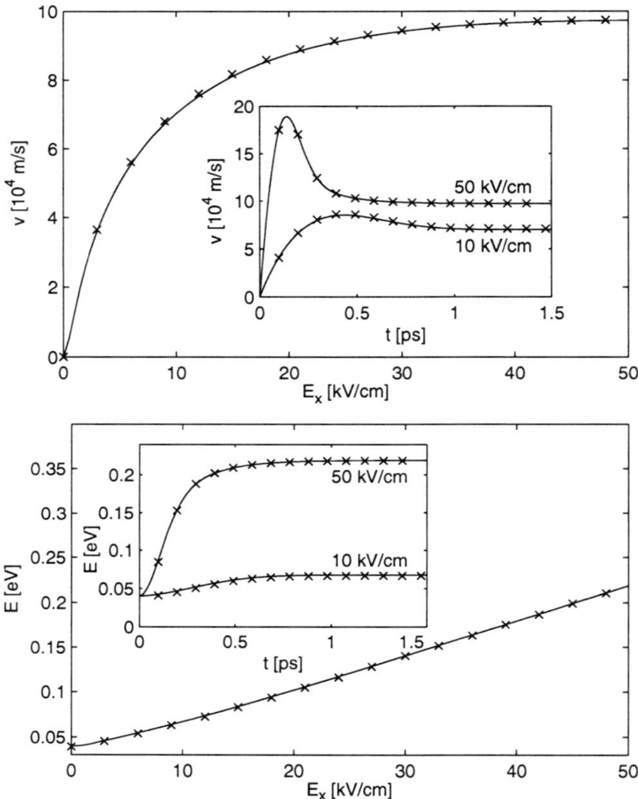

Fig. 9.1   Stationary-state drift velocity $v$ and stationary-state mean energy $E$ versus the electric field $E_x$ in silicon at the temperature $T_{\mathrm{L}}=300$ K. The insert illustrates $v$ and $E$ as a function of time $t$ in response to the onset of an electric field pulse. Solid line: Multigroup model; Crosses: WENO solver.

The symbols $K_0(\mathbf{k}, \mathbf{k}')$ and $K(\mathbf{k}, \mathbf{k}')$ are the acoustic deformation potential coupling constant and the intervalley coupling constant, respectively. According to the considerations in Sec. 2.4, they are given by

$$K_0(\mathbf{k}, \mathbf{k}') = \frac{D_{\mathrm{A}}^2 k_{\mathrm{B}} T_{\mathrm{L}}}{4\pi^2 \hbar \rho v_{\mathrm{s}}^2}, \tag{9.2a}$$

$$K(\mathbf{k}, \mathbf{k}') = \frac{D_{\mathrm{IV}}^2}{8\pi^2 \rho \omega_0}. \tag{9.2b}$$

Further, the occupation number of the optical phonons $n_{\mathrm{q}}$ is evaluated

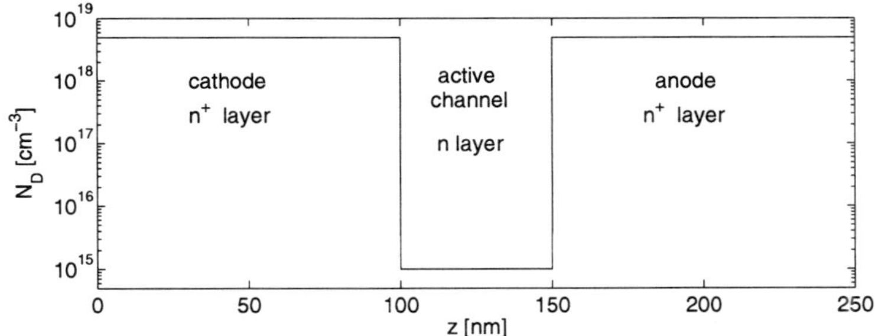


Fig. 9.2   Geometry of the one-dimensional $n^+ - n - n^+$ diode.

from

$$n_{\mathrm{q}} = \left[ \exp \frac{\hbar \omega_0}{k_{\mathrm{B}} T_{\mathrm{L}}} - 1 \right]^{-1}. \tag{9.3}$$

The appropriate material parameters for the simulation of silicon devices is given in Tab. 9.1.

## 9.2   Transport in Bulk Silicon

In Fig. 9.1, we illustrate the dependence of the stationary-state drift velocity and the mean energy on the applied electric field strength. Moreover, the inserts show the temporal evolution of these quantities in response to the onset of an electric field pulse for the field strengths $E_x = 10$ kVcm$^{-1}$ and $E_x = 50$ kVcm$^{-1}$. The parameters used in these calculations are set to $N = 100$, $M = 22$, $n_{\mathrm{mul}} = 4$. Our results are compared to those of a full WENO solver proposed in [Carrillo *et al.* (2003)a]. Here, we observe very good agreement between the results for both the steady state values and those in the transients.

The average electron energy slowly increases with the field as shown in Fig. 9.1, since much of the energy gained in the electric field is transferred to the lattice by equivalent intervalley scattering. Because the applied electric field increases the average electron energy which causes the scattering rate to increase, the momentum relaxation time and, thus, the mobility in silicon decreases with increasing field strength. Hence, the velocity field characteristics of Si is sublinear as illustrated in Fig. 9.1. Eventually, the scattering rate is so high that any further input from the field is simply lost

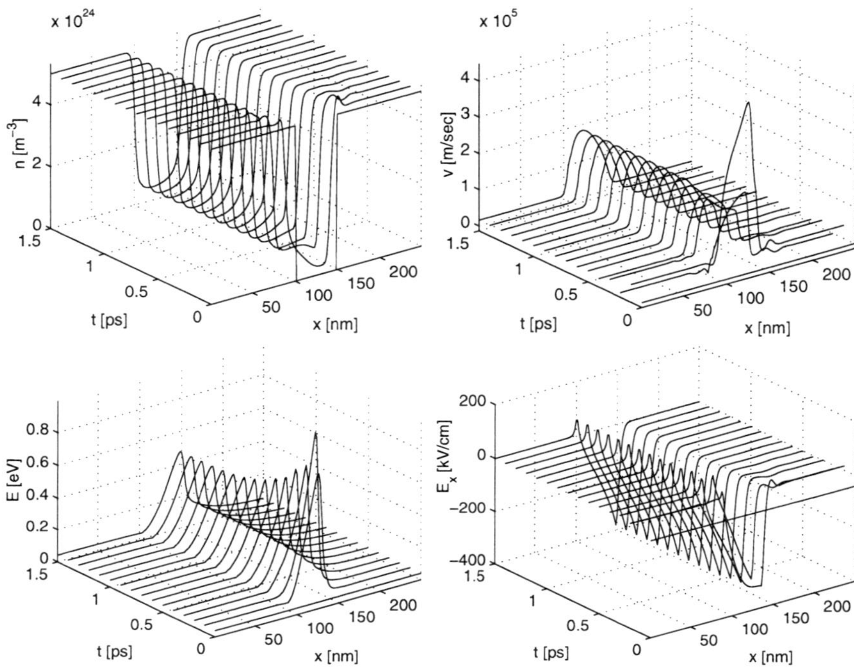

Fig. 9.3  Temporal evolution of the electron density $n$, the drift velocity $v$, the average energy $E$ and the electric field strength $E_x$ versus position $x$ in the $n^+ - n - n^+$ diode.

to collision and the drift velocity saturates.

For the time evolution of the electron drift velocity and the average energy, we find that the energy increases monotonically with time in reaching its stationary-state value, while the temporal evolution of the electron drift velocity features a velocity overshoot, which is the more pronounced, the higher the applied electric field is.

## 9.3  The Silicon $n^+ - n - n^+$ Diode

In this section, we present the results of the multigroup WENO solver for the $n^+ - n - n^+$ diode. The considered diode has a total length of 250 nm with a 50 nm active channel located in the middle of the device. The doping concentrations are set to $N_D = 5 \times 10^{18}$ cm$^{-3}$ in the $n^+$ region and $N_D = 10^{15}$ cm$^{-3}$ in the n region as it is illustrated in Fig. 9.2. The applied voltage is $V_{\text{bias}} = 1$ V and the parameters of the grid are chosen as $N = 100$,

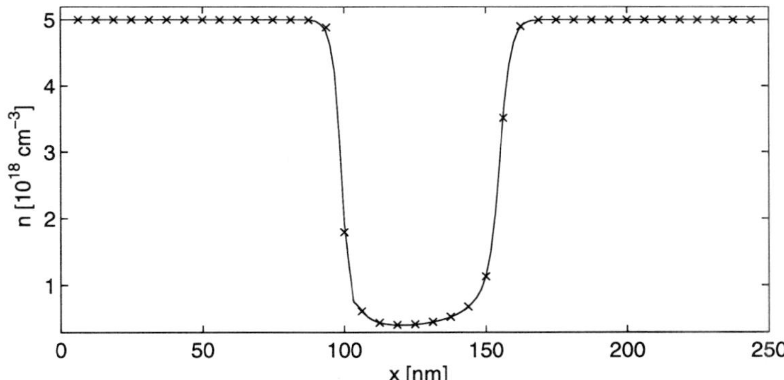

Fig. 9.4    Steady state electron density $n$ versus position $x$ in the $n^+ - n - n^+$ diode. Solid line: Multigroup WENO model; Crosses: WENO solver.

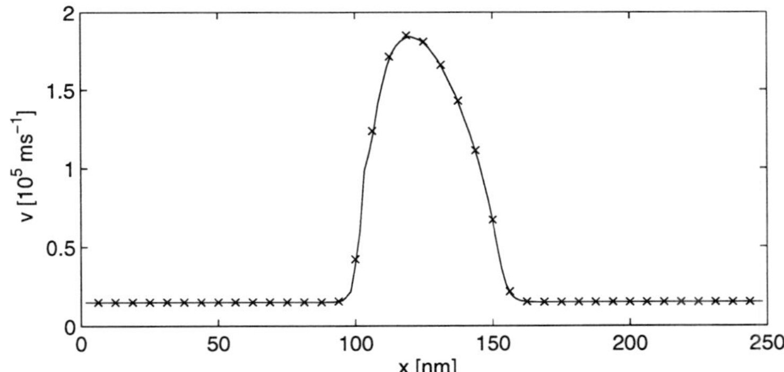

Fig. 9.5    Steady state drift velocity $v$ versus position $x$ in the $n^+ - n - n^+$ diode. Solid line: Multigroup WENO model; Crosses: WENO solver.

$M = 22$, $n_{\mathrm{mul}} = 4$ together with $P = 150$ grid points in real space. The stationary state is supposed to be reached at $t = 2$ ps. Periodic boundary conditions are used, since they are simple to impose and the results for the macroscopic quantities in the active region do hardly depend on the actual choice of the boundary conditions.

In Fig. 9.3, we show the temporal evolution of the electron density, the drift velocity, the average energy and the electric field strength versus position. We observe that electrons diffuse into the active channel right after $t_0 = 0$ ps. This diffusion is quickly counterbalanced by the increasingly

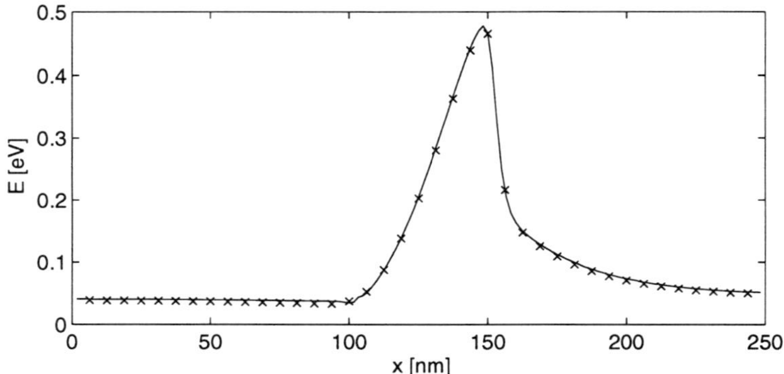

Fig. 9.6    Steady state mean energy $E$ versus position $x$ in the $n^+ - n - n^+$ diode. Solid line: Multigroup WENO model; Crosses: WENO solver.

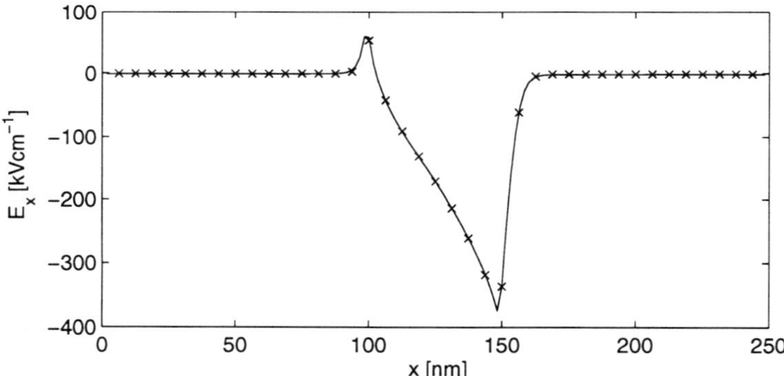

Fig. 9.7    Steady state electric field strength $E_x$ versus position $x$ in the $n^+ - n - n^+$ diode. Solid line: Multigroup WENO model; Crosses: WENO solver.

strong electric field, which causes both, a velocity overshoot and waves in the density and in the velocity plot. The maximum drift velocity reaches a value almost twice the steady state value. Later, the velocity profile becomes more and more symmetric. Moreover, it can be seen that the time needed for the relaxation of the mean energy barely differs from the stabilization time of the velocity.

Figures 9.4, 9.5, 9.6 and 9.7 display the stationary state values of the electron density, the drift velocity, the mean energy and the electric field strength as a function of position in the $n^+ - n - n^+$ diode. Moreover,

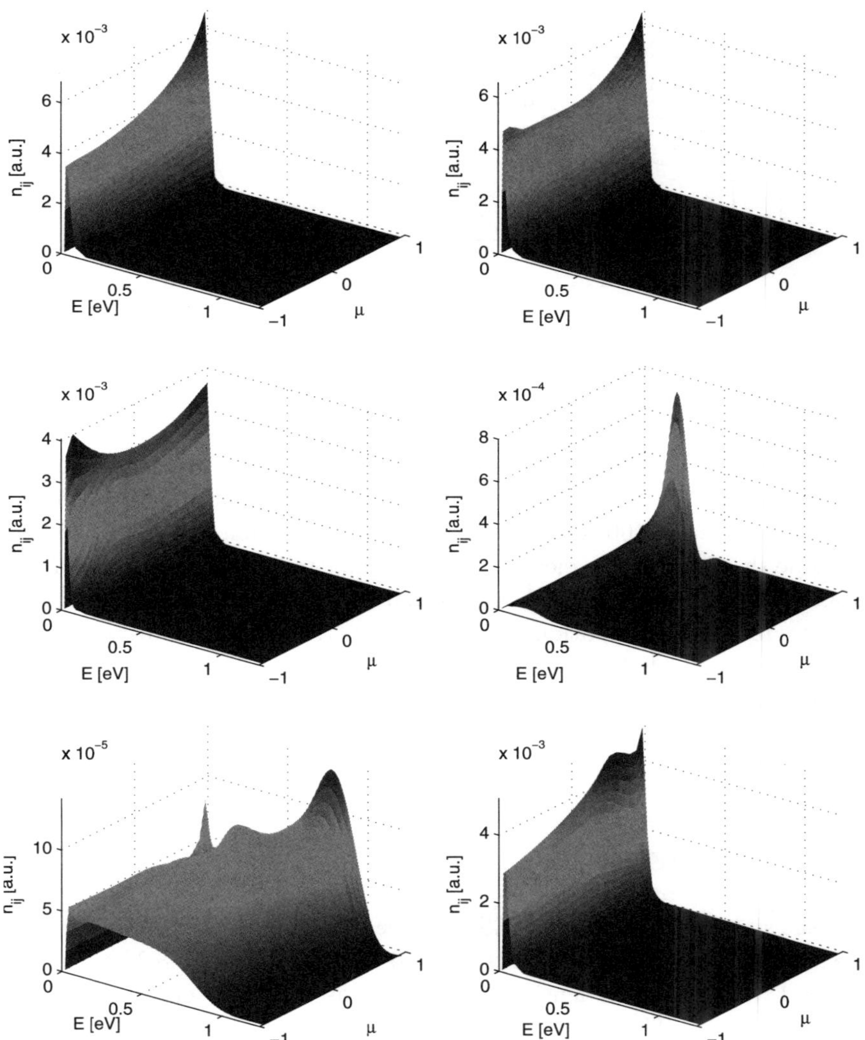

Fig. 9.8 Steady state electron distributions $n_{ij}(z_0)$ versus energy $E$ and angle $\mu$ in the $n^+-n-n^+$ diode. From left to right and from top to bottom at $z_0 = 25$ nm, $z_0 = 75$ nm, $z_0 = 100$ nm, $z_0 = 125$ nm, $z_0 = 150$ nm and $z_0 = 175$ nm.

we compare our results with those of the full WENO solver [Carrillo *et al.* (2003)a] and find that they coincide in the whole $x$ range.

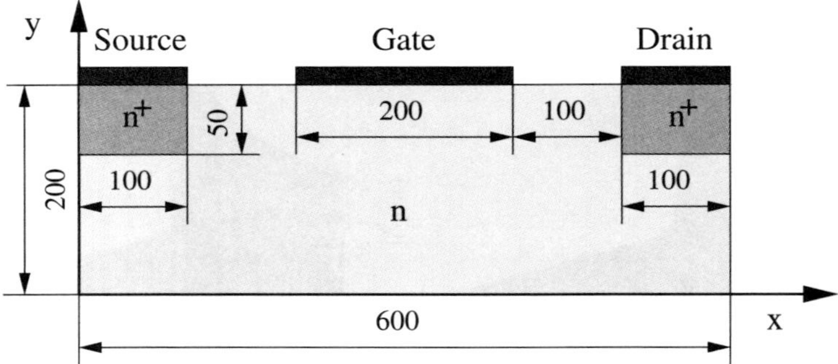


Fig. 9.9    Schematic illustration of a bidimensional MESFET with the donor densities $n = 10^{17}$ cm$^{-3}$ and $n^+ = 3 \times 10^{17}$ cm$^{-3}$. Lengths are given in nm.

Figure 9.8 illustrates the steady state electron distributions versus energy and angle in the $n^+ - n - n^+$ diode at the positions $z_0 = 25$ nm, $z_0 = 75$ nm, $z_0 = 100$ nm, $z_0 = 125$ nm, $z_0 = 150$ nm and $z_0 = 175$ nm. Outside the channel, the electron distribution can be described by a shifted Maxwell distribution in good approximation. However, inside the channel very asymmetric distribution functions occur. Hence, drift-diffusion and hydrodynamic models for the electron transport in semiconductors must fail in the correct description of this device (cf. [Carrillo *et al.* (2003)a]), since they are based on the assumption of electron distributions being shifted Maxwellians.

## 9.4    The Si-MESFET

For the simulation of the Si-MESFET, we use the geometry shown in Fig. 9.9 with the potentials at source $V_S = 0$ V, gate $V_G = -0.8$ V and drain $V_D = 1$ V. The donor densities are chosen as $n = 10^{17}$ cm$^{-3}$ and $n^+ = 3 \times 10^{17}$ cm$^{-3}$. The source and drain contacts act as particle reservoirs. Electrons may enter or exit through these contacts. The Schottky contact at the gate is assumed to be an absorbing boundary, whereas perfectly reflecting boundaries conditions are imposed at the non-contact surfaces. Concerning the boundary conditions for the Poisson equations, we apply the Neumann condition (vanishing electric field in the direction normal to the surface) on those boundary regions, where there are no contacts. These regions act as insulating boundaries, while the source, gate and drain

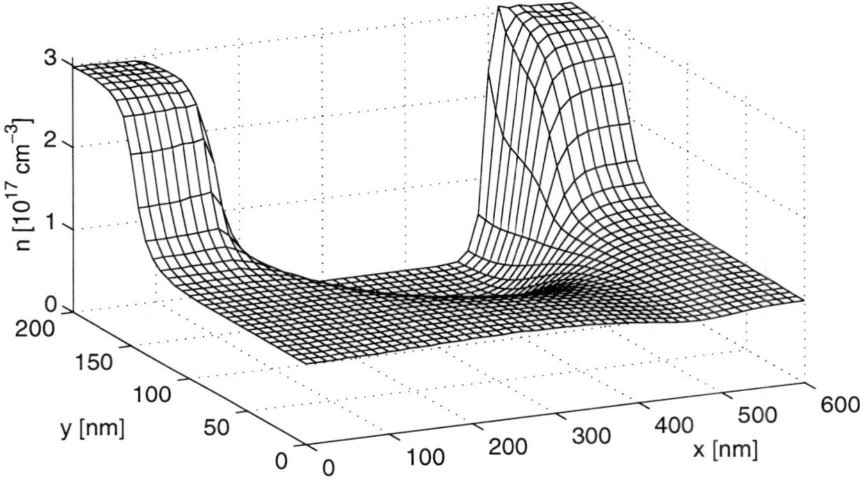

Fig. 9.10    Electron density $n$ versus position in the Si-MESFET at $t = 4$ ps.

contacts are treated as Dirichlet boundaries, where the bias voltages are applied. The parameters of the grid are chosen as $N = 75$, $M = 8$, $R = 8$ $n_{\mathrm{mul}} = 3$ together with $P = 48$ and $Q = 32$ grid points in real space.

Figure 9.10 illustrates the steady state electron density in the Si-MESFET, which is already reached at $t = 4$ ps. We observe highly accurate non-oscillatory behavior near the junctions and an asymmetric depletion region underneath the gate due to the applied drain and gate voltages. The velocity density and some stream lines are shown in Fig. 9.11. Moreover, the x and y components of the velocity density are displayed in Figs. 9.12 and 9.13. The stationary-state electric field strength and the equipotential lines are plotted in Fig. 9.14. The x- and y-components of the the electric field are illustrated in Figs. 9.15 and 9.16. Additionally, Fig. 9.17 shows the energy density in the Si-MESFET, while Fig. 9.18 illustrates the electrostatic potential.

In Figs. 9.19, 9.20, 9.21 and 9.22, we compare the cuts of the stationary-state electron density, the energy density and the x-components of the momentum and the electric field obtained with the multigroup-WENO solver with those from the full WENO solver [Carrillo *et al.* (2003)b] for several y-positions. Again, we observe good agreement between the results.

Figure 9.23 depicts the drain characteristics of the simulated MESFET

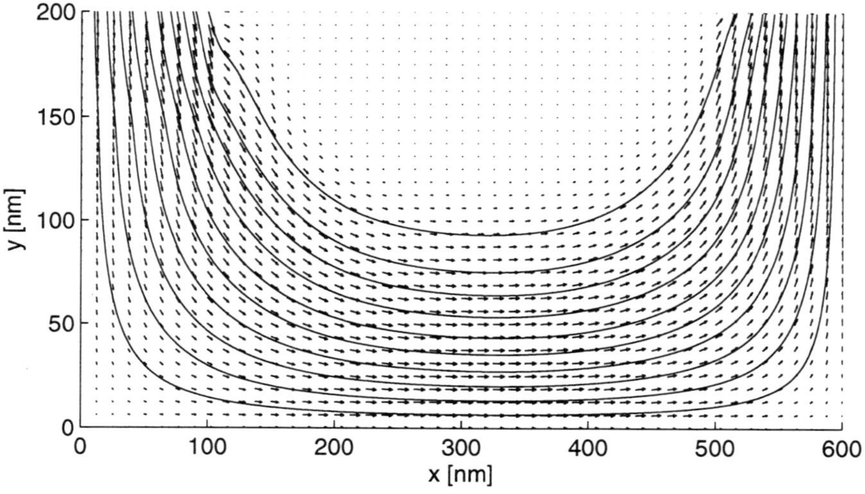

Fig. 9.11    Velocity field and streamlines versus position in the Si-MESFET at $t = 4$ ps.

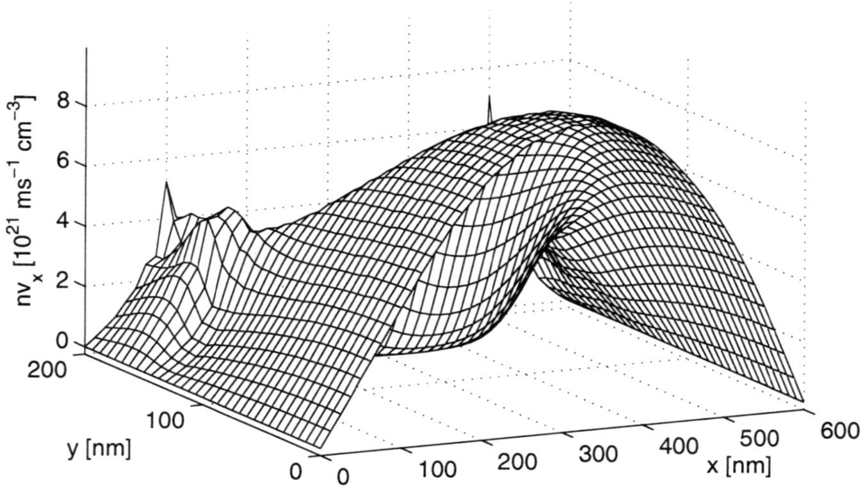

Fig. 9.12    X-component of the momentum $nv_x$ in the Si-MESFET at $t = 4$ ps.

for several gate voltages under normal operating conditions: $V_G \leq 0$ and $V_D \geq 0$. We observe that the gate voltage controls the width of the depletion

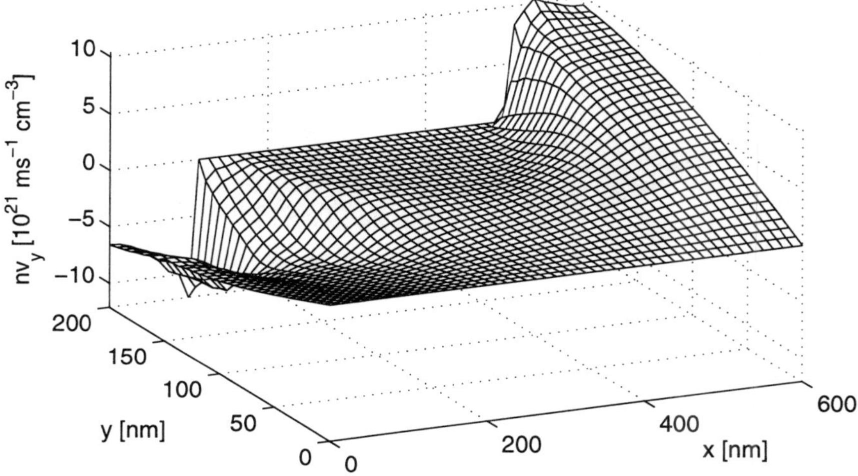

Fig. 9.13   Y-component of the momentum $nv_y$ in the Si-MESFET at $t = 4$ ps.

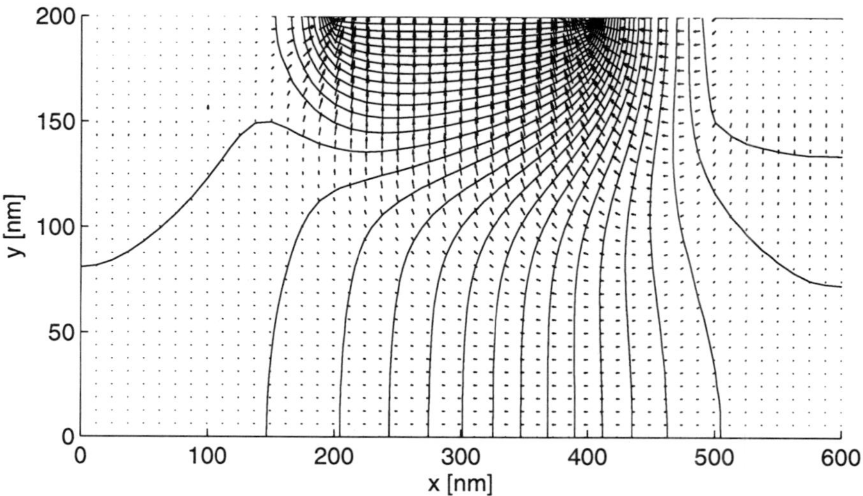

Fig. 9.14   Electric field and equipotenial lines in the Si-MESFET at $t = 4$ ps.

region and, thereby, the current through the device. Moreover, we find that the gate contact acts as an electron sink for low $V_\mathrm{D}$, when no gate voltage is

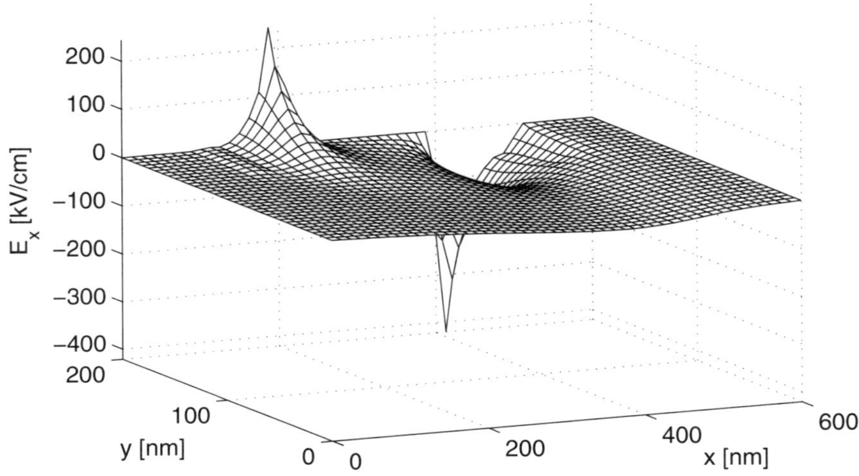

Fig. 9.15   X-component of the electric field $E_x$ in the Si-MESFET at $t = 4$ ps.

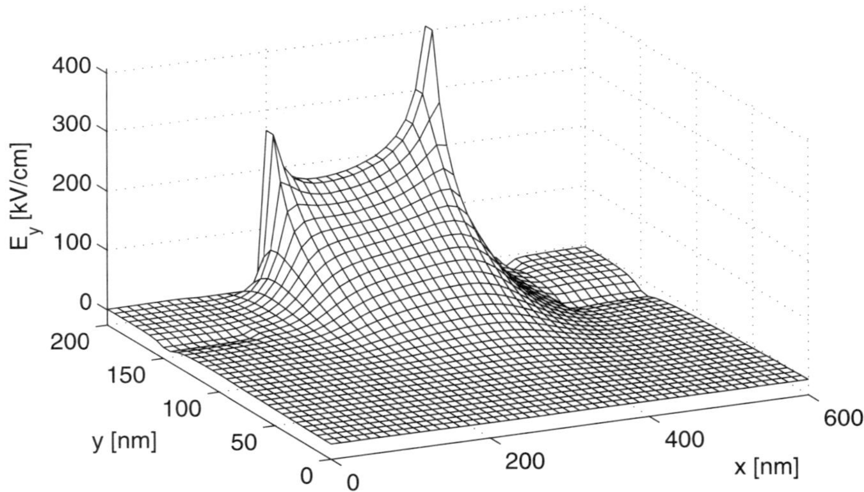

Fig. 9.16   Y-component of the electric field $E_y$ in the Si-MESFET at $t = 4$ ps.

applied. Electrons, that leave the MESFET through the gate, are replaced by ones, entering at both, the source and the drain contact. This leads to

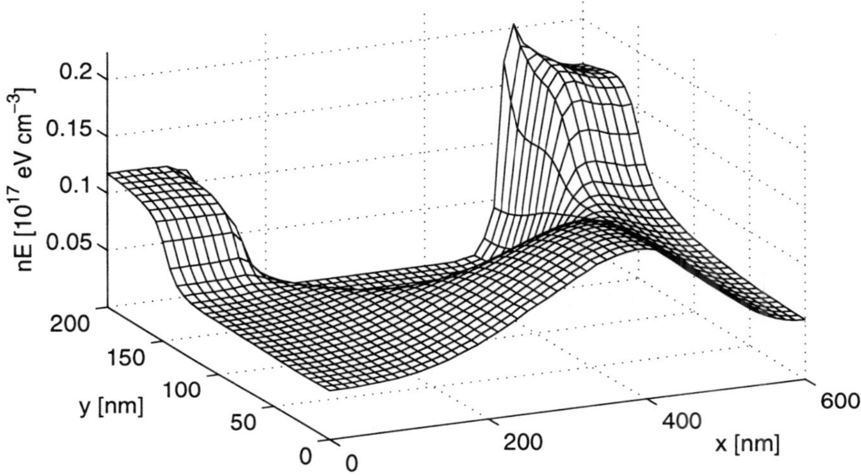

Fig. 9.17    Energy density $nE$ versus position in the Si-MESFET at $t = 4$ ps.

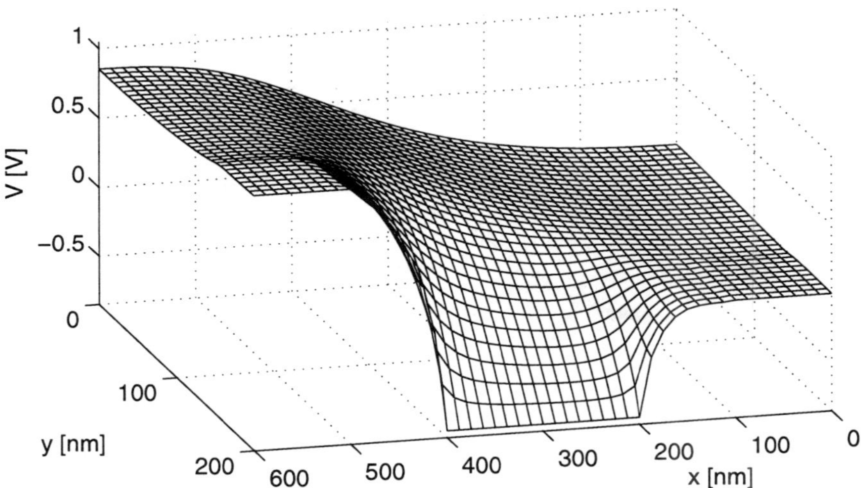

Fig. 9.18    Electrostatic potential $V$ versus position in the Si-MESFET at $t = 4$ ps.

the negative (i.e. inward) current through the drain displayed in Fig. 9.23.
For higher drain voltages, the electrons are accelerated in the resulting

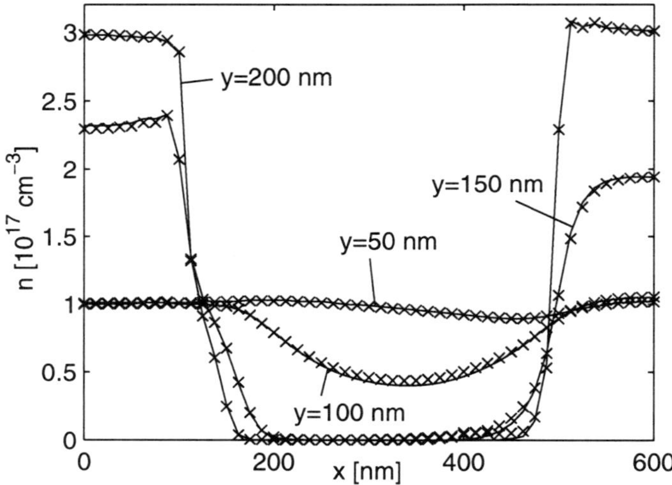


Fig. 9.19  The stationary-state electron density $n$ versus position $x$ in the Si-MESFET. (———): multigroup-WENO model; ($\times$): WENO solver.

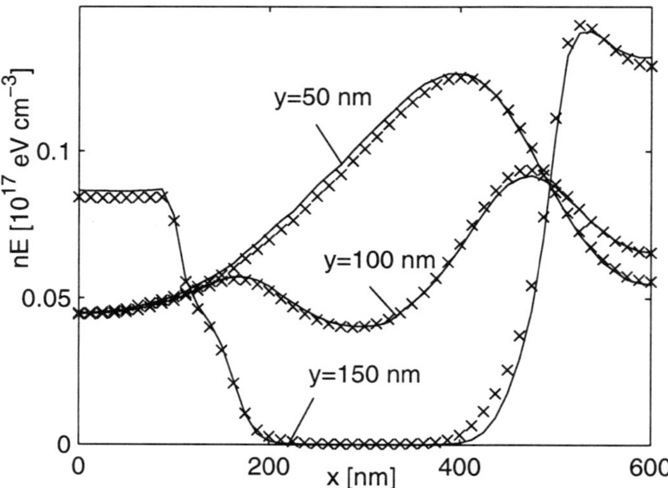


Fig. 9.20  Stationary-state energy density $nE$ versus position $x$ in the Si-MESFET. (———): multigroup-WENO model; ($\times$): WENO solver.

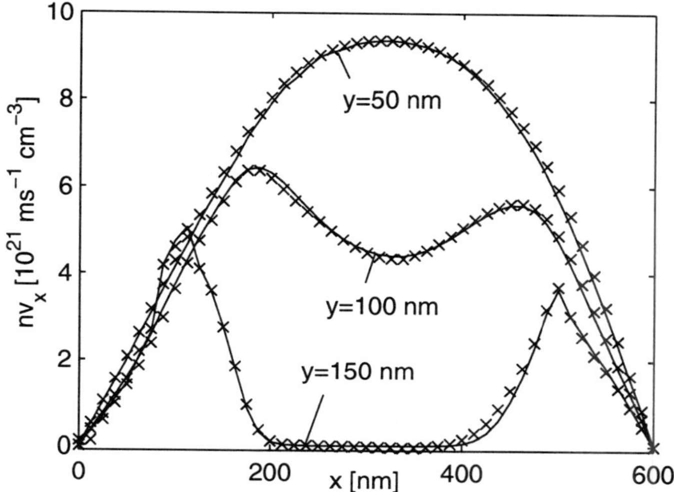

Fig. 9.21   The stationary-state x-component of the momentum $nv_x$ versus position $x$ in the Si-MESFET. (———): multigroup-WENO model; ($\times$): WENO solver.

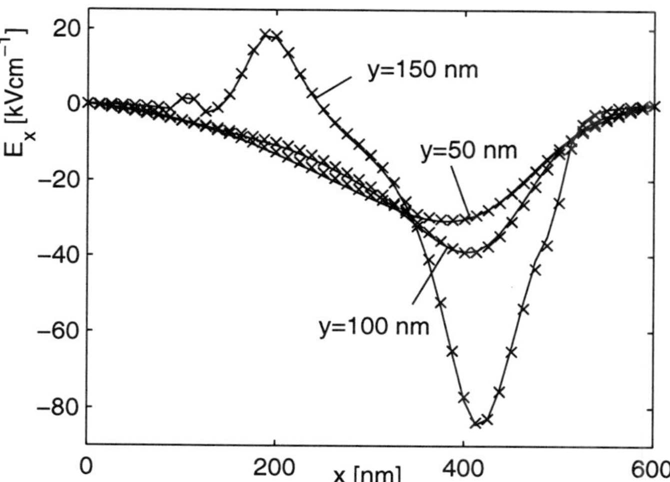

Fig. 9.22   The stationary-state the x-component of the electric field $E_x$ versus position $x$ in the Si-MESFET. (———): multigroup-WENO model; ($\times$): WENO solver.

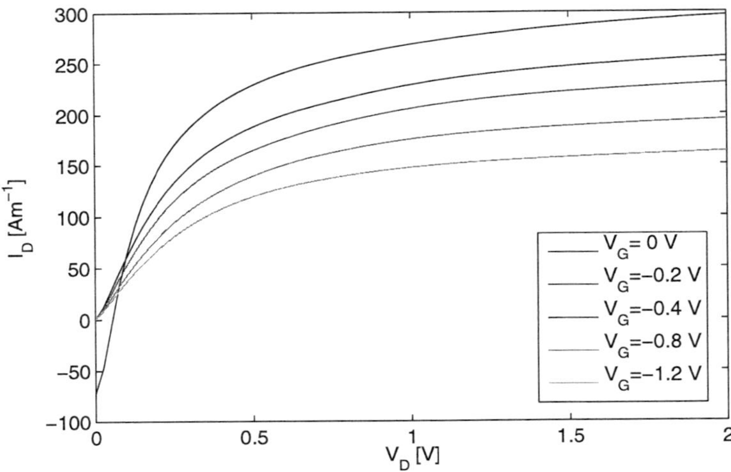

Fig. 9.23   Drain characteristics of the Si-MESFET for several gate voltages.

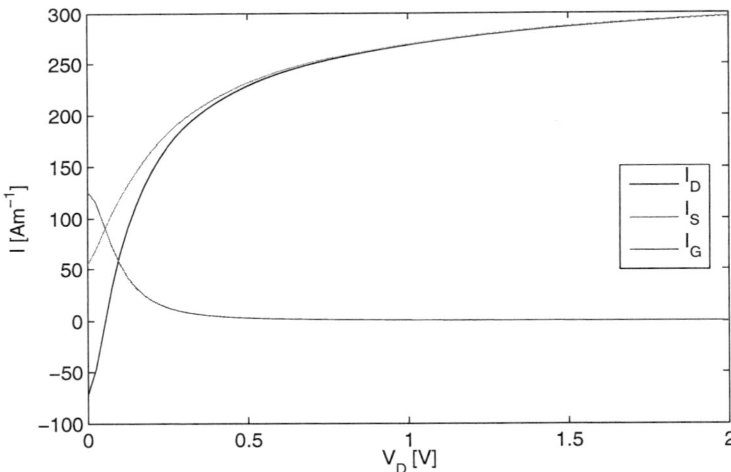

Fig. 9.24   The drain current $I_D$, the source current $I_S$ and the gate current $I_G$ versus the drain voltage $V_D$ for $V_G = 0$ V.

electric field and can pass the sink at the gate. Hence, the gate current tends to vanish for high $V_D$, as it is displayed in Fig. 9.24.

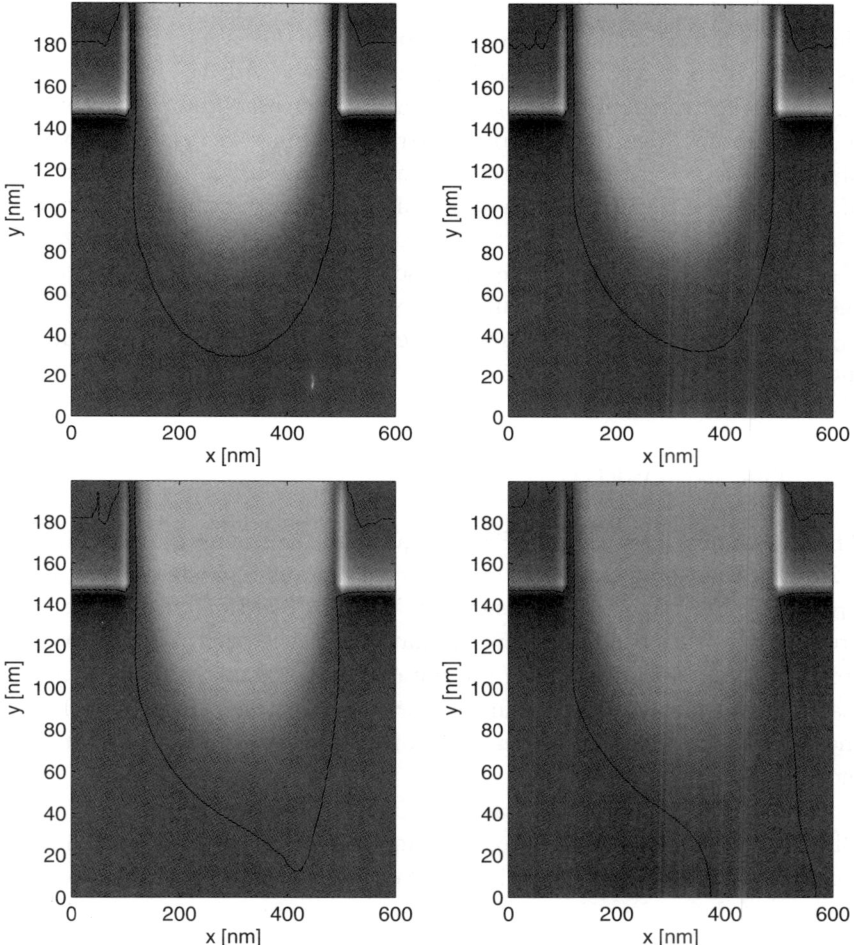

Fig. 9.25  The stationary-state carrier density $N_D - n$ in the Si-MESFET with $V_G = -1.2$ V. From top left to bottom right: $V_D = 0$ V, $V_D = 0.3$ V, $V_D = 0.4$ V and $V_D = 0.8$ V. The black lines refer to charge neutrality $n = N_D$.

The actual shape of the drain characteristics can be understood by interpreting the plots of the stationary-state carrier density in the MESFET given in Fig. 9.25. Under normal operating condition, a depletion layer is formed at the gate, and an electron flow and, therefore, a current is only possible in a conduction channel. The borderline between the conduction and the depletion region can be defined as the line of charge neutrality as it is used in Fig. 9.25. We observe that the width of the channel is hardly changed by a small drain voltage in comparison to the case $V_D = 0$ V. Hence, its resistivity is constant and $I_D$ increases linearly with $V_D$. Of course, for any given $V_D$, the voltage along the channel increases from zero at source to $V_D$ at drain. Thus, the Schottky barrier becomes increasingly reversed biased as we proceed from source to drain, leading to an asymmetric form of the depletion region and to a decreasing width of the conducting channel with increasing $V_D$. This results in a sublinear increase of $I_D$ for medium $V_D$. Finally, $V_D$ reaches a value $V_{Dsat}$ so that the depletion region touches the boundary of the device, and source and drain are pinched off. For $V_D > V_{Dsat}$, the drain current remains essentially at the value $I_{Dsat}$ and is independent of $V_D$.

## 9.5  The Si-MOSFET

The simulation of the Si-MOSFET is based on the geometry shown in Fig. 9.26 with the potentials at source $V_S = 0$ V, gate $V_G = 0.4$ V and drain $V_D = 1$ V. The donor density is chosen by $n^+ = 3 \times 10^{17}$ cm$^{-3}$. The holes are completely neglected in this simulation. Hence, the bulk substrate is treated as intrinsic silicon and the obtained results can only be seen as a starting point for the more accurate description of a MOSFET by including the holes, for instance, with the help of a simple drift-diffusion transport model.

The source and drain contacts are modeled as ohmic contacts, which allow the electrons to enter or exit the device. Perfectly reflecting boundary conditions are imposed at the Si/SiO$_2$ interface and at all of the non-contact surfaces of the device. Concerning the boundary conditions for the Poisson equation, they are the same as for the Si-MESFET. The parameters of the grid are chosen as $N = 70$, $M = 8$, $R = 8$ $n_{mul} = 3$ together with $P = 60$ and $Q = 48$ grid points in real space.

Figure 9.27 illustrates the steady state electron density in the Si-MOSFET, which is already reached at $t = 4$ ps. We observe highly accurate non-oscillatory behavior near the junctions and and the creation of a conducting channel at the Si/SiO$_2$ interface. The velocity field and some stream lines are shown in Fig. 9.28. Moreover, the x- and y-components of the drift velocity are given in Figs. 9.29 and 9.30. The stationary-state electric field strength and the equipotential lines are plotted in Fig. 9.31. The x- and y-components of the the electric field are illustrated in Figs. 9.32 and

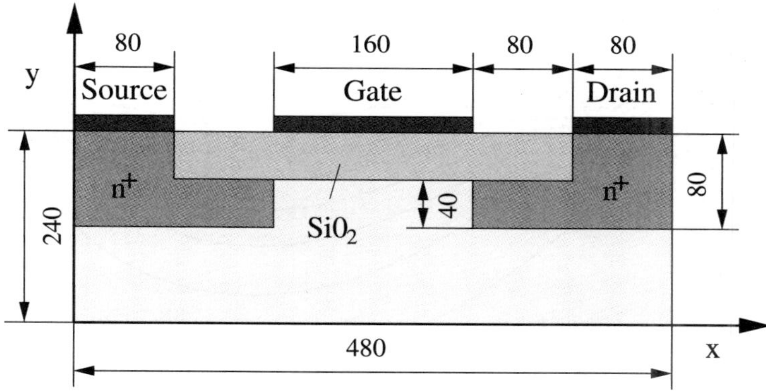



Fig. 9.26 Schematic illustration of a bidimensional MOSFET with the donor density $n^+ = 3 \times 10^{17}$ cm$^{-3}$. Lengths are given in nm.

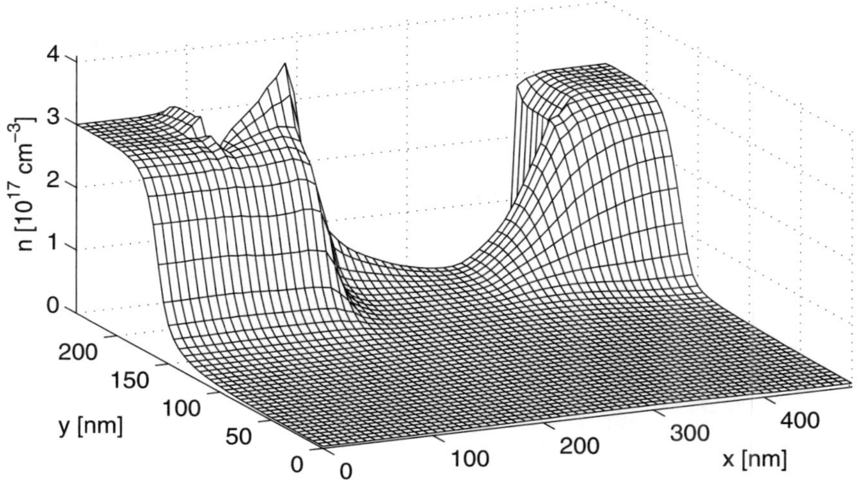

Fig. 9.27 Electron density $n$ versus position in the Si-MOSFET at $t = 4$ ps.

9.33. Additionally, Fig. 9.34 shows the mean energy in the Si-MESFET, while Fig. 9.35 illustrates the electrostatic potential.

In Figs. 9.36, 9.37, 9.38, 9.39 and 9.40, we compare the cuts of the stationary-state electron density, the x- and y-components of the drift velocity as well as the average energy and the electrostatic potential obtained

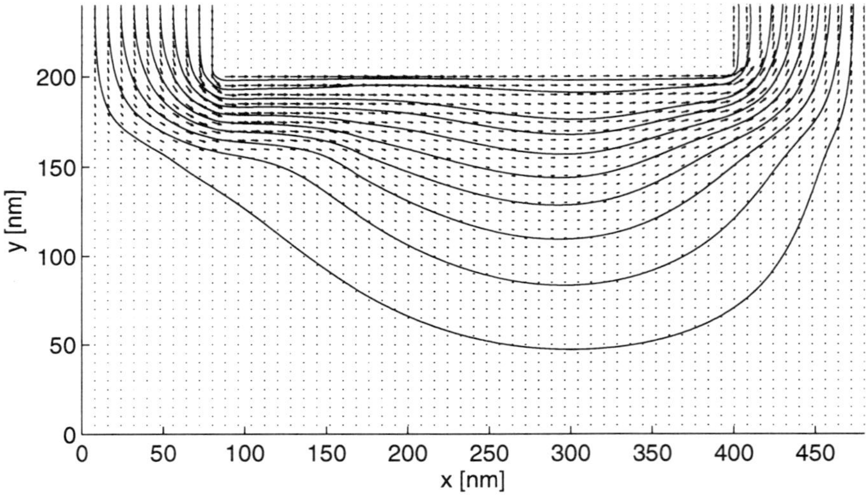

Fig. 9.28   Velocity field and current lines versus position in the Si-MOSFET at $t = 4$ ps.

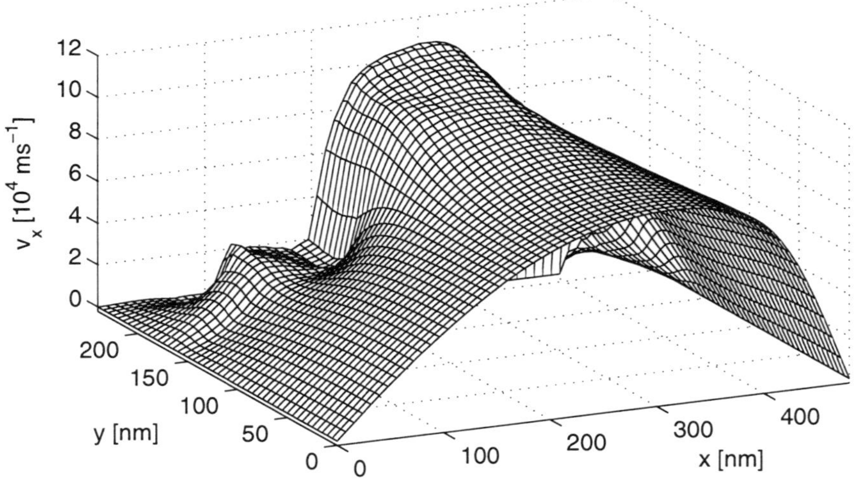

Fig. 9.29   X-component of the drift velocity $v_x$ in the Si-MOSFET at $t = 4$ ps.

with the multigroup-WENO solver with those from Monte Carlo calculations for several y-positions. Again, we observe good agreement between

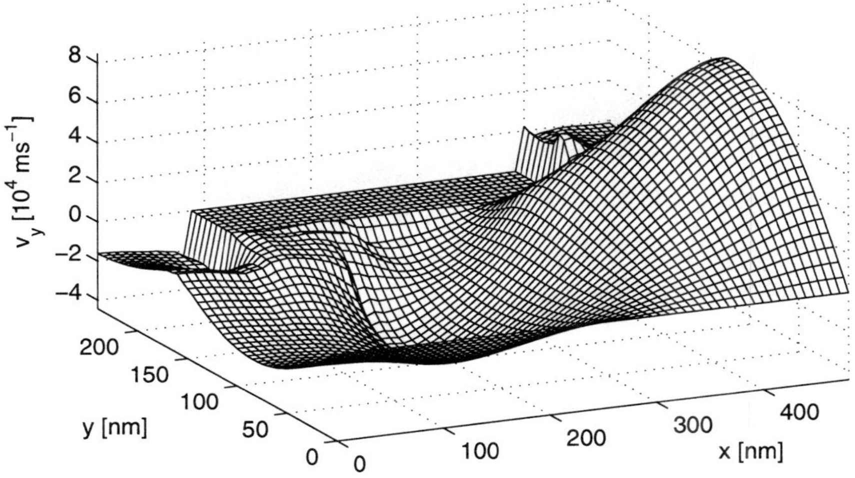

Fig. 9.30   Y-component of the drift velocity $v_y$ in the Si-MOSFET at $t = 4$ ps.

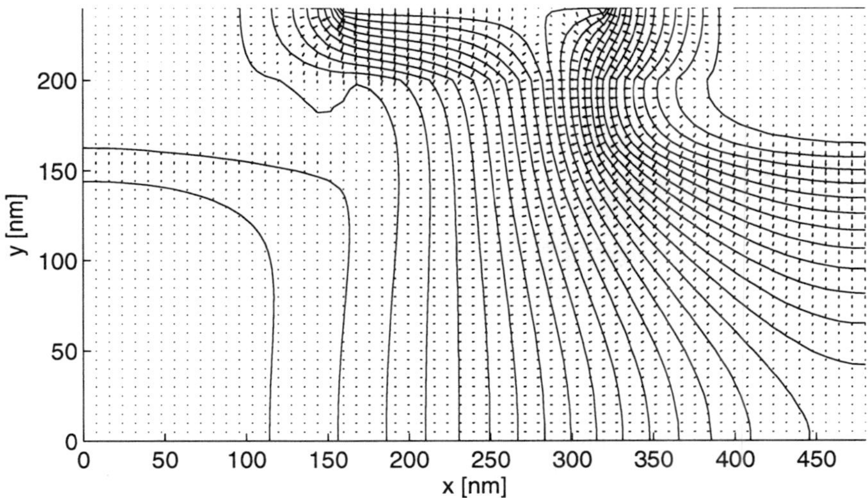

Fig. 9.31   Electric field and equipotential lines in the Si-MOSFET at $t = 4$ ps.

these results.

Hence, our studies reveal that the treatment of the momentum depen-

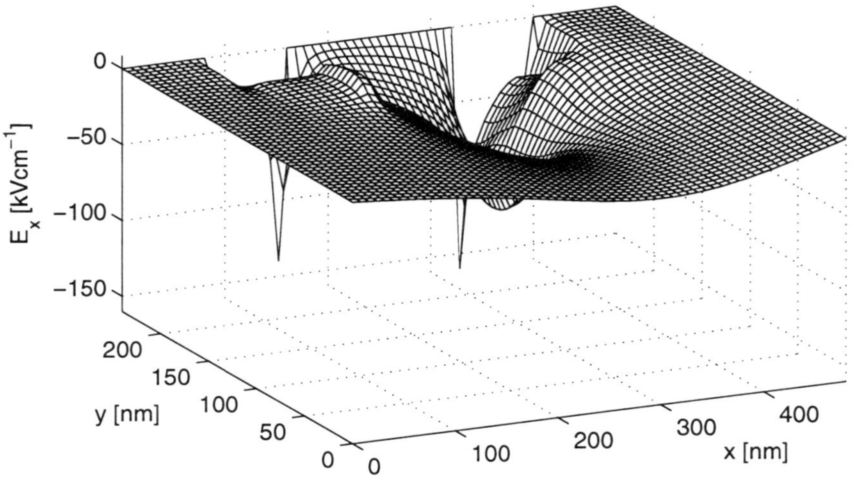

Fig. 9.32   X-component of the electric field $E_x$ in the Si-MOSFET at $t = 4$ ps.

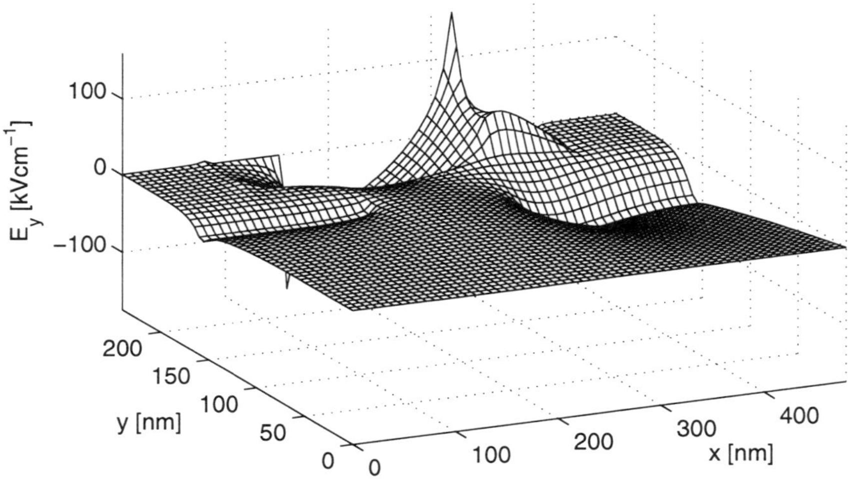

Fig. 9.33   Y-component of the electric field $E_y$ in the Si-MOSFET at $t = 4$ ps.

dence of the EDF, which does not show steep gradients, with the multigroup method and the spatial dependence with a high-order WENO scheme to

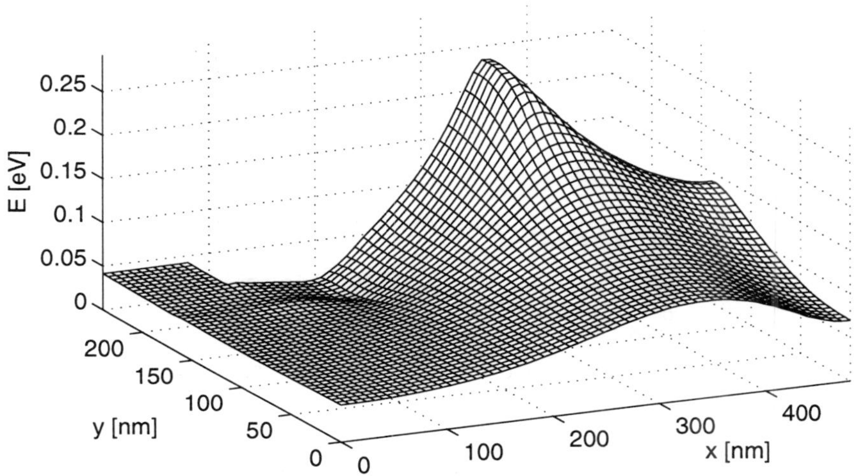

Fig. 9.34   Mean energy $E$ versus position in the Si-MOSFET at $t = 4$ ps.

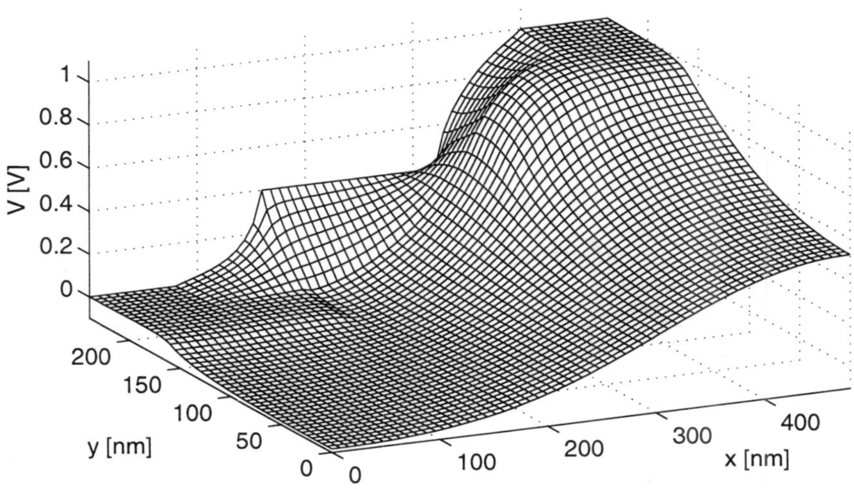

Fig. 9.35   Electrostatic potential $V$ versus position in the Si-MOSFET at $t = 4$ ps.

cope with sharp doping profiles, is an appropriate approach for the deterministic simulation of semiconductor devices.

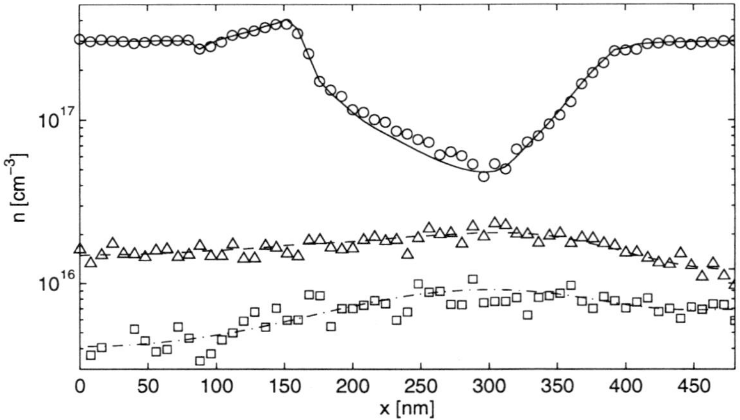

Fig. 9.36   Stationary-state electron density $n$ versus position $x$. Lines refer to present results, symbols to Monte Carlo calculations.   $(- \cdot -,\ \square)$: y=40 nm; $(- - -,\ \triangle)$: y=120 nm; $(\text{———},\ \bigcirc)$: y=200 nm.

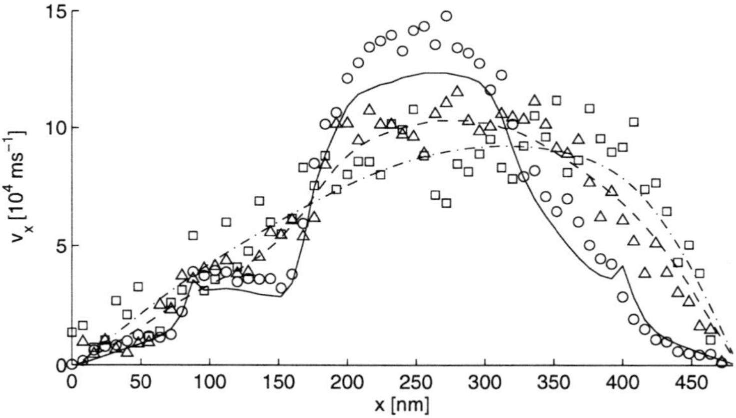

Fig. 9.37   Stationary-state velocity in x-direction $v_x$ versus position $x$. Lines refer to present results, symbols to Monte Carlo calculations. $(- \cdot -,\ \square)$: y=40 nm; $(- - -,\ \triangle)$: y=120 nm; $(\text{———},\ \bigcirc)$: y=200 nm.

In Fig. 9.41, the drain characteristics of the Si-MOSFET is displayed for several gate voltages. We observe an increasing drain current $I_D$ with increasing drain and gate voltages. The actual shape of these characteristics

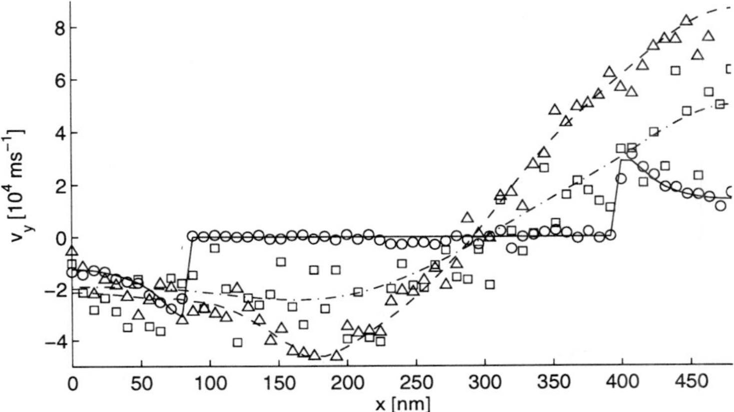

Fig. 9.38  Stationary-state velocity in y-direction $v_y$ versus position $x$. Lines refer to present results, symbols to Monte Carlo calculations. $(- \cdot -, \square)$: y=40 nm; $(- - -, \triangle)$: y=120 nm; $(\underline{\quad\quad}, \bigcirc)$: y=200 nm.

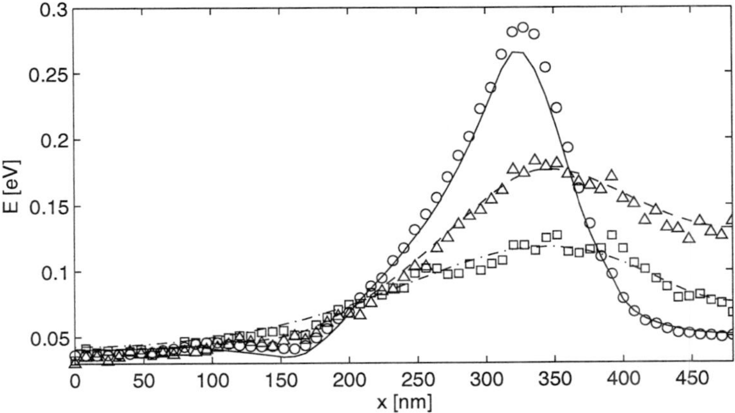

Fig. 9.39  Steady state mean energy $E$ versus position $x$. Lines refer to present results, symbols to Monte Carlo calculations. $(- \cdot -, \square)$: y=40 nm; $(- - -, \triangle)$: y=120 nm; $(\underline{\quad\quad}, \bigcirc)$: y=200 nm.

can be explained as follows. The total current through the drain consists of two portions: the current through the conduction channel formed at the Si/SiO$_2$ interface and a leakage current from drain to source via the bulk

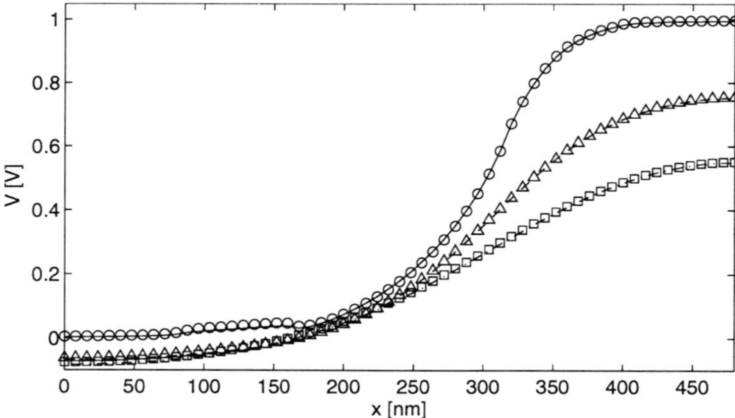

Fig. 9.40   Stationary-state electrostatic potential $V$ versus position $x$. Lines refer to present results, symbols to Monte Carlo calculations. $(- \cdot -, \square)$: y=40 nm; $(- - -, \triangle)$: y=120 nm; $(\text{———}, \bigcirc)$: y=200 nm.

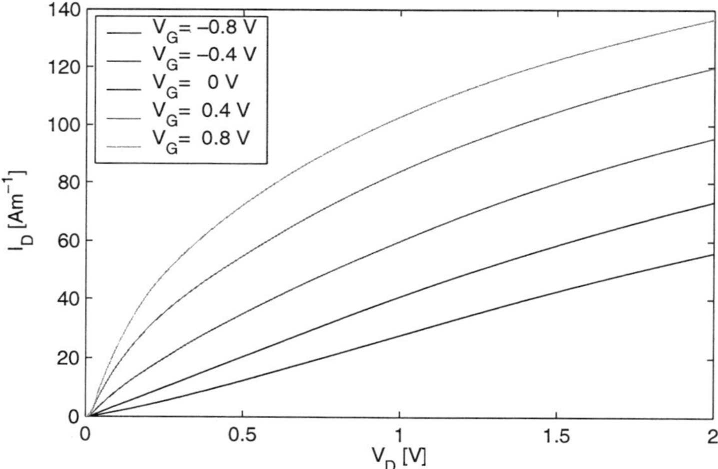



Fig. 9.41   Drain characteristics of the Si-MOSFET for several gate voltages.

substrate. Both, the channel and the substrate work as resistors, where the electric conductivity of the channel is much higher than that of the substrate. Since the amount of electrons forming the conducting channel is

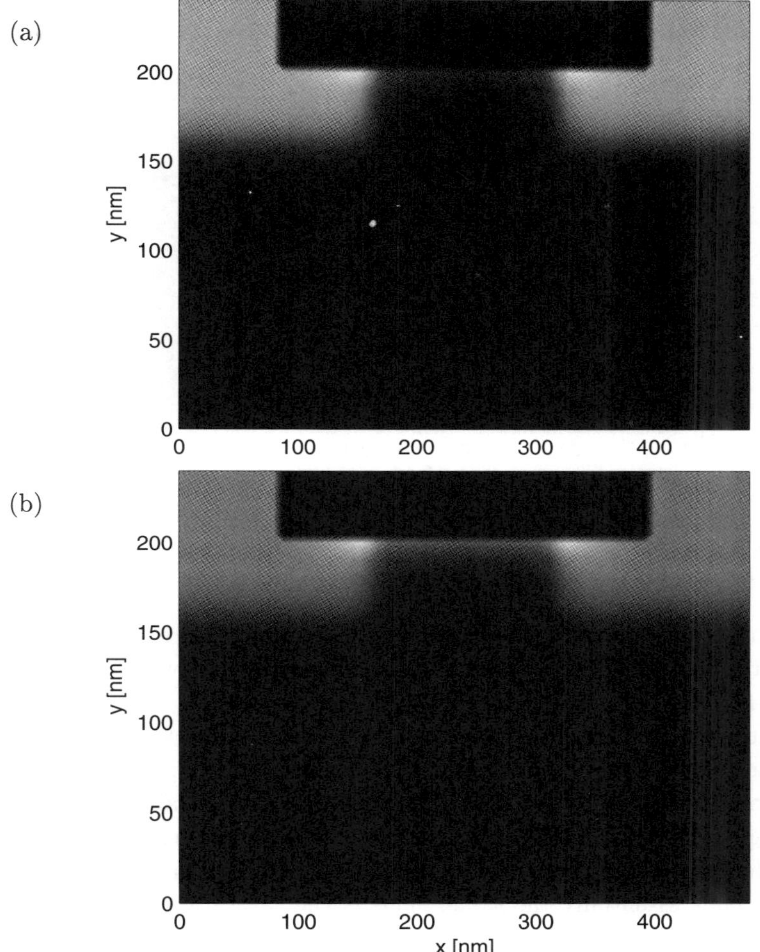

Fig. 9.42   Contour plot of the electron density in the Si-MOSFET at $t = 4$ ps with $V_\mathrm{D} = 0$ V and (a) $V_\mathrm{G} = 0.4$ V, (b) $V_\mathrm{G} = 0.8$ V.

controlled by the gate voltage as it is illustrated in Fig. 9.42, it is clear that a higher gate voltage leads to a higher channel current and consequently to a higher drain current.

Now, we consider the case that a positive gate voltage is applied so that a conducting channel is formed. If a small drain voltage is applied, most electrons flow from source to drain through the channel because of the high resistance of the substrate. Hence, the leakage current can be neglected in

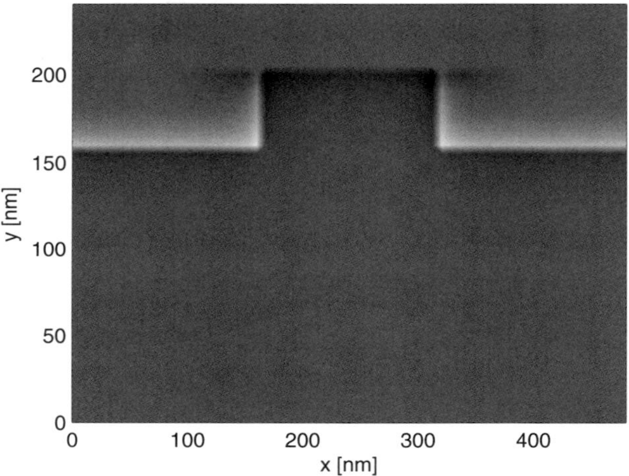

Fig. 9.43   Contour plot of the carrier density in the Si-MOSFET at $t = 4$ ps with $V_\mathrm{G} = 0.4$ V and $V_\mathrm{D} = 0$ V.

comparison to the channel current. The channel simply acts as a resistor and the drain current is proportional to the drain voltage, which is called the linear region.

When the drain voltage increases, the thickness of the conducting channel becomes reduced at the interface to the drain $n^+$ region. This trend is depicted in Figs. 9.43, 9.44 and 9.45 for rising drain voltages and a constant gate voltage. At the saturation voltage $V_\mathrm{sat}$, this thickness reduces to zero; this is called the pinch-off point. Beyond the pinch-off point, the channel current remains essentially the same. This is the saturation region, since the channel current is constant regardless of the applied drain voltage. For $V_\mathrm{D} > V_\mathrm{sat}$, the current-voltage characteristics is mainly determined by the leakage current through the substrate. In idealized characteristics of MOS-FETs, the drain current is assumed to be constant for $V_\mathrm{D} > V_\mathrm{sat}$, since the leakage current is neglected. In contrast to such models, we find a further increase of the drain current with the drain voltage in the saturation region because of the relatively high leakage current, which results from both, the little size of the simulated device (short-channel effect) and the use of an insulator instead in a p-doped material as the bulk substrate. However, the low conductivity of the substrate causes a much lower slope of the characteristics in the saturation region than that in the linear region. Hence, our results for the drain characteristics agree qualitatively with those given

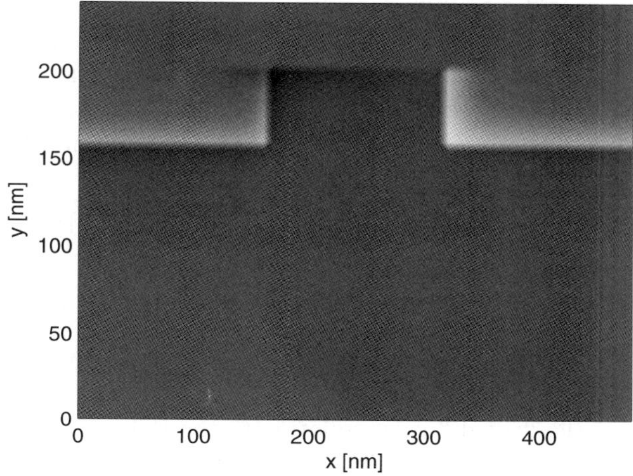

Fig. 9.44   Contour plot of the carrier density in the Si-MOSFET at $t = 4$ ps with $V_\mathrm{G} = 0.4$ V and $V_\mathrm{D} = 0.4$ V.

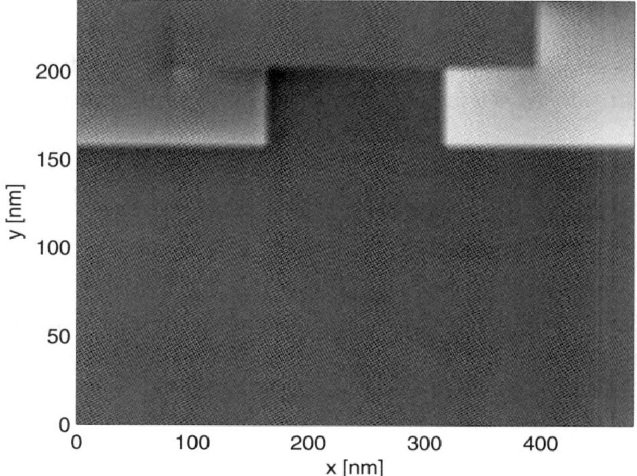

Fig. 9.45   Contour plot of the carrier density in the Si-MOSFET at $t = 4$ ps with $V_\mathrm{G} = 0.4$ V and $V_\mathrm{D} = 3$ V.

for instance in [Sze (2002)] except for the fact that the leakage current is neglected in idealized drain characteristics.

Finally, we consider an extension for the multigroup-WENO solver for a more realistic simulation of the Si-MOSFET. Therefore, we include the transport of holes through the device with the help of a drift-diffusion model. Hence, the basic equation for the evolution of the hole density $p(\mathbf{r}, t)$ depending on position $\mathbf{r}$ and time $t$ reads [Lundstrom (2000)]

$$\frac{\partial p(t, \mathbf{r})}{\partial t} + \nabla_{\mathbf{r}} \cdot \{\mu(\mathcal{E})p(t, \mathbf{r})\mathbf{E}(t, \mathbf{r}) - D(\mathcal{E})\nabla_{\mathbf{r}}p(t, \mathbf{r})\} = 0 \qquad (9.4)$$

with $\mathcal{E} = |\mathbf{E}|$, while the corresponding hole velocity density $p\mathbf{v}_{\mathrm{p}}$ is given by

$$p\mathbf{v}_{\mathrm{p}}(t, \mathbf{r}) = \mu(\mathcal{E})p(t, \mathbf{r})\mathbf{E}(t, \mathbf{r}) - D(\mathcal{E})\nabla_{\mathbf{r}}p(t, \mathbf{r}). \qquad (9.5)$$

The coupling of the drift-diffusion equation to the Boltzmann-Poisson system occurs in the Poisson equation, which must be modified in order to obtain

$$\nabla[\varepsilon_{\mathrm{r}}(\mathbf{r}) \cdot \nabla V(t, \mathbf{r})] = -\frac{e}{\varepsilon_0}[N_{\mathrm{D}}(\mathbf{r}) - n_{\mathrm{n}}(t, \mathbf{r}) - N_{\mathrm{A}}(\mathbf{r}) + p(t, \mathbf{r})] \qquad (9.6)$$

with the donor density $N_{\mathrm{D}}$, the acceptor density $N_{\mathrm{A}}$ and the electron density $n_{\mathrm{n}}$.

To begin with, we summarize the approximations used for the hole mobility and diffusion coefficient in silicon. The hole mobility $\mu$ is approximated by

$$\mu(\mathcal{E}) = \mu^0 \left[1 + \left(\frac{\mathcal{E}}{\mathcal{E}_{\mathrm{cr}}}\right)^2\right]^{-\frac{1}{2}} \qquad (9.7)$$

with the critical electric field $\mathcal{E}_{\mathrm{cr}}$ and the low-field mobility $\mu^0$ as suggested in [Lundstrom (2000)]. The quantity $\mathcal{E}_{\mathrm{cr}}$ is related to the saturation velocity $v_{\mathrm{sat}}^{\mathrm{p}}$ of holes via

$$\mathcal{E}_{\mathrm{cr}} = \frac{v_{\mathrm{sat}}^{\mathrm{p}}}{\mu^0}. \qquad (9.8)$$

The low-field mobility in pure silicon can be fitted by an expression of the form

$$\mu^0 = \frac{A}{T_{\mathrm{L}}^a}. \qquad (9.9)$$

Table 9.2   Material parameters for the hole transport in silicon.

| Quantity | Symbol | Unit | Value |
|---|---|---|---|
| Saturation velocity[a] | $v_{\text{sat}}^{\text{P}}$ | $10^4$ ms$^{-1}$ | 9.6 |
| Fitting parameter[b] | A | $10^4$ m$^2$K$^{2.2}$V$^{-1}$s$^{-1}$ | 1.30 |
|  | a |  | 2.2 |
| Lattice temperature | $T_{\text{L}}$ | K | 300 |
| Low-field mobility (9.9) | $\mu_{\text{p}}^0$ | $10^{-2}$ m$^2$V$^{-1}$s$^{-1}$ | 4.616 |
| Critical field (9.8) | $\mathcal{E}_{\text{cr}}$ | kVcm$^{-1}$ | 20.80 |

[a] Ref. [Smith *et al.* (1981)],  [b] Ref. [Ali Omar and Reggiani (1987)]

Concerning the diffusion coefficient of holes $D$, we apply the Einstein relation

$$D(\mathcal{E}) = \frac{k_{\text{B}} T_{\text{p}}(\mathcal{E})}{e} \mu(\mathcal{E}) \qquad (9.10)$$

with the field dependent hole temperature [Lundstrom (2000)]

$$T_{\text{p}}(\mathcal{E}) = T_{\text{L}} \left[ 1 + \left( \frac{\mathcal{E}}{\mathcal{E}_{\text{cr}}} \right)^2 \right]. \qquad (9.11)$$

Numerical values for the quantities used to calculate $\mu$ and $D$ are given in Tab. 9.2.

For the spatially two-dimensional case, the drift-diffusion equation reads

$$\frac{\partial p}{\partial t} + \frac{\partial}{\partial x}(\mu p E_x) + \frac{\partial}{\partial y}(\mu p E_y) - \frac{\partial}{\partial x}\left( D \frac{\partial p}{\partial x} \right) - \frac{\partial}{\partial y}\left( D \frac{\partial p}{\partial y} \right) = 0. \qquad (9.12)$$

This equation is discretized on the same spatial grid as used in the multigroup-WENO solver; the spatial derivatives are approximated by means of standard finite difference schemes. Hence, we obtain

$$\frac{\partial}{\partial t} p_{xy} + \frac{1}{\Delta x}([\mu p E^x]_{xy}^+ - [\mu p E^x]_{xy}^-) + \frac{1}{\Delta y}([\mu p E^y]_{xy}^+ - [\mu p E^y]_{xy}^-) \qquad (9.13)$$

$$- \frac{1}{\Delta x^2}[\mu_{x+1/2,y} p_{x+1,y} - (\mu_{x+1/2,y} + \mu_{x-1/2,y}) p_{xy} + \mu_{x-1/2,y} p_{x-1,y}]$$

$$- \frac{1}{\Delta y^2}[\mu_{x,y+1/2} p_{x,y+1} - (\mu_{x,y+1/2} + \mu_{x,y-1/2}) p_{xy} + \mu_{x,y-1/2} p_{x,y-1}] = 0$$

with the abbreviations

$$p_{xy} = p(x_y, y_y), \quad E_{xy}^x = E^x(x_y, y_y), \quad E_{xy}^y = E^y(x_y, y_y), \qquad (9.14\text{a})$$

$$\mu_{xy} = \mu(\mathcal{E}_{xy}), \quad D_{xy} = D(\mathcal{E}_{xy}), \quad \mathcal{E}_{xy} = \left[ E_{xy}^{x\,2} + E_{xy}^{y\,2} \right]^{\frac{1}{2}}. \qquad (9.14\text{b})$$

Moreover, we apply

$$[\mu p E^x]_{xy}^- = \begin{cases} \mu_{x-1,y} p_{x-1,y} E_{x-1,y}^x, & \text{if } E_{xy}^x > 0, \\ \mu_{xy} p_{xy} E_{xy}^x, & \text{if } E_{xy}^x < 0, \end{cases} \tag{9.15a}$$

$$[\mu p E^x]_{xy}^+ = \begin{cases} \mu_{x+1,y} p_{x+1,y} E_{x+1,y}^x, & \text{if } E_{xy}^x < 0, \\ \mu_{xy} p_{xy} E_{xy}^x, & \text{if } E_{xy}^x > 0, \end{cases} \tag{9.15b}$$

$$[\mu p E^y]_{xy}^- = \begin{cases} \mu_{x,y-1} p_{x,y-1} E_{x,y-1}^y, & \text{if } E_{xy}^y > 0, \\ \mu_{xy} p_{xy} E_{xy}^y, & \text{if } E_{xy}^y < 0, \end{cases} \tag{9.15c}$$

$$[\mu p E^y]_{xy}^+ = \begin{cases} \mu_{x,y+1} p_{x,y+1} E_{x,y+1}^y, & \text{if } E_{xy}^y < 0, \\ \mu_{xy} p_{xy} E_{xy}^y, & \text{if } E_{xy}^y > 0 \end{cases} \tag{9.15d}$$

and

$$\mathcal{E}_{x\pm1/2,y} = \frac{1}{2}\left[(E_{xy}^x + E_{x\pm1/2y}^x)^2 + (E_{xy}^y + E_{x\pm1/2,y}^x)^2\right]^{\frac{1}{2}}, \tag{9.16a}$$

$$\mathcal{E}_{x,y\pm1/2} = \frac{1}{2}\left[(E_{xy}^x + E_{x,y\pm1/2}^x)^2 + (E_{xy}^y + E_{x,y\pm1/2}^y)^2\right]^{\frac{1}{2}}. \tag{9.16b}$$

The components $n v_{\mathrm{p}}^x$ and $n v_{\mathrm{p}}^y$ of the hole velocity density in x and y directions are evaluated via

$$n v_{\mathrm{p}}^x(x_x, y_y) = \mu_{xy} p_{xy} E_{xy}^x - \frac{D_{xy}}{2\Delta x}(p_{x+1,y} - p_{x-1,y}), \tag{9.17a}$$

$$n v_{\mathrm{p}}^y(x_x, y_y) = \mu_{xy} p_{xy} E_{xy}^y - \frac{D_{xy}}{2\Delta y}(p_{x,y+1} - p_{x,y-1}). \tag{9.17b}$$

Concerning the boundary conditions for the hole density, we proceed as follows. For modeling a reflecting boundary, for instance, at $x = 0$, we set $[\mu p E^x]_{xy}^- = 0$ and $\mu_{-1/2,y} p_{-1,y} = \mu_{1/2,y} p_{1,y}$. For an ohmic contact at $y = Q$, we set

$$[\mu p E^y]_{xy}^+ = \begin{cases} \mu_{x,y} N_{\mathrm{A}}(x_x, y_y) E_{x,y}^y, & \text{if } E_{xy}^y < 0, \\ \mu_{xy} p_{xy} E_{xy}^y, & \text{if } E_{xy}^y > 0 \end{cases} \tag{9.18}$$

and $\mu_{x,Q+1/2} p_{x,Q+1} = \mu_{x,Q} N_A(x_x, y_Q)$.

The time integration of (9.13) is performed by means of an explicit scheme for the advection term for conserving the hole density and a fully implicit algorithm for treating the diffusion term, which allows us to obtain non-oscillating solutions with large time steps. In other words, we can use the time step $\Delta t$ evaluated from (8.30) for the time integration of both, the Boltzmann equation and the drift-diffusion equation.

The geometry used for simulating the coupled electron-hole transport in the Si-MOSFET is displayed in Fig. 9.46. It agrees with that shown in Fig.

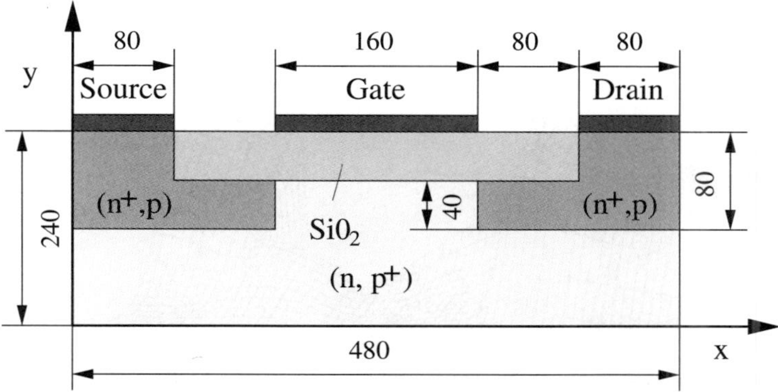



Fig. 9.46   Schematic illustration of a bidimensional MOSFET used to simulate electron and hole transport. Lengths are given in nm.

9.26 besides the changes in the doping profiles. The donor densities are set to $n^+ = 3 \times 10^{17}$ cm$^{-3}$ and $n = 10^{12}$ cm$^{-3}$, while $p^+ = 10^{17}$ cm$^{-3}$ and $p = 10^{12}$ cm$^{-3}$ are used as the acceptor densities. Moreover, the applied voltages are set to $V_S = 0$ V at source, $V_G = 0.4$ V at the gate and $V_D = 1$ V at drain in coincidence with the previous simulation. The model parameters used are $N = 60$, $M = 6$, $R = 6$, $n_{\mathrm{mul}} = 2$, $P = 60$, $Q = 48$.

Figures 9.47 and 9.48 display the electron and the hole densities in the considered Si-MOSFET at $t = 6$ ps. In addition, Fig. 9.49 depicts the velocity field of the electrons; the electric field is shown in Fig. 9.50.

We observe that the electric fields around the drain $n^+$ region are stronger in this simulation than they are for considering only the electron transport (cf. Fig. 9.31), because of the formation of a hole depletion region in this area (cf. Fig. 9.48). This strong electric field, that acts repulsive on the holes, accelerates the electrons in a way, so that they hardly take their ways in the device through the substrates, as it is clearly illustrated by the low density of current lines and by the low electron density in the right half of the substrate in Figs. 9.49 and 9.47. Therefore, the electron current through the considered Si-MOSFET is mainly determined by the electron flow through the conducting channel formed at the Si/SiO$_2$ interface and the current voltage characteristics feature more agreement with the idealized drain characteristics mentioned above.

For giving an example about this fact, we consider Fig. 9.51, which shows the drain characteristics based on the simulation of the coupled

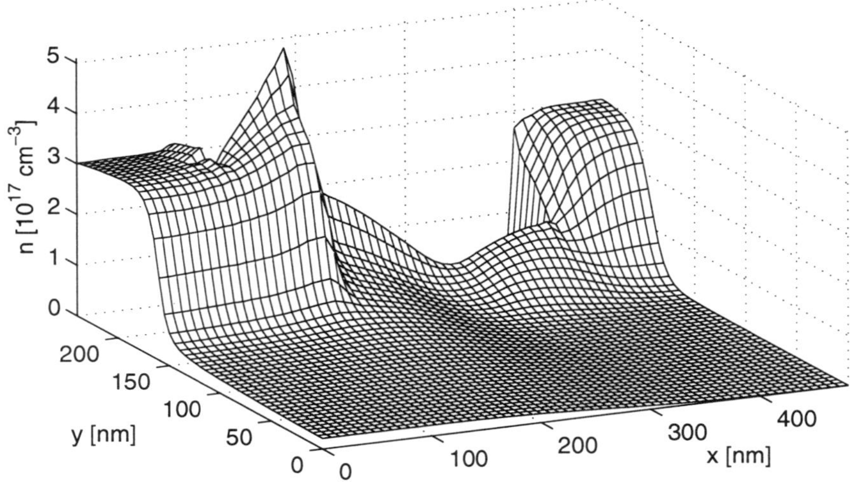

Fig. 9.47   Electron density $n$ versus position for the coupled electron-hole transport in the Si-MOSFET at $t = 6$ ps.

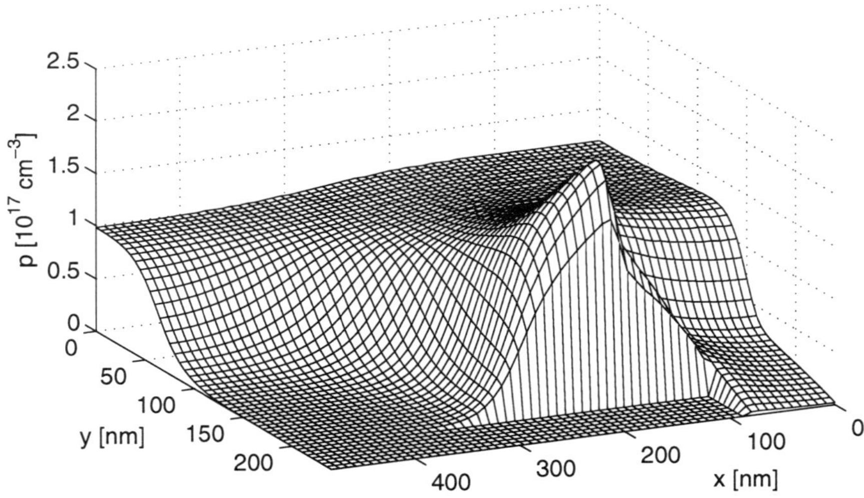

Fig. 9.48   Hole density $p$ versus position for the coupled electron-hole transport in the Si-MOSFET at $t = 6$ ps.

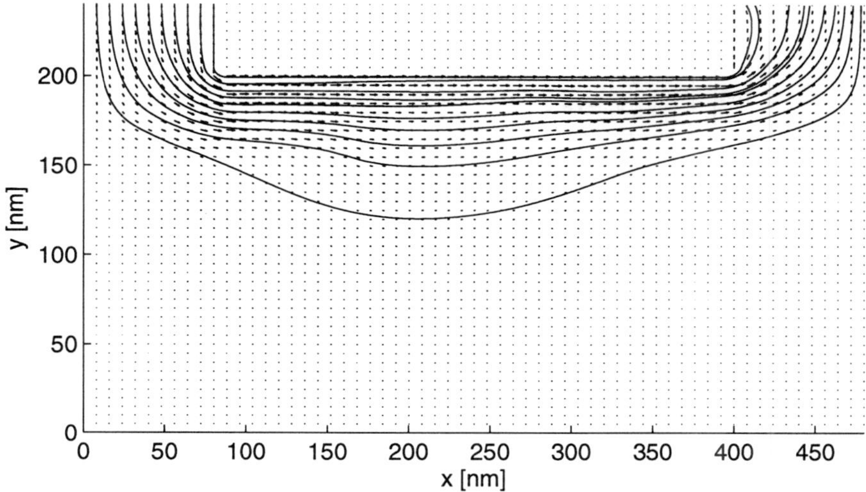

Fig. 9.49  Electron velocity field and current lines versus position for the coupled electron-hole transport in the Si-MOSFET at $t = 6$ ps.

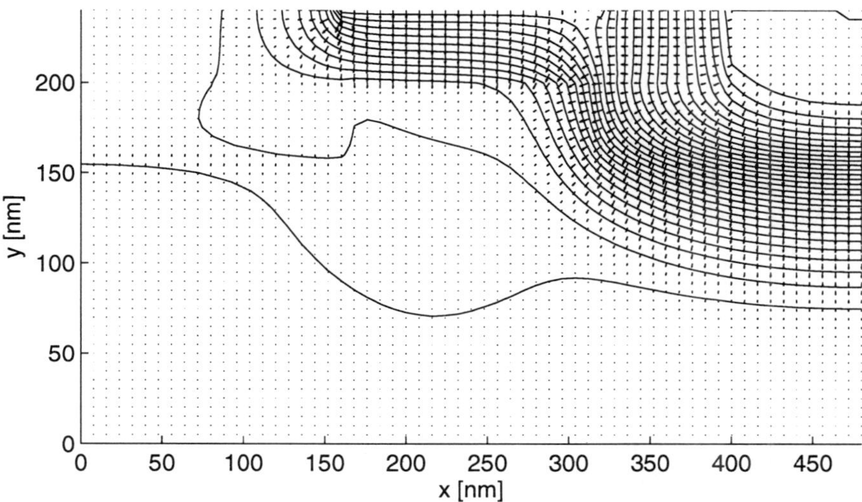

Fig. 9.50  Electric field and equipotential lines versus position for the coupled electron-hole transport in the Si-MOSFET at $t = 6$ ps.

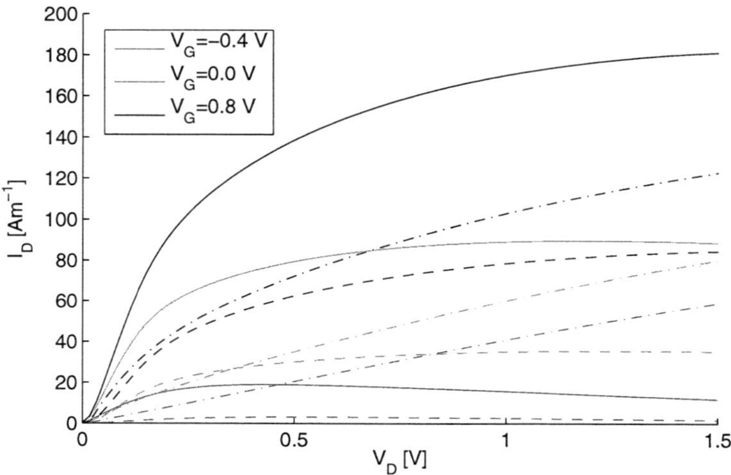

Fig. 9.51 Drain characteristics in the Si-MOSFET for several gate voltages. Solid lines: total current for coupled electron-hole system; dashed lines: electron current for the coupled electron-hole system; dashed-dotted lines: pure electron transport.

electron-hole transport through the Si-MOSFET for several gate voltages. In this figure, the solid lines refer to the total, i.e., the sum of the electron and the hole currents, whereas the dashed lines refer to the electron current. The dashed-dotted curves illustrate the electron current for considering only the electron transport through the device as given in Fig. 9.41. We find the split of the drain characteristics into the linear region and the saturation region typically for MOSFETs. Moreover, we observe that the hole current is of the same order as the electron current for the simulated configuration of the device, which is a consequence of the high chosen acceptor density. The comparison between the electron currents obtained by considering the coupled electron-hole system and by dealing with the pure electron system is of special interest. Here, we find that these quantities exhibit almost the same dependence on the drain voltage for small $V_D$, while they differ significantly for high drain voltages due to the much smaller leakage current in the electron-hole simulation.

# Chapter 10

# Simulation of Gallium Arsenide Devices

## 10.1 Introduction

In this chapter, we present numerical results for the transport of electrons in bulk GaAs, in a GaAs $n^+ - n_i - n^+$ diode and in a GaAs-MESFET obtained by means of our multigroup approach (8.17). All of these calculations are performed for the lattice temperature $T_L = 300$ K. We approximate the conduction band of GaAs by taking into account the $\Gamma$-valley in the center of the first Brillouin zone and four equivalent $L$-valleys along the crystallographic directions $\langle 1, 1, 1 \rangle$. The zero energy level is set to the energy of the bottom of the $\Gamma$-valley, in other words, $\Delta_{0\Gamma} = 0$. In our transport model for GaAs, we regard the acoustic deformation potential scattering (ADP) with the transition rate

$$S_{\nu\nu}^{\mathrm{ADP}}(\mathbf{k}, \mathbf{k}') = \frac{D_A^2 k_B T_L}{4\pi^2 \hbar \rho v_s^2} \, \delta[E_\nu(\mathbf{k}') - E_\nu(\mathbf{k})] \tag{10.1}$$

and the ionized impurity scattering (IMP)

$$S_{\nu\nu}^{\mathrm{IMP}}(\mathbf{k}, \mathbf{k}') = \frac{N_I}{\hbar} \left( \frac{e^2}{2\pi\varepsilon_{st}\varepsilon_0} \right)^2 \frac{\delta[E_\nu(\mathbf{k}') - E_\nu(\mathbf{k})]}{(|\mathbf{k} - \mathbf{k}'|^2 + q_D^2)^2} \tag{10.2}$$

with the screening parameter $q_D^2 = e^2 N_I / \varepsilon_{st}\varepsilon_0 k_B T_L$. The most important intravalley scattering mechanism for GaAs is the polar optical interaction (POP) with the transition rate

$$S_{\nu\nu}^{\mathrm{POP}}(\mathbf{k}, \mathbf{k}') = \frac{\omega_p e^2}{8\pi^2} \left( \frac{1}{\varepsilon_{hf}} - \frac{1}{\varepsilon_{st}} \right) \frac{I_\nu(\mathbf{k}, \mathbf{k}')}{|\mathbf{k} - \mathbf{k}'|^2} \tag{10.3}$$

$$\times \{ n_p \, \delta[E_\nu(\mathbf{k}') - E_\nu(\mathbf{k}) - \hbar\omega_p] + (n_p + 1) \, \delta(E_\nu(\mathbf{k}') - E_\nu(\mathbf{k}) + \hbar\omega_p] \},$$

191

Table 10.1  Material parameters for GaAs.

| Quantity | Symbol | Unit | Value | |
|---|---|---|---|---|
| Mass density | $\rho$ | kgm$^{-3}$ | 5360 | |
| Sound velocity | $v_{\rm s}$ | ms$^{-1}$ | 5240 | |
| Static dielectric constant | $\varepsilon_{\rm st}$ | | 12.90 | |
| High-frequency dielectric constant | $\varepsilon_{\rm hf}$ | | 10.92 | |
| Acoustic deformation potential | $D_{\rm A}$ | eV | 7 | |
| Optical deformation potential | $D_{\rm O}$ | eVm$^{-1}$ | $10^{11}$ | |
| Non-polar optical phonon energy | $\hbar\omega_{\rm o}$ | eV | 0.032 | |
| Polar optical phonon energy | $\hbar\omega_{\rm p}$ | eV | 0.032 | |
| Impurity concentration | $N_{\rm I}$ | cm$^{-3}$ | $10^{14}$ | |
| Quantity | Symbol | Unit | $\Gamma$-valley | $L$-valley |
| Effective mass ratio | $m_\nu^*/m_0$ | | 0.067 | 0.35 |
| Non-parabolicity factor | $\alpha_\nu$ | eV$^{-1}$ | 0.611 | 0.242 |
| Valley bottom energy | $\Delta_{0\nu}$ | eV | 0 | 0.32 |
| Number of equivalent valleys | $Z_\nu$ | | 1 | 4 |

where $n_{\rm p} = [\exp\left(\hbar\omega_{\rm p}/k_{\rm B}T_{\rm L}\right) - 1]^{-1}$ is the polar phonon occupation number. The overlap factor is defined as

$$I_\nu(\mathbf{k}, \mathbf{k}') = a_\nu(\mathbf{k})a_\nu(\mathbf{k}') + c_\nu(\mathbf{k})c_\nu(\mathbf{k}')\frac{\mathbf{k}\cdot\mathbf{k}'}{|\mathbf{k}||\mathbf{k}'|} \tag{10.4}$$

with

$$a_\nu = \sqrt{\frac{1 + \alpha_\nu[E_\nu(\mathbf{k}) - \Delta_{0\nu}]}{1 + 2\alpha_\nu[E_\nu(\mathbf{k}) - \Delta_{0\nu}]}}, \quad c_\nu = \sqrt{\frac{\alpha_\nu[E_\nu(\mathbf{k}) - \Delta_{0\nu}]}{1 + 2\alpha_\nu[E_\nu(\mathbf{k}) - \Delta_{0\nu}]}}. \tag{10.5}$$

Besides these intravalley scattering mechanisms, we also consider the intervalley scattering (IV) by non-polar optical phonons transferring electrons between the $\Gamma$- and $L$-valleys (non-equivalent intervalley scattering) and between the $L$-valleys (equivalent intervalley scattering). Here, the transition rate reads

$$S_{\nu\mu}^{\rm IV}(\mathbf{k}, \mathbf{k}') = \frac{Z_\mu D_{\rm O}^2}{8\pi^2\rho\omega_{\rm o}}\{n_{\rm o}\,\delta[E_\mu(\mathbf{k}') - E_\nu(\mathbf{k}) - \hbar\omega_{\rm o}] \tag{10.6}$$

$$+(n_{\rm o} + 1)\,\delta[E_\mu(\mathbf{k}') - E_\nu(\mathbf{k}) + \hbar\omega_{\rm o}]\}$$

with the number of equivalent final valleys $Z_\mu$ and the occupation number of non-polar phonons $n_{\rm o} = [\exp\left(\hbar\omega_{\rm o}/k_{\rm B}T_{\rm L}\right) - 1]^{-1}$. For more details, we refer to Tab. 10.1 and [Ziman (2001)].

The material parameters used are given in Tab. 10.1. Concerning the model parameters, we set $n_{\rm mul} = 3$, $N_\Gamma = 105$, $N_L = 75$, $M = 12$, which imply that $E_{\rm max}^\Gamma = E_{\rm max}^L = 1.12$ eV. For the validation of our numerical

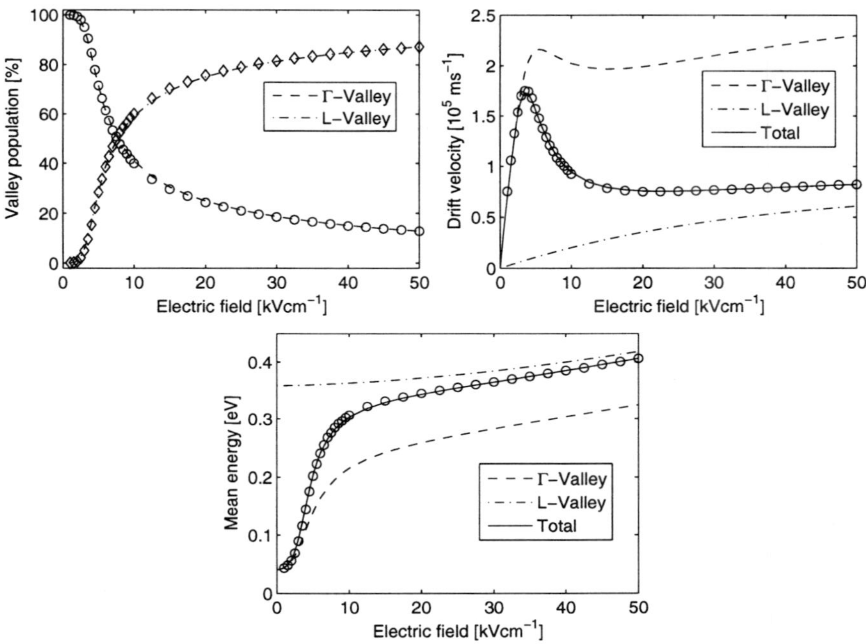

Fig. 10.1  Stationary-state valley population, drift velocity and mean energy as functions of the electric field in bulk GaAs. Lines refer to present results, symbols refer to MC calculations.

scheme, we compare our results with those of Monte Carlo simulations, which are performed with a code based on [Tomizawa (1993)].

## 10.2  Bulk GaAs

The stationary-state valley population, the drift velocity and the mean electron energy in GaAs as a function of the applied electric field strength is displayed in Fig. 10.1 for the impurity density $N_\mathrm{I} = 10^{14}$ cm$^{-3}$. We observe that the electrons populate mainly the $L$-valleys for increasing electric field strength. Moreover, the negative differential conductivity, i.e., the decreasing drift velocity with increasing field, is clearly visible. It is caused by the fact that the drift velocity of electrons in the $L$-valleys is much smaller than that in the $\Gamma$-valley due to the high effective mass and the high scattering rate for equivalent intervalley scattering of $L$ electrons. The mean electron

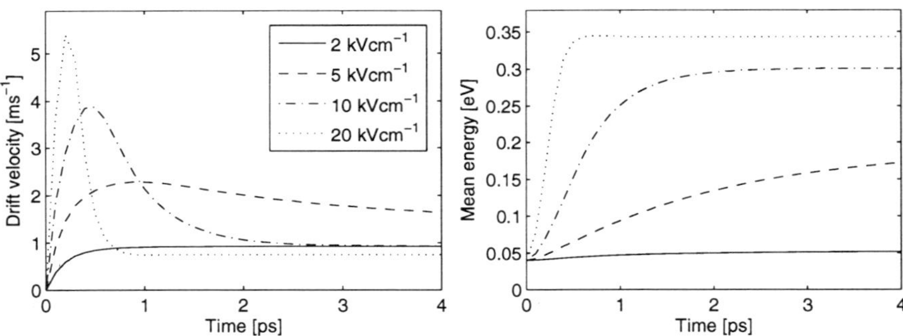

Fig. 10.2    Drift velocity and mean energy versus time after the onset of an electric field pulse in bulk GaAs for several applied voltages.

energy tends to saturate to a constant low slope for high electric fields, since most of the energy input is transfered to the lattice by $\Gamma-L$ and $L-L$ transitions.

All of these features are typical for the common III-V compound semiconductors such as GaAs and well distinct from those of the elementary semiconductors like silicon. Finally, we point out the excellent agreement between the results of the deterministic solver and the stochastic Monte Carlo scheme.

Figure 10.2 illustrates the temporal evolution of the electron drift velocity and the mean energy in GaAs from an initial equilibrium value towards a stationary-state value under the influence of the onset of an electric field at $t = 0$ ps for several field strengths. In these simulations, the impurity density is set to $N_I = 10^{17}$ cm$^{-3}$. We observe the occurrence of velocity overshoots for high fields as a consequence of the ballistic motion of electrons right after the onset of the electric field. Additionally, we allude to the fact that the higher value of $N_I$ in comparison to that used above leads to significantly lower stationary-state values for the drift velocity at 2 kVcm$^{-1}$ and 5 kVcm$^{-1}$, while all of the steady state values of the other drift velocities as well as all of the energies displayed in Fig. 10.2 are hardly influenced by the change of this parameter.

One of the advantages of deterministic methods for solving the Boltzmann transport equation is the availability of the electron distribution function for all times and positions in noise-free resolution. Concerning the bulk case, we present the distributions of $\Gamma$ and $L$ electrons for $N_I = 10^{17}$ cm$^{-3}$ at $t = 5$ ps after the onset of an electric field of 10 kVcm$^{-1}$ in Fig. 10.3.

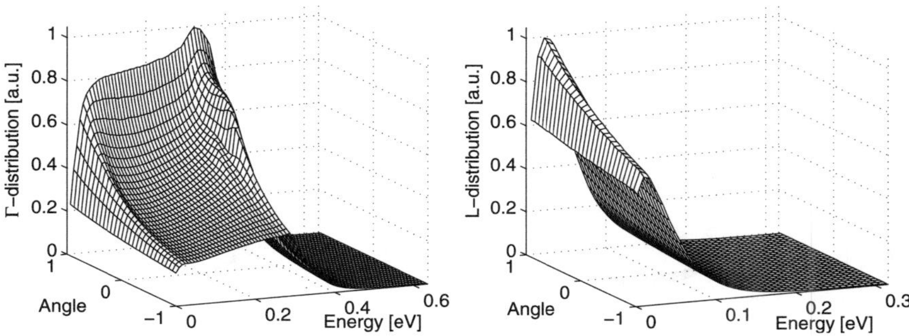

Fig. 10.3  Distribution of $\Gamma$ electrons and $L$ electrons in bulk GaAs for $N_I = 10^{17}$ cm$^{-3}$ at $t = 5$ ps after the onset of an electric field of 10 kVcm$^{-1}$.

The distribution of $L$ electrons can be interpreted as a shifted Maxwellian. In contrast, the distribution of electrons in the $\Gamma$-valley is featured by a complicated structure. We observe high asymmetry of the distribution function as well as a drastic decline for energies higher than $\Delta E_{0L}$, since electrons with such energies preferably populate the $L$-valleys. Moreover, the wall effect, i.e., the accumulation of electrons at energies right below $\Delta E_{0L}$ with a maximum in the direction normal to the electric field, is clearly observable.

## 10.3  The GaAs $n^+ - n_i - n^+$ Diode

The considered GaAs $n^+ - n_i - n^+$ diode has a total length of 750 nm. The active channel is situated in the middle of the device with a thickness of 250 nm. The cathode and anode $n^+$-layers of 250 nm are highly doped with $n^+ = 2 \times 10^{17}$ cm$^{-3}$, while the $n_i$-region is assumed to be undoped. Hence, the donor density in this layer equals the intrinsic value $n_i = 1.79 \times 10^6$ cm$^{-3}$. For simplicity, the impurity concentration important for impurity scattering is not set to the position-dependent donor density but to $N_I = 10^{14}$ cm$^{-3}$. In the simulations presented in this section, we use 101 grid points in space, which implies that $\Delta x = 7.5$ nm.

Figures 10.4 and 10.5 depict the electron density, the drift velocity, the mean energy and the electric field strength in the GaAs $n^+ - n_i - n^+$ diode as a function of the position at $t = 10$ ps. This long simulation time guarantees that the stationary-state regime is almost reached. Plots on the left refer to the applied voltage $V_{\text{bias}} = 0.25$ V; those on the right

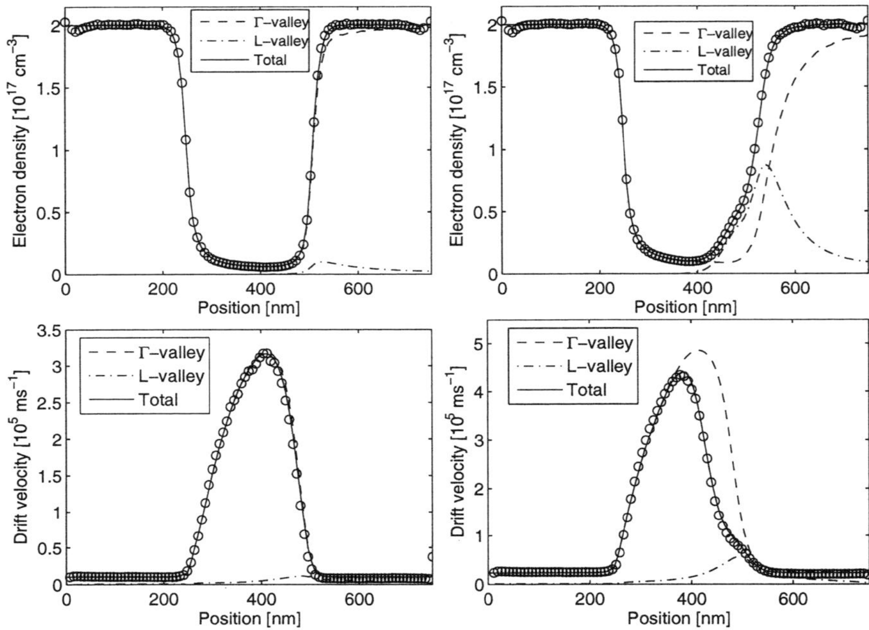

Fig. 10.4　Electron density and drift velocity in the GaAs $n^+ - n_{\rm i} - n^+$ diode at $t = 10$ ps: $V_{\rm bias} = 0.25$ V (left), $V_{\rm bias} = 0.75$ V (right). Lines refer to present results, symbols refer to MC calculations.

are obtained for $V_{\rm bias} = 0.75$ V. Moreover, the contributions of $\Gamma$ and $L$ electrons to the total macroscopic quantities are displayed for delivering some additional information.

Concerning the case $V_{\rm bias} = 0.25$ V, we find that the electron transport takes place mainly in the $\Gamma$-valley. In other words, most of the electrons do not gain enough energy to be scattered into the energetically high $L$-valleys. However, the influence of the $L$-valleys cannot be neglected even for this case, since we observe that though $L$ electrons are rare, they contribute significantly to the mean energy (cf. the solid and the dashed lines in Fig. 10.5, top-left). None the less, the simulated case can be seen as an almost one-valley transport. Hence, the shapes of macroscopic quantities versus position agree qualitatively with those obtained for simulating a silicon $n^+ - n - n^+$ diode.

Some new transport features occur when studying the considered $n^+ - n_{\rm i} - n^+$ diode for $V_{\rm bias} = 0.75$ V applied voltage. We note an accumulation

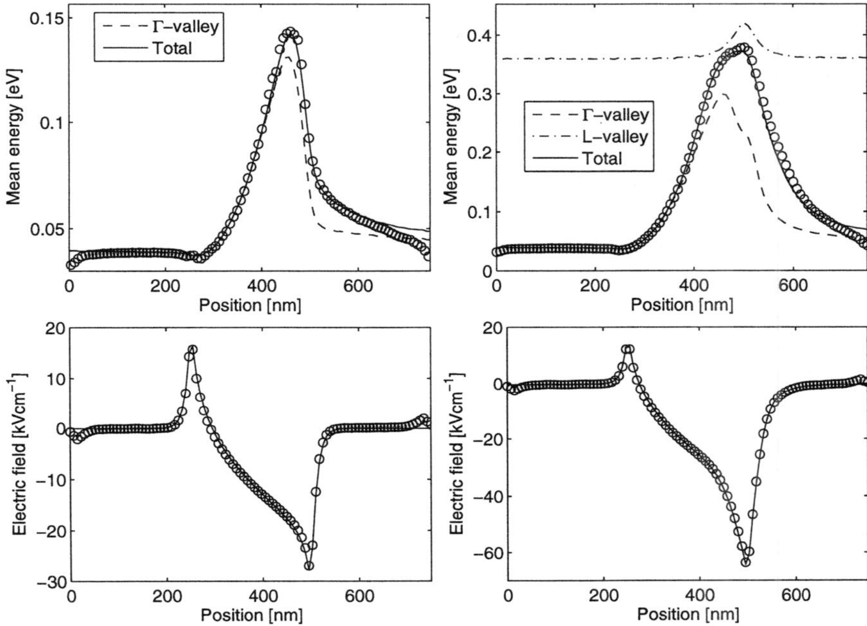

Fig. 10.5   Mean energy and electric field in the GaAs $n^+ - n_\mathrm{i} - n^+$ diode at $t = 10$ ps: $V_\mathrm{bias} = 0.25$ V (left), $V_\mathrm{bias} = 0.75$ V (right). Lines refer to present results, symbols refer to MC calculations.

of electrons between about 400 nm and 500 nm in Fig. 10.4 top-right, which is not visible for $V_\mathrm{bias} = 0.25$ V (Fig. 10.4, top-left). As discussed in [Tomizawa (1993)], this charge accumulation is not a result of the scattering in the active layer, but it is caused by the back-scattering of electrons from the anode $n^+$-layer. The back-scattering of these electrons takes place, since the electrons which move almost ballistically through the active layer gain high enough energies to enter the $L$-valleys in the $n^+$ anode layer, where they are scattered isotropically with high scattering probabilities and are subsequently retransferred to the $\Gamma$-valley. A great portion of these just created $\Gamma$ electrons possesses a component of the wave vector in direction towards the $n_\mathrm{i}$-layer and, therefore, are able to reenter the active layer. Hence, when designing a submicron GaAs device, the anode region cannot be considered simply as a drain of the electrons, since this region has a strong influence on the distribution of electrons in the active zone.

Concerning the drift velocity and the mean electron energy for $V_\mathrm{bias} =$

0.75 V, we observe that they feature some kinks in their dependence on position caused by the charge accumulation at the $n_i - n^+$ anode interface and, hence, differ notably in their shapes in comparison to the results obtained for $V_{bias} = 0.25$ V. Moreover, we find that the contributions of the $\Gamma$ and of the $L$ electrons to the total macroscopic quantity are of the same order in the right half of the diode. Thus, only many-valley models can yield accurate results for the considered situation. Finally, we note that the maximum value of the drift velocity reached in the active layer exceeds the maximum stationary-state bulk velocity greatly for both $V_{bias} = 0.25$ V and $V_{bias} = 0.75$ V .

Finally, we note that the results obtained with the help of the multigroup-WENO solver agree very well with those of the Monte Carlo calculations. Slight differences appear at the boundaries of the device, which result from the difficulties in imposing corresponding boundary conditions for the deterministic method and the stochastic scheme. However, the more interesting region around the active channel is described by both techniques in excellent consistency.

In Fig. 10.6, we show the temporal evolution of the electron density, the drift velocity, the mean energy and the electric field in the GaAs $n^+ - n_i - n^+$ diode for $V_{bias} = 0.75$ V. Although the chosen initial distribution is not relevant from a physical point of view, it is interesting to study this time-accurate relaxation process. First of all, we remark that all of the displayed curves feature non-oscillatory behavior for all times and positions. Thus, our multigroup-WENO solver is able to cope with the abrupt changes in the doping concentration at 250 nm and 500 nm, which constitutes an important property for reliable device simulation.

In the plot of the electron density, we find a package of electrons, which are firstly transported ballistically through the active layer and later backscattered from the anode to form the charge accumulation in the stationary state. A corresponding wave is found in the drift velocity, which also features regions of negative velocities due to the diffusion of electrons from the anode into the $n_i$-layer right after the beginning of the simulation. For the electric field, we can clearly study the formation of the back-drawing field for inhibiting electrons entering the active channel. While the electron density, the drift velocity and the electric field have almost reached their stationary-state values in the time domain plotted in Fig. 10.6, the mean electron energy still changes significantly, especially in the anode region, since the energy relaxation time is much longer than the momentum relaxation time.

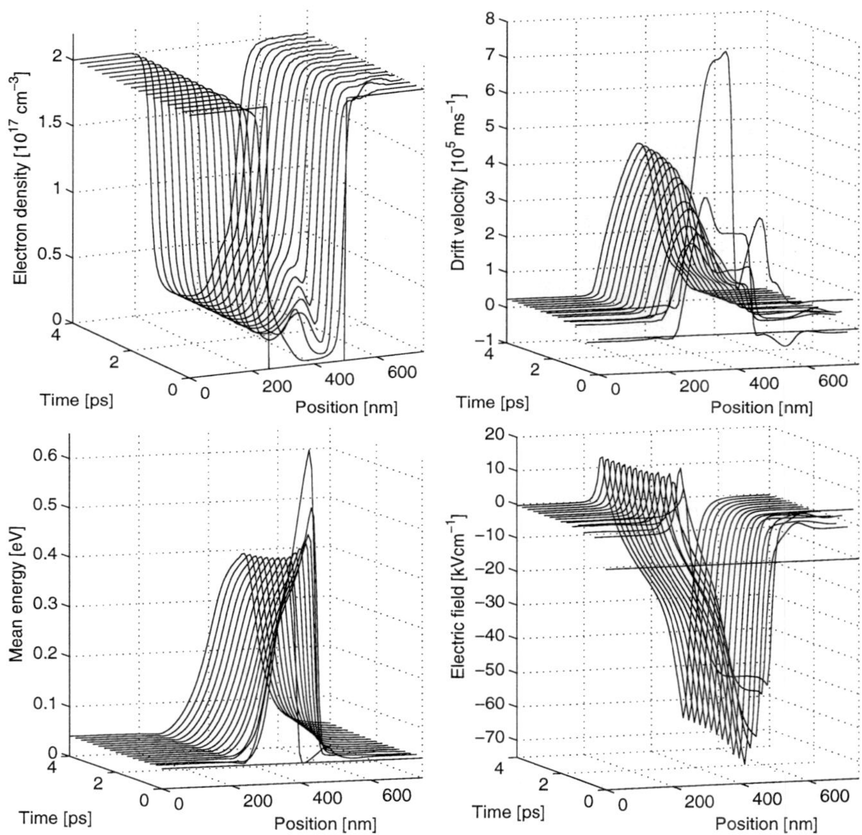

Fig. 10.6  Temporal evolution of electron density, drift velocity, mean energy and electric field in the GaAs $n^+ - n_i - n^+$ diode with $V_{\text{bias}} = 0.75$ V.

Figure 10.7 illustrates the $\Gamma$ and the $L$ electron distributions in the GaAs $n^+ - n_i - n^+$ diode with $V_{\text{bias}} = 0.75$ V at $t = 10$ ps and at the $n_i - n$ anode interface ($x = 500$ nm). Similar to the bulk case, we find that the $L$ electron distribution is almost a shifted Maxwellian, while the $\Gamma$ distribution is assigned to a complicated structure far away from the equilibrium distribution. The sharp drop at $\Delta E_{0L}$ found for the bulk case is washed out, i.e., electron states above 0.32 eV are likely to be populated due to diffusion processes. In addition, we can see a region of isotropically distributed electrons around $\Delta E_{0L}$ caused by $L$ electrons reentering the $\Gamma$-valley by isotropic $L-\Gamma$ intervalley scattering, which are the reason for the

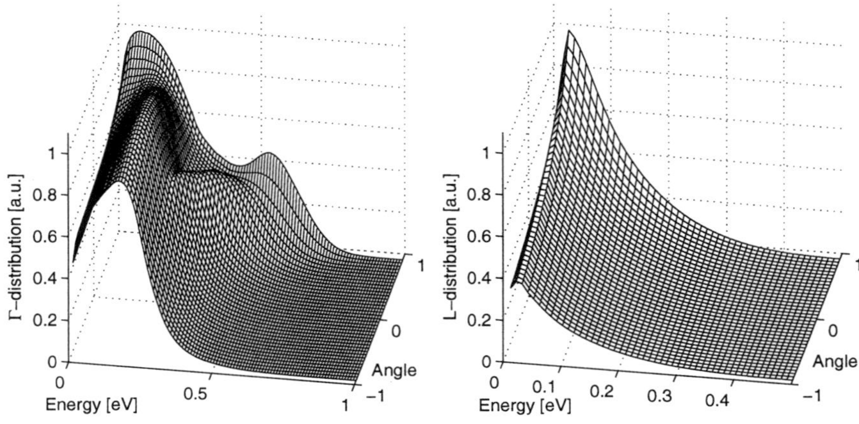

Fig. 10.7  $\Gamma$ and $L$ electron distributions in the GaAs $n^+ - n_{\mathrm{i}} - n^+$ diode with $V_{\mathrm{bias}} = 0.75$ V at $t = 10$ ps and $x = 500$ nm.

charge accumulation mentioned above. Based on these results, we can state that the presented simulation can only be performed accurately by means of kinetic transport equations, since the occurring distribution functions are far from being shifted Maxwellians, which is the fundamental demand that hydrodynamic models work well.

## 10.4  The GaAs-MESFET

In this section, we present the results for the electron transport through a GaAs-MESFET obtained by means of our multigroup-WENO solver (8.17), (8.22). The geometry of the investigated GaAs-MESFET is shown in Fig. 10.8. The donor densities are chosen as $n^+ = 7 \times 10^{16}$ cm$^{-3}$ and $n = 10^{16}$ cm$^{-3}$. The source and drain contacts act as particle reservoirs. Electrons may enter or exit through these contacts, which is modelled according to (8.28c). The Schottky contact at the gate is assumed to be a totally absorbing boundary (8.28b), whereas perfectly reflecting boundary conditions are imposed at the non-contact surfaces (8.28a). Concerning the boundary conditions for the Poisson equation, we apply the Neumann condition (vanishing electric field in the direction normal to the surface) on the non-contact boundaries to simulate insulating boundaries. The source, gate and drain contacts are treated as Dirichlet boundaries, where the bias voltages are applied. They equal $V_{\mathrm{S}} = 0$ V at the source, $V_{\mathrm{G}} = -0.4$ V at

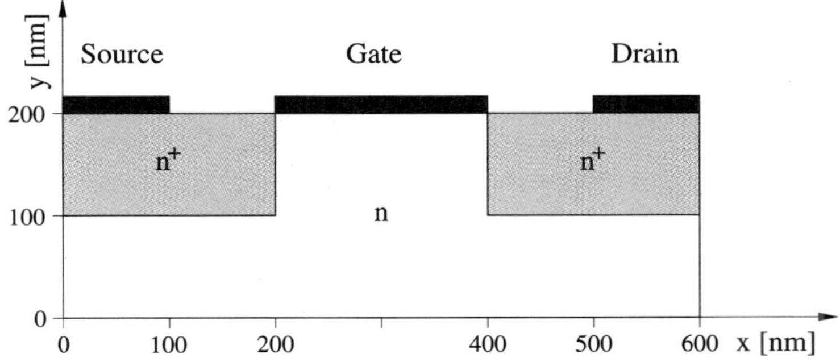


Fig. 10.8    Schematic illustration of the considered MESFET geometry.

the gate and $V_D = 0.8$ V at the drain contact.

The parameters of the grid are chosen as $N^\Gamma = 50$, $N^L = 30$, $n_{\mathrm{mul}} = 2$, $M = R = 10$ for the momentum space together with $P = 48$, $Q = 24$ grid points in real space. The time integration is performed from the initial time $t_0 = 0$ ps up to the final time $t = 4$ ps, when the stationary state is approximately reached.

In Figs. 10.9, 10.10, 10.11 and 10.12, the most important hydrodynamic

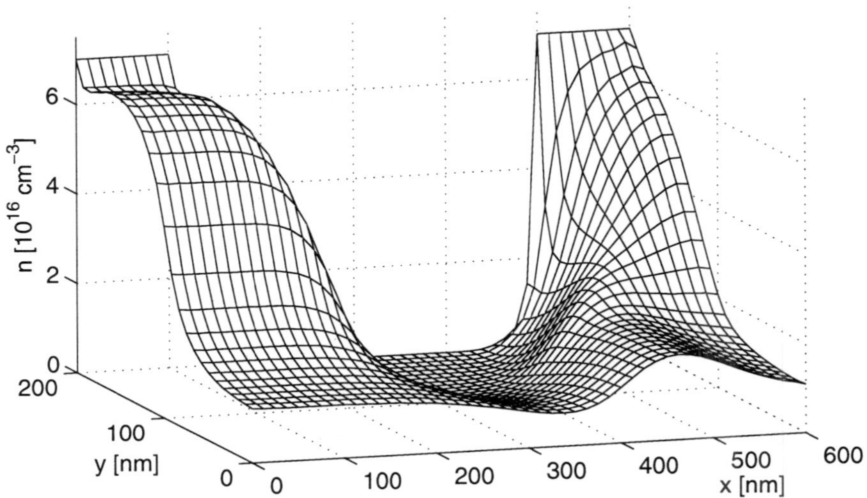


Fig. 10.9    Electron density $n$ versus position $(x, y)$ in the GaAs-MESFET at $t = 4$ ps.

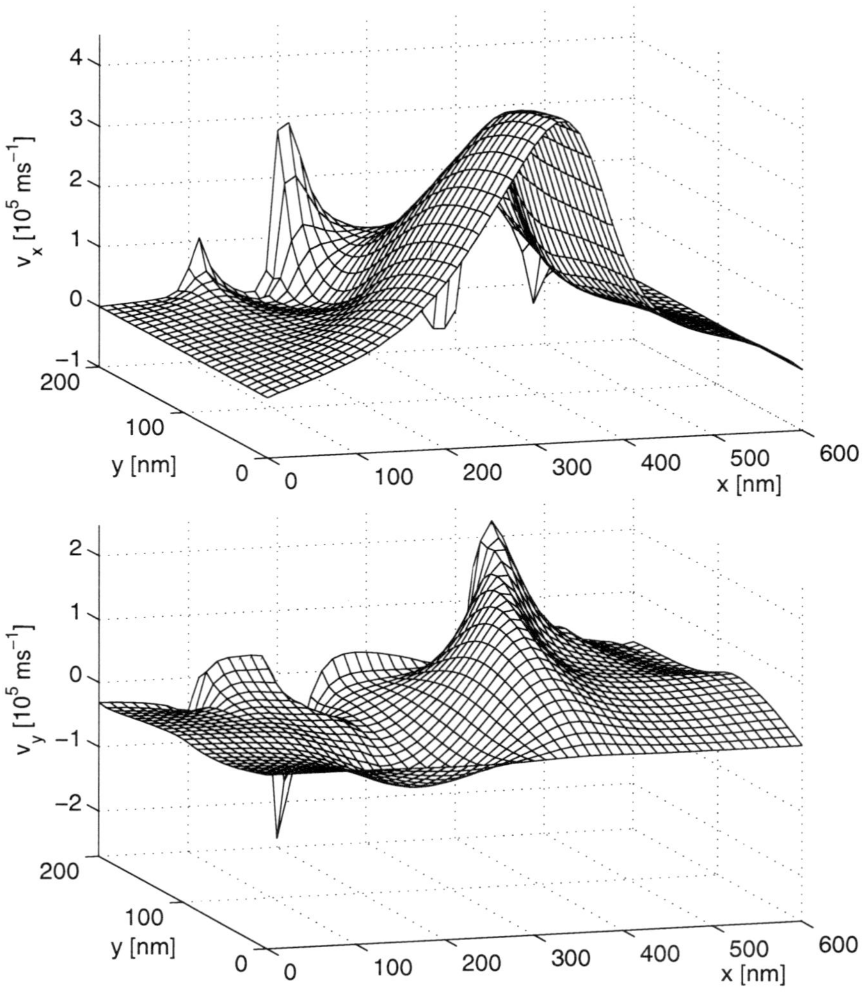

Fig. 10.10　X-component $v_x$ and y-component $v_y$ of the drift velocity versus position $(x, y)$ at $t = 4$ ps.

transport quantities, namely the electron density, the components of the drift velocity, the velocity field and the mean electron energy are displayed. First of all, we point out the highly regular, non-oscillatory behavior of these quantities throughout the whole device. This behavior implies that the proposed numerical technique is able to resolve the spatial dependence

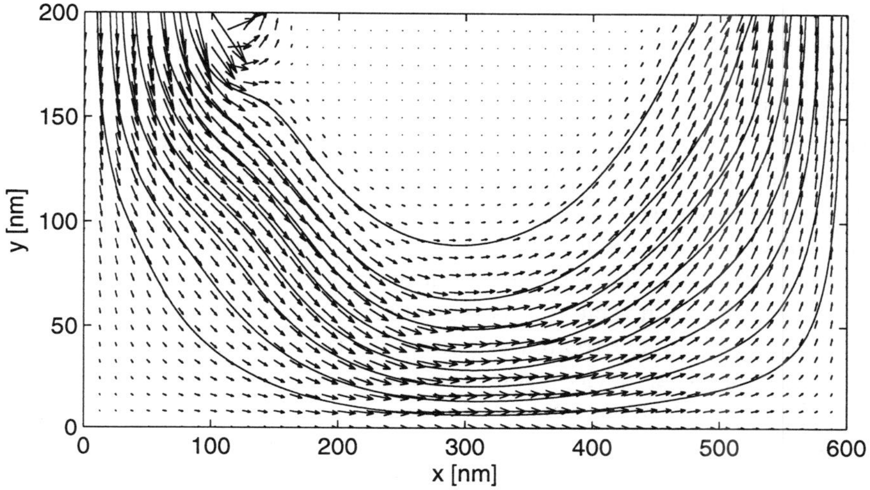

Fig. 10.11   Velocity field and current lines at $t = 4$ ps.

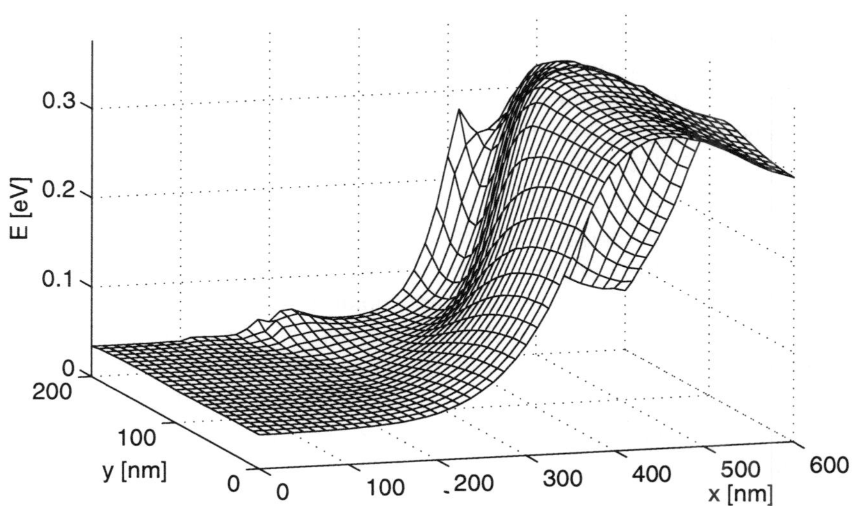

Fig. 10.12   Mean electron energy $E$ versus position $(x, y)$ at $t = 4$ ps.

of transport quantities even in regions of very low electron densities and at abrupt changes of the donor density.

Concerning the electron density in Fig. 10.9, we observe the formation of a depletion layer at the gate contact, since strong electric fields repulse the carriers from this region. In this respect, the behavior of the GaAs-MESFET agrees with that of the silicon MESFET, studied in detail in chapter 9. On the other hand, the charge accumulation, which is formed around the drain $n - n^+$ interface, is not found in the corresponding silicon device. As discussed in [Tomizawa (1993)] for the GaAs $n^+ - n - n^+$ diode, this enhanced electron density near the drain results from the backscattering of carriers into the active region after having already entered the high doping drain region. Due to this typical multi-valley effect, the anode contact cannot be simply treated as a drain for the electrons in the designing of submicron GaAs devices, since this region has a strong influence on the distribution of electrons in the active zone.

In the x- and the y-component of the electron drift velocity in Fig. 10.10, we find values which exceed the maximum drift velocity of bulk GaAs (about $1.8 \times 10^5$ ms$^{-1}$ at room temperature) significantly. Thus, ballistic transport plays an important role in the considered device. A more descriptive illustration of the electron drift velocity is given in Fig. 10.12. Here, we display the electron velocity field and some current lines. Although this representation gives only qualitative information, it offers a simple way for studying the transport of the electrons through the device.

The mean electron energy in Fig. 10.11 equals approximately the mean lattice energy $3k_{\mathrm{B}}T_{\mathrm{L}}/2$ in the source region. In the active region, the electrons are accelerated by strong electric fields and gain energy, leading to a maximum electron energy at the drain $n - n^+$ interface. At the drain contact, the mean energy is low again. This is caused by the fact that the electrons lose their energies in several scattering events and that hot, i.e., high energetic outflowing electrons are replaced by cold inflowing carriers.

For illustrating the electrical properties of the considered GaAs-MESFET, we display the x- and the y-components of the electric field as well as the electrostatic potential versus position in Figs. 10.13 and 10.14. The potential is very regular, while the components of the electric fields feature some spikes due to their singularities in the points of changing boundary conditions. Additionally, we find strong y-components of the field repulsing the electrons from the gate, as well as the back-drawing fields at the changes in the doping concentration. An illustrative summary of the Figs. 10.13 and 10.14 is given in Fig. 10.15 depicting the electric field and the equipotential lines in the MESFET at $t = 4$ ps.

For the validation of our numerical scheme, we compare our results

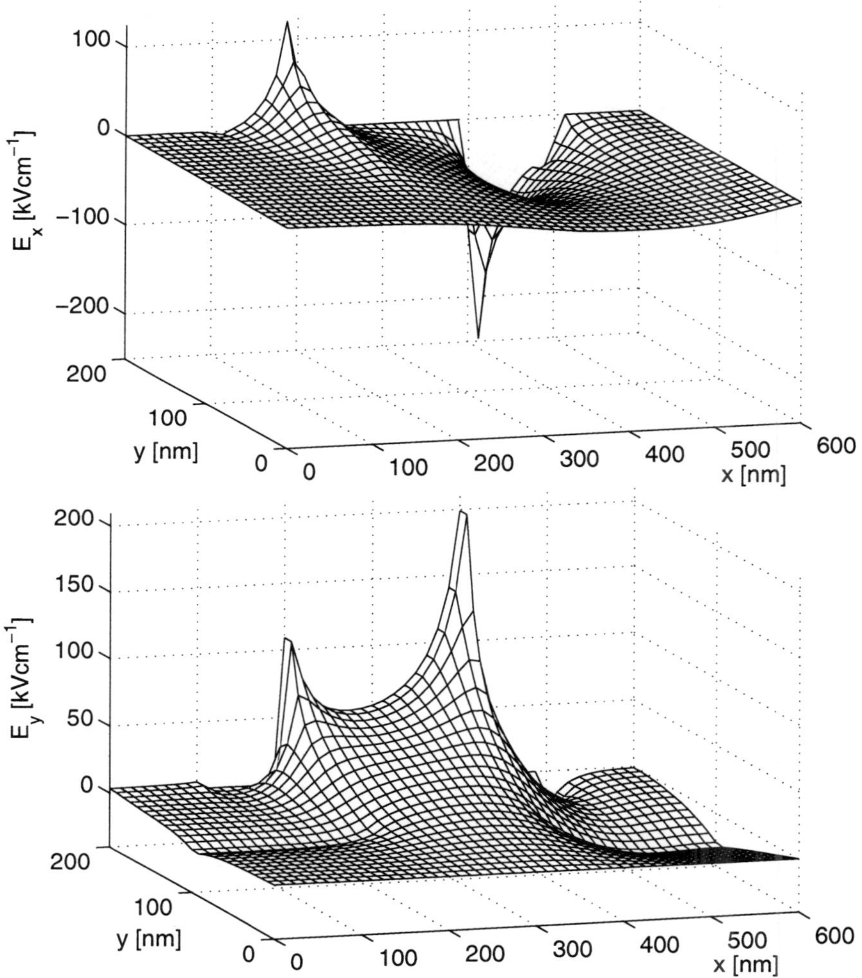

Fig. 10.13  X-component $E_x$ and y-component $E_y$ of the electric field versus position $(x, y)$ at $t = 4$ ps.

with those of Monte Carlo calculations, which are obtained by a slightly modified version of the code of Tomizawa [Tomizawa (1993)]. Therefore, we show some cuts of the electron density, the drift velocity, the mean energy and the electrostatic potential in the Figs. 10.16, 10.17, 10.18 and 10.19. We find that all of the displayed cuts of the electrostatic potential as well

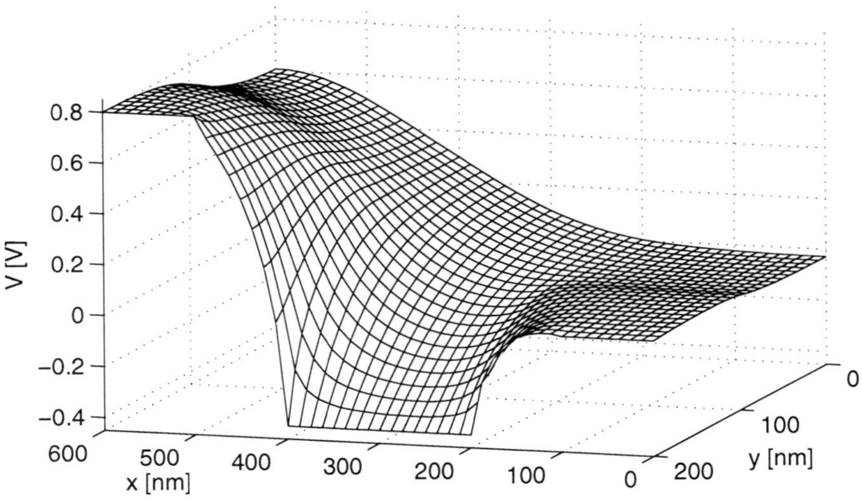

Fig. 10.14   Electrostatic potential $V$ versus position $(x, y)$ at $t = 4$ ps.

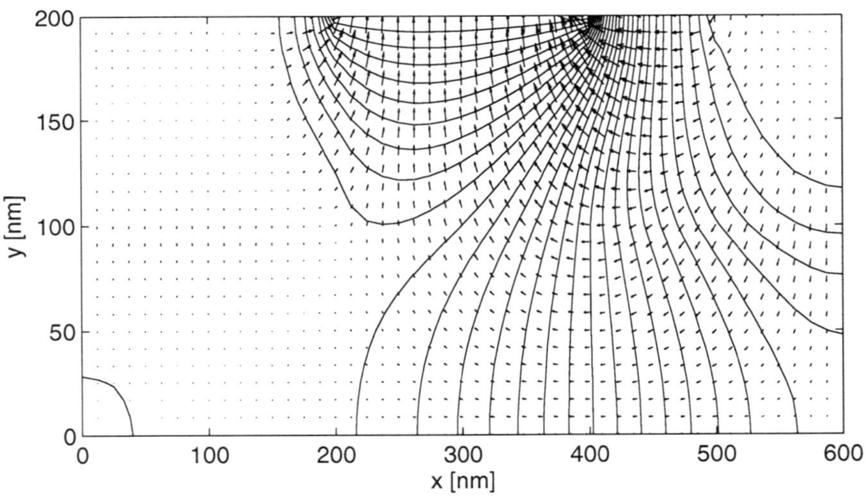

Fig. 10.15   Electric field and equipotential lines at $t = 4$ ps.

as all of the hydrodynamic quantities at $y = 50$ nm agree very well. In addition, the cuts at $y = 150$ nm exhibit the same good consistency in

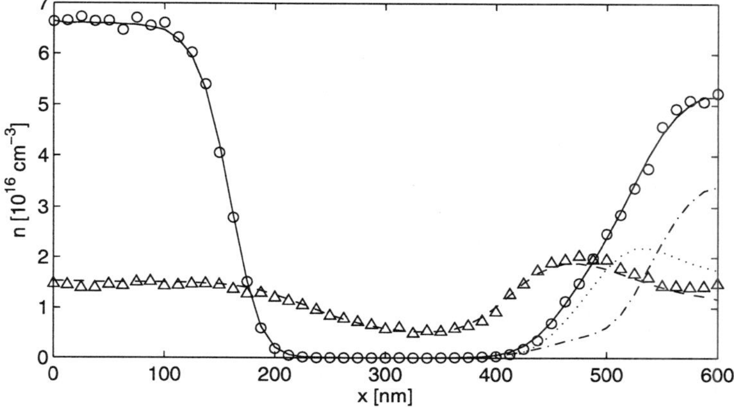

Fig. 10.16   Electron density $n$ versus position $x$ in the GaAs-MESFET at $t = 4$ ps. The lines refer to present results, the symbols depict Monte Carlo data. $(---, \triangle)$ electron density at $y = 50$ nm; $(\text{———}, \bigcirc)$ electron density at $y = 150$ nm; $(-\cdot-)$ $\Gamma$-electron density at $y = 150$ nm; $(\cdots)$ $L$-electron density at $y = 150$ nm.

the regions, where the electron density is not too small. We also observe that the Monte Carlo method is not able to determine the macroscopic quantities in the depletion layer in a sufficiently accurate order, while our deterministic technique allows us the investigation of this region.

Besides the total electron density, velocity and energy, we also display the portions of these quantities due to $\Gamma$- and $L$-electrons for $y = 150$ nm. We find that the transport is dominated by $\Gamma$-electrons between $x = 0$ nm and $x = 300$ nm. For $x$-values higher than 300 nm, electrons have gained high enough energies so that a significant portion of them is scattered into the $L$-valleys and the total macroscopic quantities set up by a complicated interplay of $\Gamma$- and $L$-electrons. We observe that there are more $L$-electrons than $\Gamma$-electrons between $x = 400$ nm and $x = 550$ nm. The $L$-drift velocity is low in comparison to the $\Gamma$-velocity as a consequence of the high effective mass of $L$-electrons and the very efficient equivalent intervalley scattering in these valleys. Finally, we find that both the $\Gamma$- and the $L$-electrons are significantly heated in the drain region.

In Fig. 10.20, we display the temporal evolution of the boundary $y_{\mathrm{D}}$ of the depletion layer, which we implicitly define as

$$n[t, x, y_{\mathrm{D}}(t, x)] = 0.75 \times 10^{22}\,\mathrm{m}^{-3}. \tag{10.7}$$

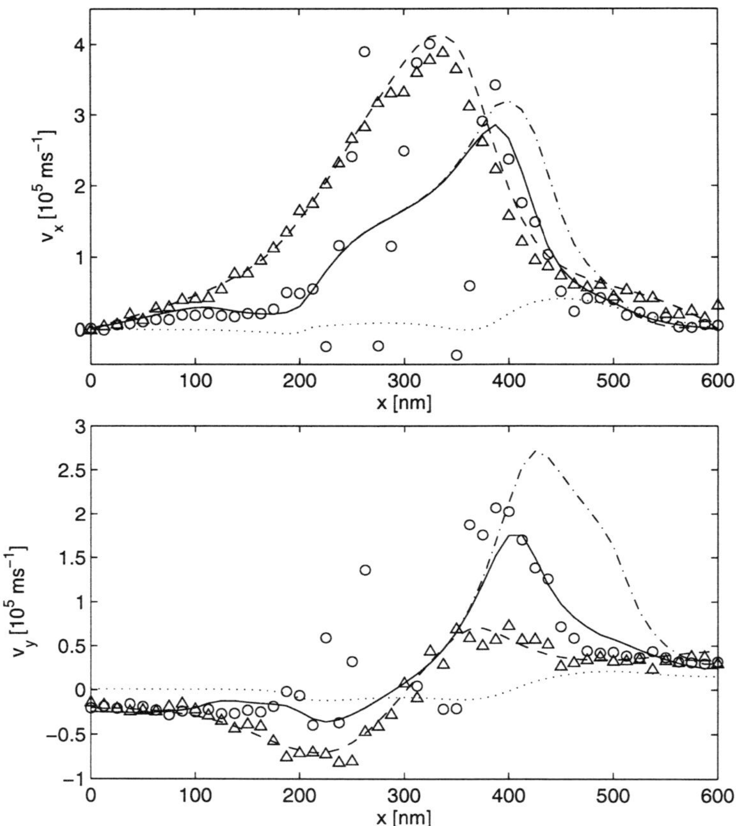

Fig. 10.17   X-component $v_x$ and y-component $v_y$ of the drift velocity versus position $x$ at $t = 4$ ps. The lines refer to present results, the symbols depict Monte Carlo data. $(-\,-\,-,\ \triangle)$ drift velocity at $y = 50$ nm; $(\text{———},\ \bigcirc)$ drift velocity at $y = 150$ nm; $(-\,\cdot\,-)$ $\Gamma$-drift velocity at $y = 150$ nm; $(\cdots)$ $L$-drift velocity at $y = 150$ nm.

Of course, the studied evolution process is not physically relevant, since the initial conditions are not. However, Fig. 10.20 allows us to cherish hopes that the applicability of our numerical scheme for investigating time dependent problems, and it shows some interesting features of the depletion layer, which should also influence the behavior of real MESFETs in switching processes. We observe that the depletion layer is not formed by a simple evacuation of electrons, but its formation exhibits a temporally periodic structure. Rapid increasing of the depletion layer at $t = 0.3$ ps

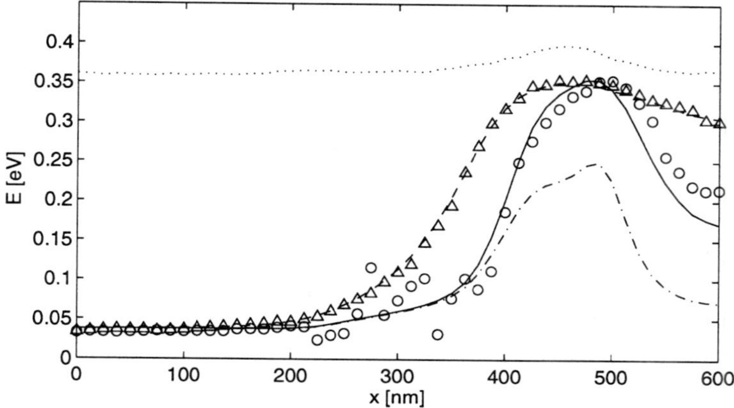

Fig. 10.18   Mean electron energy $E$ versus position $x$ at $t = 4$ ps.  The lines refer to present results, the symbols depict Monte Carlo data.  $(- - -, \triangle)$ electron energy at $y = 50$ nm; $(\text{------}, \bigcirc)$ electron energy at $y = 150$ nm; $(- \cdot -)$ $\Gamma$-electron energy at $y = 150$ nm; $(\cdots)$ $L$-electron energy at $y = 150$ nm.

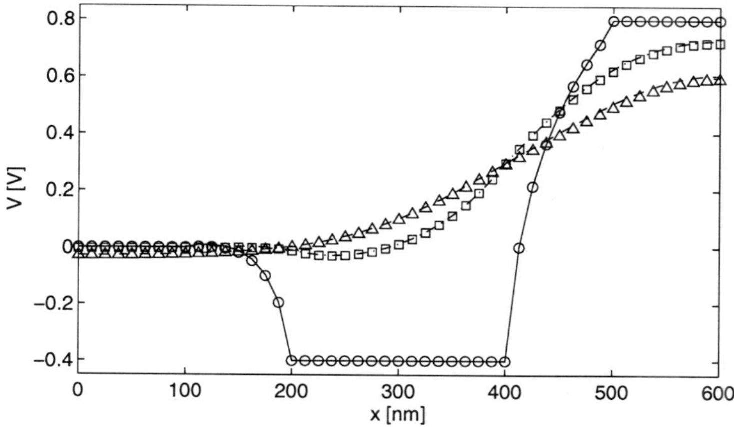

Fig. 10.19   Electrostatic potential $V$ versus position $x$ at $t = 4$ ps.  The lines refer to present results, the symbols depict Monte Carlo data.  $(- - -, \triangle)$ potential at $y = 0$ nm; $(- \cdot -, \square)$ potential at $y = 100$ nm; $(\text{------}, \bigcirc)$ potential at $y = 200$ nm.

and $t = 0.6$ ps is followed by a much smaller growth or even a reduction of the $y_D$.  Similarly, an oscillation between more and less symmetric depletion layers is visible.  After $t = 1$ ps, the depletion layer evolves quite slowly to

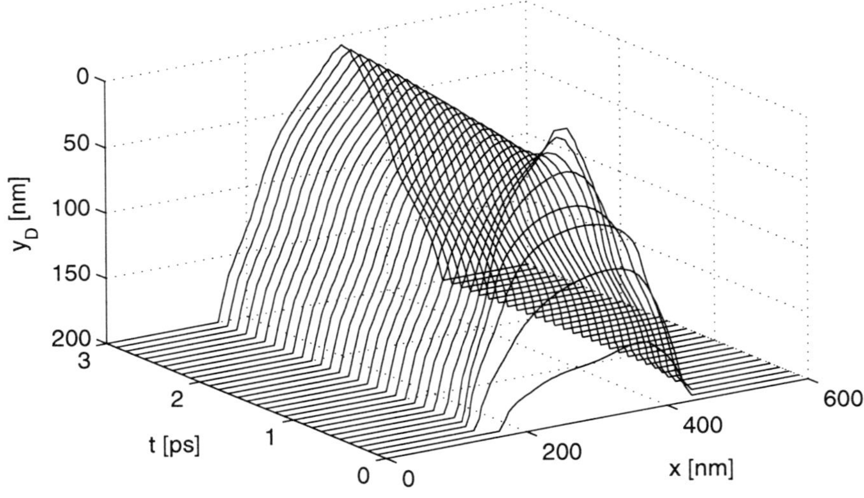

Fig. 10.20   Temporal evolution of the boundary of the depletion layer $y_\mathrm{D}(t,x)$.

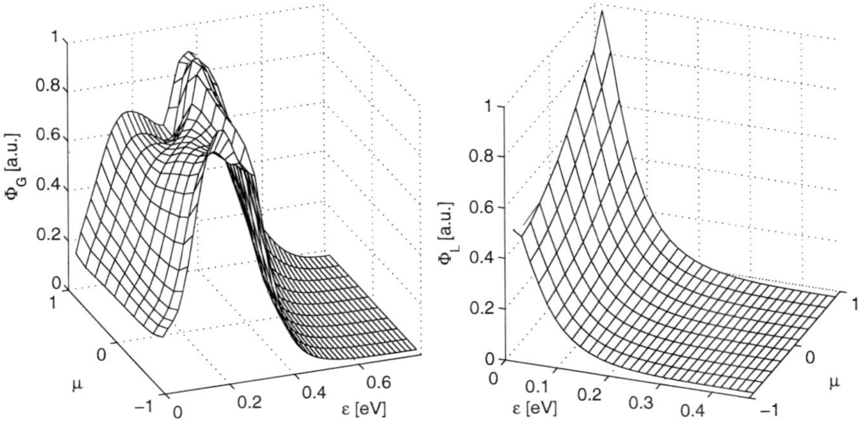

Fig. 10.21   Normalized electron distribution $\Phi_G$ in the $\Gamma$-valley and $\Phi_L$ in the $L$-valley at $t = 4$ ps, $x = 500$ nm and $y = 150$ nm.

its stationary state shape, which is almost reached at $t = 3$ ps.

One of the advantages of deterministic methods for solving the Boltzmann transport equation is the availability of the electron distribution function for all times and positions in noise-free resolution. For instance, we

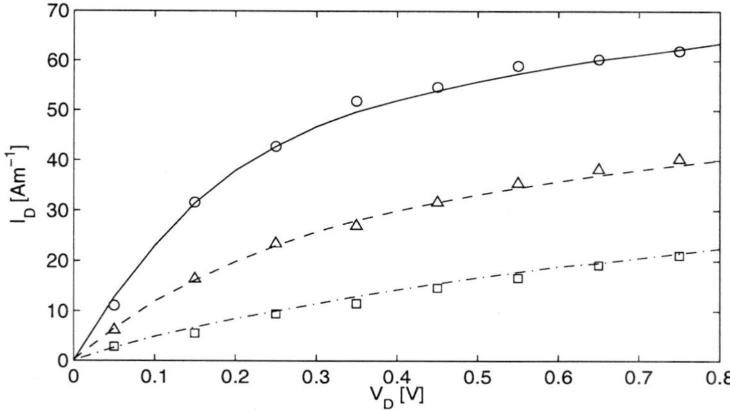

Fig. 10.22  Steady state drain current $I_\mathrm{D}$ versus drain voltage $V_\mathrm{D}$ for several gate voltages. Lines refer to present results, symbols denote Monte Carlo data. (——, $\bigcirc$) $V_\mathrm{G} = -0.2$ V, $(- - -,\ \triangle)$ $V_\mathrm{G} = -0.4$ V, $(- \cdot -,\ \square)$ $V_\mathrm{G} = -0.6$ V.

depict the distribution functions for electrons in the $\Gamma$- and in the $L$-valley versus energy and polar angle at the point $x = 500$ nm, $y = 150$ nm at $t = 4$ ps in Fig. 10.21. It should be noted that these functions are obtained by averaging the coefficients $n^\nu_{ijk}$ with respect to the azimuth angle $\varphi$, which allows the convenient representation. We find that both, the $\Gamma$- and the $L$-distribution function, are very smooth. The distribution of $L$-electrons can be seen as a shifted Maxwell distribution with a temperature much higher than the lattice temperature. On the other hand, the $\Gamma$-distribution function is assigned to a complicated structure, and it is far from equilibrium. Consequently, the application of hydrodynamic transport equations for the investigation of the presented submicron GaAs-MESFET cannot yield accurate results.

Finally, we report the results for the drain characteristics of the considered GaAs-MESFET in Fig. 10.22 in comparison to the results of Monte Carlo simulations. The drain current $I_\mathrm{D}$ is obtained from the momentum $nv_y$ at the drain via

$$I_\mathrm{D}(t) = -\mathcal{Q} \sum_{x_n \in \mathrm{drain}} nv_y(t, x_n, y_Q)\Delta x. \qquad (10.8)$$

The stationary state is assumed to be reached, when the relative change of $I_\mathrm{D}$ and the source current $I_\mathrm{S}$ is less than 1 percent within a time interval of 0.2 ps. All of the calculations for determining the drain characteristics are performed with the grid $N^\Gamma = 25$, $N^L = 15$, $n_{\mathrm{mul}} = 1$, $M = R = 4$ in momentum space and $P = 18$, $Q = 10$ in real space. Concerning the shape of the drain characteristics, we find the typical behavior of a MESFET [Tomizawa (1993); Sze (1990)]. Moreover, we observe a very good agreement of the deterministic and the stochastic results. This implies that the presented numerical technique is able to give sufficiently accurate information on the main transport features of a device even when a very coarse grid is used.

The presented numerical results on the main transport quantities in a GaAs-MESFET as well as their comparison to the data of Monte Carlo simulations prove the applicability and the validity of the proposed numerical scheme. The proposed numerical method is not competitive with Monte Carlo simulations on the level of computational costs for determining stationary states with the fine grid (about two days on two AMD Athlon MP 2000+ processors, 1666 MHz, 2 GB RAM). However, it provides detailed, highly accurate information on distribution functions and on macroscopic quantities in the whole device. Therefore, the multigroup-WENO solver can usefully be applied for creating benchmarks for all of the solvers for semiconductor device simulation. On the other hand, the presented numerical scheme gives sufficiently accurate results even for coarse grids, which allows the obtaining of a noise-free overview of the main transport properties of a device within justifiable computation times (about 20 minutes for determining the 17 points of one of the current-voltage curves in Fig. 10.22 with one of the mentioned processors).

Hence, the multigroup-WENO solver is found to be a powerful tool for GaAs device simulation and an interesting alternative to the usually applied Monte Carlo techniques.

# Chapter 11

# Conclusion

This book is aimed at presenting new deterministic solution methods for the Bloch-Boltzmann-Peierls equations governing the carrier transport in semiconductors. Therefore, we present a multigroup model to the Boltzmann transport equations for polar semiconductors. Special effort is invested in the proper formulations of the force term and the polar optical interaction terms. In addition, expressions for handling all the other relevant scattering mechanisms are deduced. We prove that this multigroup model fulfills the related conservation laws for the electron density and the total energy density. As concerns the numerical properties of our model, we find two advantages in comparison to other mesoscopic methods: the collision coefficients are found to be analytical expressions; the evaluation of the collision terms is performed in a very efficient way, even the nonlinear POP interaction term is simply given as the product of the unknowns with a constant collision coefficient. Consequently, our method combines high numerical accuracy and affordable computation time.

The developed method is used to study the transient transport regime in InP in response to a step-like dc electric field pulse. The dependence of the computation time on the applied electric field strength and the demanded relative accuracy is presented. The results for the average drift velocity as a function of the applied electric field are in good agreement with several experimental and other theoretical studies. Moreover, we discuss the phenomenon of the velocity overshot for high electric field strengths and demonstrate that the influence of hot phonons on the average drift velocity of electrons cannot be neglected for high electron densities. Several figures display the electron and the LO phonon distributions under typical velocity overshot conditions. Additionally, the dependence of some macroscopic quantities on time after the onset of the electric field pulse is discussed.

In addition, this deterministic solution method for the BTEs is used for investigating the transport properties of GaAs in response to a time-depending external electric field. Overshoot and undershoot phenomena as well as the Rees effect are studied. The obtained results are in good agreement with related MC calculations. The influence of the electron density on the drift velocity and the average electron energy is discussed. Moreover, we demonstrate that calculations containing the usual assumption of equilibrium phonon distributions fail for sufficiently high electron densities. The stationary state distribution of the electrons in response to an external electric field is discussed and the enlarged electron density in the $\Gamma$ valley with energies just below the bottom of the $L$ valleys is clarified.

A direct solution approach to the BBP equations for degenerated carrier gases is presented, containing the full quantum statistics of carriers and phonons. This transport model is based on a general expression for the carrier band structure. Moreover, it allows the investigation of particle distributions of arbitrary anisotropy with respect to a main direction. Concerning the mathematical properties of the deduced transport model equations, we show the boundedness of the solution according to the Pauli principle. The conservational properties of the multigroup model are discussed and a Boltzmann H-Theorem for the obtained evolution equations is proved. The equilibrium solution to the multigroup model is a set of discretized versions of the Fermi-Dirac and Bose-Einstein distributions. Numerical results are given for the stationary-state distributions of a coupled electron-longitudinal optical phonon system in GaAs. To sum up, we find that the major mathematical features of the multigroup model equations coincide with those of the original BBP equations. Hence, we regard our transport model as an interesting starting point for the formulation of direct solution methods for the device simulation.

Furthermore, the two-dimensional electron transport in AlGaN/GaN heterostructures in presence of strain polarization fields is simulated with the help of a multigroup model. The envelope wave functions for the confined electrons are calculated using a self-consistent Poisson-Schrödinger solver. The electron gas degeneracy and hot phonons are included into these transport equations. Numerical results are given for the field and time dependence of macroscopic quantities and for the electron and phonon distribution functions. Moreover, we compare our results to those of Monte Carlo simulations and with experiments.

A multigroup-WENO solver for the non-stationary Boltzmann-Poisson system is applied for simulating the electron transport in silicon and GaAs based devices, which are the spatially one-dimensional $n^+-n-n^+$ diode and the spatially two-dimensional MESFET and MOSFET. The comparison of these results with those obtained by full WENO schemes and Monte Carlo calculations clarifies that the proposed multigroup-WENO solver is at least an interesting alternative to the other methods for the accurate simulation of the carrier transport in semiconductor devices.

We regard our multigroup method as a very powerful tool for dealing with nonlinear kinetic transport equations (e.g, the Bloch-Boltzmann-Peierls equations for the coupled carrier-phonon system of polar semiconductors and the Boltzmann-Poisson system for self-consistent device simulation), whose solution is essential for the deeper understanding of modern highly integrated semiconductor devices.

# Bibliography


ALAM M. A., STETTLER M. S. and LUNDSTROM M. S., Formulation of the Boltzmann equation in terms of scattering matrices, *Solid-State Electronics*, **36** (1993), 263–271.

ALI OMAR M. and REGGIANI L., Drift and diffusion of charge carriers in silicon and their empirical relation to the electric field, *Solid-state Electronics*, **30** (1987), 693–697.

AMBACHER O., SMART J., SHEALY J. R., WEIMANN N. G., CHU K., MURPHY M., DIMITROV R., WITTMER L., STUTZMANN M., RIEGER W. and HILSENBECK J., Two-dimensional electron gases induced by spontaneous and piezoelectric polarization in N- and Ga-face AlGaN/GaN heterostructures, *J. Appl. Phys.*, **85** (1999), 3222–3233.

ANDO T., FOWLER A. . and STERN F., Electronic properties of two-dimensional systems, *Rev. Mod. Phys.*, **54** (1982), 437–672.

ASHCROFT N. W. and MERMIN N. D., **Solid-State Physics**, Saunders College, Philadelphia, (1976).

AUBERT J. P., VAISSIERE J. V. and NOUGIER J. P., Matrix determination of the stationary solution of the Boltzmann equation for hot carriers in semiconductors, *J. Appl. Phys.*, **56** (1984), 1128–1132.

AUER C., SCHÜRRER F. and KOLLER W., A semicontinuous formulation of the Bloch-Boltzmann-Peierls equations, *SIAM J. Appl. Math.*, **64** (2004), 1457–1475.

BARDEEN J. and SHOCKLEY W., Deformation potentials and mobilities in nonpolar crystals, *Phys. Rev.*, **80** (1950) 72–80.

BARONI S., DE GIRONCOLI S., CORSO A. D. and GIANOZZI P., Phonons and related crystal properties from density-functional perturbation theory, *Rev. Mod. Phys.*, **73** (2001), 515–562.

BOHM D., **Quantum theory**, Prentice-Hall, Englewood Cliffs, (1951).

BROOKS H. and HERRING C., Scattering by ionized impurities in semiconductors, *Phys. Rev.*, **83** (1951), 879.

BUDD H., Path variable formulation of the hot carrier problem, *Phys. Rev.*, **158** (1967), 798–804.

BULUTAY C., RIDLEY B. K. and ZAKHLENIUK N. A., Full band polar optical phonon scattering analysis and negative differential conductivity in wurtzite GaN, *Phys. Rev. B*, **62** (2000), 15754–15763.

CÁCERES M. J., CARRILLO J. A. and MAJORANA A., Deterministic solution of the Boltzmann-Poisson system in GaAs-based semiconductors, *HYKE preprint HYKE2004-111*, www.hyke.org (2004).

CARAFFINI G. L., GANAPOL B. and SPIGA G. A., Multigroup approach to the non-linear extended Boltzmann equation, *Il Nouvo Cimento*, **17** (1995), 129–142.

CARRILLO J. A., GAMBA I.M., MAJORANA A. and SHU C.-W., A WENO-solver for 1D non-stationary Boltzmann-Poisson systems for semiconductor devices, *J. Comp. Electr.*, **1** (2002), 365–375.

CARRILLO J. A., GAMBA I.M., MAJORANA A. and SHU C.-W., A WENO-solver for the transients of Boltzmann-Poisson system for semiconductor devices: performance and comparisons with Monte Carlo methods, *J. Comput. Phys.*, **184** (2003) 498–525.

CARRILLO J. A., GAMBA I.M., MAJORANA A. and SHU C.-W., A direct solver for 2D non-stationary Boltzmann-Poisson systems for semiconductor devices: a MESFET simulation by WENO-Boltzmann schemes, *J. Comp. Electr.*, **2** (2003), 375–380.

COHEN M. L. and BERGSTRESSER T. K., Band structures and pseudo-potential form factors for fourteen semiconductors of the diamond and zinc-blende structures, *Phys. Rev.*, **141** (1966), 789–796.

CONSTANT E., Non-Steady-State Carrier Transport in Semiconductors in Perspective with Submicrometer Devices, **Hot-Electron Transport in Semiconductors**, Springer, New York, (1985).

CONWELL E. M. and VASSEL M. O., High-field transport in n-type GaAs, *Phys. Rev.*, **166** (1968), 797–821.

CZYCHOLL G., **Theoretische Festkörperphysik**, Vieweg, Braunschweig, (2000).

DAMEN T. C., LEITE R. C. C. and SHAH J., **Proc. Tenth International Conference on the Physics of Semiconductors**, U.S. Atomic Energy Commission, Cambridge, (1970).

DATTA S., **Quantum Phenomena**, Addison-Wesley, New York, (1989).

ERTLER C. and SCHÜRRER F., A multicell matrix solution to the Boltzmann equation applied to the anisotropic electron transport in silicon, *J. Phys. A: Math. Gen.*, **36** (2003), 8759–8774.

FATEMI E. and ODEH E., Upwind finite difference solution of Boltzmann equations applied to electron transport in semiconductor devices, *J. Comput. Phys.*, **108** (1993), 209–217.

FAWCETT W., BOARDMAN A. D. and SWAIN S., Monte Carlo determination of electron transport properties in Gallium Arsenide, *J. Phys. Chem. Solids*, **31** (1970), 1963–1990.

FAWCETT W. and REES H. D., Calculation of hot electron diffusion rate in GaAs, *Phys. Lett. A*, **29** (1969), 578–579.

FERRY D. K., **Semiconductors**, Maxwell Macmillian Editions, New York, (1991).

FERRY D. K. and GOODNICK S. M., **Transport in Nanostructures**, Cambridge University Press, Cambridge, (1997).

FETTER A. L. and WALECKA J. D., **Quantum Theory of Many-Particle Systems**, McGraw Hill, New York, (1971).

FISCETTI M. V., Monte Carlo simulation of transport in technologically significant semiconductors the diamond and zinc-blende structures. Part I: homogeneous transport, *IEEE Trans. Electron. Devices*, **38** (1991), 634–649.

FRÖHLICH H. and MOTT N. F., The mean free path of electrons in polar crystals, *Proc. Roy. Soc. A*, **171** (1939), 496.

GALLER M., SCHÜRRER F. and ROSSANI A., Moment equations of a conservative multigroup approximation to the nonlinear 3D Boltzmann equation, *Trans. Theo. Stat. Phys.*, **33** (2004), 203–221.

GALLER M., ROSSANI A. and SCHÜRRER F., A variable multigroup approach to the nonlinear Boltzmann equation based on the method of weighted residuals, **Rarefied Gas Dynamics: Proceedings of the 23rd International Symposium on Rarefied Gas Dynamics**, Springer, New York, (2003).

GASKA R., YANG J. W., OSINSKY A., CHEN Q., KHAN M. A., ORLOV A. O., SNIDER G. L. and SHUR M. S., Electron transport in AlGaN-GaN heterostructures grown on 6H-SiC substrates, *Appl. Phys. Lett.*, **72** (1998), 707–709.

GIANNOZZI P., DE GIRONCOLI S., PAVONE P. and BARONI S., Ab initio calculation of phonon dispersions in semiconductors, *Phys. Rev. B*, **43** (1991), 7231-7242.

GLOVER G. H., Microwave measurements of the velocity-field characteristic of n-type InP, *Appl. Phys. Lett.*, **20** (1972), 224–225.

GNUDI A., VENTURA D., BACCARINI G. and ODEH F., Two-dimensional MOSFET simulation by means of a multidimensional spherical harmonics expansion of the Boltzmann transport equation, *Solid-State electronics*, **35** (1993), 575–581.

GONZÁLEZ SÁNCHEZ T., VELÁZQUEZ PÉREZ J. E., GUTIÉRREZ CONDE O. M. and PARDO COLLANTES D., Electron transport in InP under high electric field conditions, *Semicond. Sci. Technol.*, **7** (1991), 31–36.

HENNACY K. A. and GOLDSMAN N., A generalized Legendre polynomials/spare matrix approach for determining the distribution function in non-polar semiconductors, *Solid-State electronics*, **36** (1993), 869–877.

HENNACY K. A., WU Y.-J., GOLDSMAN N. and MAYERGOYZ I. D., Deterministic MOSFET simulation using a generalized spherical harmonics expansion of the Boltzmann equation, *Solid-State electronics*, **38** (1995), 1485–1495.

HUTSON A. R., Piezoelectric scattering and phonon drag in ZnO and CdS, *J. Appl. Phys.*, **32** (1961), 2287–2292.

JACOBONI C. and LUGLI P., **The Monte Carlo Method for Semiconductor Device Simulation**, Springer, New York, (1989).

JACOBONI C. and REGGIANI L., The Monte Carlo method for the solution of the charge transport in semiconductors with application to covalent materials, *Rev. Mod. Phys.*, **55** (1983), 645–705.

JIANG G. and SHU C.-W., Efficient implementation of weighted ENO schemes,

*J. Comput. Phys.*, **126** (1996), 202–228.

JUNGEMANN C. and MEINZERHAGEN B., **Hierarchial Device Simulation**, Springer, Vienna, (2003).

KITTEL C., **Quantum Theory of Solids**, Wiley, New York, (1963).

KLEMENS P. G., Anharmonic decay of optical phonons, *Phys. Rev.*, **148** (1966), 845–848.

KOBAYASHI T., TAKAHARA K., KIMURA T. and ABE K., High-field properties of n-InP under high pressure, *Solid-State Electronics*, **21** (1978), 79–82.

KOLNIK J., OGUZMAN I. H., BRENNAN K. F., WANG R., RUDEN P. P. and WANG Y., Electronic transport studies of bulk zincblende and wurtzite phases of GaN based on an ensemble Monte Carlo calculation including a full zone bandstructure, *J. Appl. Phys.*, **78** (1995), 1033–1038.

LAPIDUS L. and PINDER G. F., **Numerical Solutions of partial differential equations in science and engineering**, Wiley, New York, (1982).

LEVEQUE R. J., **Numerical Methods for Conservation Laws**, Birkhäuser, Basel, (1992).

LIFSCHITZ E. M. and PITAEVSKII L. P., **Physical Kinetics**, Pergamon Press, Oxford, (1981).

VON DER LINDE D., KUHL J. and KLINGENBERG H., Raman scattering from nonequilibrium LO phonons with picosecond resolution, *Phys. Rev. Lett.*, **44** (1980), 1505–1508.

LUGLI P., BORDONE P., REGGIANI L., RIEGER M., KOCEVAR P., GOODNICK S. M., Monte Carlo studies of nonequilibrium phonon effects in polar semiconductors and quantum wells. I. Laser photoexcitation, *Phys. Rev. B*, **39** (1989), 7852–7865.

LUNDSTROM M., **Fundamentals of carrier transport**, Cambridge University Press, Cambridge, (2000).

MAJORANA A. and PIDATELLA R. M., A finite difference scheme solving the Boltzmann-Poisson system for semiconductor devices, *J. Comput. Phys.*, **174** (2001), 649–668.

MALONEY T. J. and FREY J., Transient and steady-state electron transport properties of GaAs and InP, *J. Appl. Phys.*, **48** (1977), 781–787.

MARKOWICH P., RINGHOFER C. and SCHMEISER C., **Semiconductor Equations**, Springer, Vienna, (1990).

MASSIDDA S., CONTINENZA A., FREEMAN A. J., PASCALE T. M., MELONI F. and SERRA M., Structural and electronic properties of narrow-band-gap semiconductors: InP, InAs, InSb, *Phys. Rev. B*, **41** (1990), 12079–12085.

MATULIONIS A., LIBERIS J., ARDARAVIČUS L., RAMONAS M., MATULIONIENĖ UI. and SMART J., Hot-electron energy relaxation time in AlGaN/GaN, *Semicond. Sci. Technol.*, **17** (2002), L9–L14.

MATULIONIS A., LIBERIS J., ARDARAVIČUS L., RAMONAS M., ZUBKUTĖ T., MATULIONIENĖ I., EASTMAN F., SHEALY J. R. and SMART J., Fast and ultrafast processes in AlGaN/GaN channels, *Phys. Stat. Sol. b*, **234** (2002), 826–829.

MEISTER A. and STRUCKMEIER J., **Hyperbolic Partial Differential Equations**, Vieweg, Braunschweig, (2002).

MENENDEZ J. and CARDONA M., Temperature dependence of the first order

Raman scattering by phonons in Si, Ge and $\alpha$-Sn: anharmonic effects, *Phys. Rev. B*, **29** (1984), 2051–2059.

NAG B., **Electron Transport in Compound Semiconductors**, Springer, New York, (1980).

NICLOT B., DEGOND P. and POUPAUD F., Deterministic simulations of the Boltzmann transport equation of semiconductors, *J. Comput. Phys.*, **78** (1988), 313–349.

NIELSEN L. D., Microwave measurement of electron drift velocity in Indium Phosphide for electric fields up to 50 kV/cm, *Phys. Lett. A*, **38** (1972), 221–222.

NOUGIER J. P. and ROLLAND M., Mobility, noise temperature and diffusivity of hot holes in germanium, *Phys. Rev. B*, **8** (1973), 5728–5737.

O'LEARY S. K., FOUTZ B. E., SHUR M. S., BHAPKAR U. V., and EASTMAN L. F., Electron transport in wurtzite indium nitride, *J. Appl. Phys.*, **83** (1998), 826–829.

PÖTZ W. and VOGL P., Theory of optical-phonon deformation potentials in tetrahedral semiconductors, *Phys. Rev. B*, **24** (1981), 2025–2037.

PRICE P. J., Two-dimensional electron transport in semiconductor layers. I: phonon scattering, *Ann. Phys.*, **133** (1981), 217–239.

RAMONAS M., MATULIONIS A. and ROTA L., Monte Carlo simulation of hot-phonon and degeneracy effects in the AlGaN/GaN two-dimensional electron gas channel, *Semicond. Sci. Technol.*, **18** (2003), 118–123.

REGGIANI L., **Hot-Electron transport in Semiconductors**, Springer, Berlin, (1985).

RIDLEY B. K., **Quantum Processes in Semiconductors**, Oxford University Press, Oxford, (1982).

RIDLEY B. K., Hot electrons in low-dimensional structures, *Rep. Prop. Phys.*, **54** (1991), 169–256.

RIDLEY B. K., FOUTZ B. E. and EASTMAN L. F., Mobility of electrons in bulk GaN and $Al_xGa_{1-x}N$/GaN heterostructures, *Phys. Rev. B*, **61** (2000), 16862–16869.

RINGHOFER C., Computational methods for semiclassical and quantum transport in semiconductor devices, *Acta Num.*, **3** (1997), 485–521.

RINGHOFER C., Space-time discretization methods for series expansion solutions of the Boltzmann equation for semiconductors, *SIAM J. Num. Anal.*, **38** (2000), 442–465.

RINGHOFER C. SCHMEISER C. and ZWIRCHMAYER A., Moment methods for the semiconductor Boltzmann equation in bounded position domains, *J. Num. Anal.*, **39** (2001), 1078–1095.

ROSSANI A., Generalized kinetic theory of electrons and phonons, *Physica A*, **305** (2002), 323–329.

ROSSANI A. and KANIADAKIS G., A generalized quasi-classical Boltzmann equation, *Physica A*, **277** (2000), 349–358.

RUCH J., Electron Dynamics in short channel field-effect transistors, *IEEE Trans. Electron Devices*, **ED-19** (1972), 652–654.

SHU C.-W. and OSHER S., Efficient implementation of essentially non-oscillatory shock capturing schemes, *J. Comp. Phys.*, **77** (1988), 439–471.

SHUR M. S., Influence of non-uniform field distribution in the channel on the frequency performance of GaAs FETs, *Electron. Lett.*, **12** (1976), 615–616.

SMITH P. M., FREY J. and CHATTERJEE P., High-field transport of holes in silicon, *Appl. Phys. Lett.*, **39** (1981), 332–333.

SZE S. M., **High-speed Semiconductor Devices**, Wiley, New York, (1990).

SZE S. M., **Modern Semiconductor Devices**, Wiley, New York, (1998).

SZE S. M., **Semiconductor Devices**, Wiley, New York, (2002).

TOMIZAWA K., **Numerical Simulation Of Submicron Semiconductor Devices**, Artech House, Boston, (1993).

VAISSIERE J. C., NOUGIER J. P., FADEL M., HLOU L. and KOCEVAR P., Numerical solution of coupled steady state hot-phonon-hot-electron Boltzmann equations in InP, *Phys. Rev. B*, **46** (1992), 13082–13099.

VAISSIERE J. C., NOUGIER J. P., VARANI L., HOULET P., HLOU L., REGGIANI L. and KOCEVAR P., Nonequilibrium phonon effects on the transient high-field transport regime in InP, *Phys. Rev. B*, **53**, 9886–9894 (1996).

VENTURA D., GNUDI A. and BACCARANI G., A deterministic approach to the solution of the BTE in semiconductors, *Rivista del Nuovo Cimento*, **18** (1995), 1–33.

WEISSMANTEL C. and HAMANN C., **Grundlagen der Festkörperphysik**, Johann Ambrosius Barth Verlag, Heidelberg, (1995).

WINDHORN T. H., COOK L. W., HAASE M. A. and STILLMAN G. E., Electron transport in InP at high electric field, *Appl. Phys. Lett.*, **42** (1983), 725–727.

YU P. Y. and CARDONA M., **Fundamentals of Semiconductors**, Springer, Berlin, (2001).

ZIMAN J. M., **Electrons and Phonons**, Clarendon Press, Oxford, (2001).

ZOLLNER S., SCHMID U., CHRISTENSEN N. E. and CARDONA M., Conduction band minima of InP: Ordering and absolute energies, *Appl. Phys. Lett.*, **57** (1990), 2339–2341.

# Related Publications of the Author


GALLER M. and SCHÜRRER F., A deterministic solution method for the coupled system of transport equations for the electrons and phonons in polar semiconductors, *J. Phys. A: Math. Gen.*, **37** (2004), 1479–1497.

GALLER M. and SCHÜRRER F., A multigroup approach to the coupled electron-phonon Boltzmann equations in InP, *Trans. Theo. Stat. Phys.*, **33** (2004), 485–501.

GALLER M. and SCHÜRRER F., Multigroup equations to the hot-electron hot-phonon system in III-V compound semiconductors, *Comp. Methods Appl. Mech. Engrg.*, **194** (2005), 2806–2818.

GALLER M. and SCHÜRRER F., A deterministic solver for the transport of the AlGaN/GaN 2D electron gas including hot-phonon and degeneracy effects, *J. Comput. Phys.*, (2004), in press.

GALLER M. and SCHÜRRER F., Mathematical properties of a kinetic transport model for carriers and phonons in semiconductors, submitted to *ZAMP*, (2004).

GALLER M. and SCHÜRRER F., Stability analysis of a multigroup model for the Boltzmann transport equations of carriers and phonons, **Rarefied Gas Dynamics: Proceedings of the 24rd International Symposium on Rarefied Gas Dynamics**, Springer, New York, (2005).

GALLER M., MAJORANA A. and SCHÜRRER F., A multigroup WENO solver for the non-stationary Boltzmann-Poisson system for semiconductor devices submitted to **Proc. 5$^{\text{th}}$ International Workshop on Scientific Computing in Electrical Engineering**, (2004).

GALLER M. and MAJORANA A., Deterministic and stochastic simulations of electron transport in semiconductors, submitted to *Trans. Theo. Stat. Phys.*, (2004).

GALLER M. and SCHÜRRER F., A deterministic solver for the 1D non-stationary Boltzmann-Poisson system for GaAs devices: bulk GaAs and GaAs $n^+ - n_i - n^+$ diode, submitted to *J. Comput. Electronics*, (2004).

GALLER M. and SCHÜRRER F., A Direct Multigroup-WENO Solver for the 2D Non-Stationary Boltzmann-Poisson System for GaAs Devices: GaAs-MESFET, submitted to *J. Comput. Phys.*, (2005).

# Index


acoustic deformation potential
scattering
in 2D, 112
in 3D, 16
acoustic phonons, 9
AlGaN/GaN heterojunction, 132
approximative band model, 7

ballistic motion, 74, 194
band structure, 6
Bloch functions, 6
Bloch theorem, 6
Bloch-Boltzmann-Peierls equations,
28, 88
in 2D, 122
Boltzmann transport equation, 27,
148
of electrons, 40
of LO phonons, 45
Boltzmann-Poisson system, 148
Bose-Einstein distribution, 10, 34, 99
boundary conditions, 153
boundedness of the solution
of the BBP equations, 32
of the multigroup equations, 93
box method to the BTE, 2
Brillouin zone, 6
Brooks-Herring model, 22

CFL condition, 154
charge accumulation, 197
collision coefficient, 51

collision term, 27
in 2D, 123
condition of balance, 27
conducting channel, 181
conduction band, 7
confining potential, 135
conservation
of carrier density, 33, 55, 95
of total energy density, 33, 59, 95
of total momentum, 33
conservative form, 149
CPU time, 64, 212
crystal momentum, 8

Debye approximation, 11
Debye length, 101
deformation potential interaction, 15
degeneracy effects, 100, 136, 142
density of states, 39, 149
in 2D, 110
of carriers, 90
of phonons, 90
depletion layer, 208
dielectric function, 118
diffusion coefficient, 185
diffusion term, 27, 147
Dirac distribution, 41
direct matrix method to the BTE, 2
Dirichlet condition, 163
dispersion law
of electrons, 6
of phonons, 9

distribution function
  of carriers, 26
  of electrons, 100
  of electrons in 2D, 120, 145
  of electrons in GaAs, 82, 103, 194,
      199, 211
  of electrons in InP, 66, 73
  of electrons in silicon, 162
  of phonons, 27, 100
  of phonons in 2D, 121, 146
  of phonons in GaAs, 103
  of phonons in InP, 68, 75
drain characteristics
  of GaAs MESFET, 211
  of silicon MESFET, 163
  of silicon MOSFET, 178, 187
drift term, 27
drift-diffusion model, 184

Einstein approximation, 11
elastic coefficients, 138
electromechanical coupling coefficient,
  133
electron-hole system, 186
electron-phonon scattering, 15
  in 2D, 111
electron-phonon system, 63
electrostatic interaction, 15
energy gap, 7
energy subbands, 109, 135
equilibrium solution
  to multigroup model equations, 98
  to the BBP equations, 34
equipartition of phonons, 10

Fermi energy, 136
Fermi's golden rule, 12
Fermi-Dirac distribution, 34, 99
finite differences approach to the
  BTE, 2
Fröhlich model, 19

Galerkin method to the BTE, 2
gallium arsenide, 77, 191
GaN-based HFET, 107
group velocity, 26, 123

H-theorem
  to multigroup model equations, 97
  to the BBP equations, 33
Hamilton's equations of motion, 25
hole mobility, 184
Hook's law, 137
hot electrons, 47
hot phonons, 47, 72, 100, 142

indium phosphide, 61
initial data, 154
intervalley scattering, 18
intervalley transitions, 12
intravalley transitions, 12
ionized impurity scattering, 24
iterative technique to the BTE, 2

k·p perturbation theory, 8
Kane model, 8, 155

leakage current, 183
longitudinal phonons, 9
low density approximation, 29, 45,
  149

macroscopic quantities
  in 2D, 128
  in multigroup formulation, 41, 92,
      151
  of carriers, 31
  of phonons, 31
material parameters
  for silicon holes, 185
  of AlGaN/GaN, 133
  of GaAs, 78, 192
  of InP, 62, 71
  of silicon, 155
mathematical properties, 87
Maxwell distribution, 35
method of weighted residuals, 41, 90,
  151
MinMod slope limiter, 129, 151
Monte Carlo methods to the BTE, 1
multigroup model equations, 38, 87,
  91
multigroup-WENO solver, 147

multivalley-model, 147

negative differential resistivity, 75, 193
Neumann condition, 162
non-equilibrium phonons, 69
non-parabolic band approximation, 8, 39, 148

ohmic contact, 154
optical deformation potential scattering, 17
optical phonons, 9
overlap factor, 14, 192
overshoot phenomena, 78

parabolic band approximation, 7
path integral solution to the BTE, 2
Pauli principle, 28, 103
periodic crystal potential, 5
phonon relaxation time, 63, 101
phonon-phonon interaction, 45, 77, 101
phonons, 9
piezoelectric polarization, 132
piezoelectric scattering
    in 2D, 114
    in 3D, 22
pinch-off point, 182
Poisson equation, 135, 148
polar optical phonon interaction term, 47
polar optical scattering
    in 2D, 115
    in 3D, 21
polar semiconductor, 38
pseudo-Newtonian law, 26, 43

quantum statistics, 88
quantum well, 107
    square, 109

random phase approximation, 118
reciprocal lattice, 6
Rees effect, 78
reflecting boundary, 153

relaxation time, 45

scattering matrix approach to the BTE, 2
scattering mechanisms
    in GaAs, 77, 191
    in GaN, 132
    in InP, 61
    in silicon, 155
scattering rate
    in 3D, 14
    in 2D, 112
    of acoustic deformation potential scattering, 16
    of intervalley scattering, 18
    of ionized impurity scattering, 24
    of optical deformation potential scattering, 17
    of polar optical scattering, 21
scattering sources, 12
Schottky contact, 154
Schrödinger equation, 6, 108, 136
screened Coulomb potential, 23
screening, 118
screening parameter, 23
    in 2D, 119
semiclassical dynamics of carriers, 24
short-channel effect, 182
silicon, 155
silicon MOSFET, 107
spherical harmonics expansion to the BTE, 2
spontaneous polarization, 132
strain vector, 137
subband envelope functions, 136
successive over-relaxation scheme, 154
symmetry relations, 95, 96

three-valley model, 71
threshold field, 73
transfer
    of energy, 100
    of momentum, 100
transport
    at AlGaN/GaN heterojunction, 139
    in bulk GaAs, 193

in bulk silicon, 157
in GaAs $n^+ - n - n^+$ diode, 195
in GaAs MESFET, 200
in silicon $n^+ - n - n^+$ diode, 158
in silicon MESFET, 162
in silicon MOSFET, 172
transverse phonons, 9
TVD forward Euler scheme, 154
two-dimensional electron gas, 107
two-valley model, 61, 191

umklapp processes, 13
undershoot phenomena, 78
upwind scheme, 43, 129, 151

valence bands, 7
velocity overshoot, 78
velocity undershoot, 79

wall effect, 85
WENO scheme, 152
WENO solver to the BTE, 3
wurtzite crystal, 134

# Series on Advances in Mathematics for Applied Sciences

## Aims and Scope

This Series reports on new developments in mathematical research relating to methods, qualitative and numerical analysis, mathematical modeling in the applied and the technological sciences. Contributions related to constitutive theories, fluid dynamics, kinetic and transport theories, solid mechanics, system theory and mathematical methods for the applications are welcomed.

This Series includes books, lecture notes, proceedings, collections of research papers. Monograph collections on specialized topics of current interest are particularly encouraged. Both the proceedings and monograph collections will generally be edited by a Guest editor.

High quality, novelty of the content and potential for the applications to modern problems in applied science will be the guidelines for the selection of the content of this series.

## Instructions for Authors

Submission of proposals should be addressed to the editors-in-charge or to any member of the editorial board. In the latter, the authors should also notify the proposal to one of the editors-in-charge. Acceptance of books and lecture notes will generally be based on the description of the general content and scope of the book or lecture notes as well as on sample of the parts judged to be more significantly by the authors.

Acceptance of proceedings will be based on relevance of the topics and of the lecturers contributing to the volume.

Acceptance of monograph collections will be based on relevance of the subject and of the authors contributing to the volume.

Authors are urged, in order to avoid re-typing, not to begin the final preparation of the text until they received the publisher's guidelines. They will receive from World Scientific the instructions for preparing camera-ready manuscript.

# SERIES ON ADVANCES IN MATHEMATICS FOR APPLIED SCIENCES

Vol. 17 The Fokker–Planck Equation for Stochastic Dynamical Systems
and Its Explicit Steady State Solution
*by C. Soize*

Vol. 18 Calculus of Variation, Homogenization and Continuum Mechanics
*eds. G. Bouchitté et al.*

Vol. 19 A Concise Guide to Semigroups and Evolution Equations
*by A. Belleni-Morante*

Vol. 20 Global Controllability and Stabilization of Nonlinear Systems
*by S. Nikitin*

Vol. 21 High Accuracy Non-Centered Compact Difference Schemes for
Fluid Dynamics Applications
*by A. I. Tolstykh*

Vol. 22 Advances in Kinetic Theory and Computing: Selected Papers
*ed. B. Perthame*

Vol. 23 Waves and Stability in Continuous Media
*eds. S. Rionero and T. Ruggeri*

Vol. 24 Impulsive Differential Equations with a Small Parameter
*by D. Bainov and V. Covachev*

Vol. 25 Mathematical Models and Methods of Localized Interaction Theory
*by A. I. Bunimovich and A. V. Dubinskii*

Vol. 26 Recent Advances in Elasticity, Viscoelasticity and Inelasticity
*ed. K. R. Rajagopal*

Vol. 27 Nonstandard Methods for Stochastic Fluid Mechanics
*by M. Capinski and N. J. Cutland*

Vol. 28 Impulsive Differential Equations: Asymptotic Properties of
the Solutions
*by D. Bainov and P. Simeonov*

Vol. 29 The Method of Maximum Entropy
*by H. Gzyl*

Vol. 30 Lectures on Probability and Second Order Random Fields
*by D. B. Hernández*

Vol. 31 Parallel and Distributed Signal and Image Integration Problems
*eds. R. N. Madan et al.*

Vol. 32 On the Way to Understanding The Time Phenomenon: The
Constructions of Time in Natural Science — Part 1.
Interdisciplinary Time Studies
*ed. A. P. Levich*

Vol. 33 Lecture Notes on the Mathematical Theory of the Boltzmann Equation
*ed. N. Bellomo*

Vol. 34 Singularly Perturbed Evolution Equations with Applications to
Kinetic Theory
*by J. R. Mika and J. Banasiak*

# SERIES ON ADVANCES IN MATHEMATICS FOR APPLIED SCIENCES

Vol. 35    Mechanics of Mixtures
*by K. R. Rajagopal and L. Tao*

Vol. 36    Dynamical Mechanical Systems Under Random Impulses
*by R. Iwankiewicz*

Vol. 37    Oscillations in Planar Dynamic Systems
*by R. E. Mickens*

Vol. 38    Mathematical Problems in Elasticity
*ed. R. Russo*

Vol. 39    On the Way to Understanding the Time Phenomenon: The Constructions of Time in Natural Science — Part 2. The "Active" Properties of Time According to N. A. Kozyrev
*ed. A. P. Levich*

Vol. 40    Advanced Mathematical Tools in Metrology II
*eds. P. Ciarlini et al.*

Vol. 41    Mathematical Methods in Electromagnetism Linear Theory and Applications
*by M. Cessenat*

Vol. 42    Constrained Control Problems of Discrete Processes
*by V. N. Phat*

Vol. 43    Motor Vehicle Dynamics: Modeling and Simulation
*by G. Genta*

Vol. 44    Microscopic Theory of Condensation in Gases and Plasma
*by A. L. Itkin and E. G. Kolesnichenko*

Vol. 45    Advanced Mathematical Tools in Metrology III
*eds. P. Cianlini et al.*

Vol. 46    Mathematical Topics in Neutron Transport Theory — New Aspects
*by M. Mokhtar-Kharroubi*

Vol. 47    Theory of the Navier–Stokes Equations
*eds. J. G. Heywood et al.*

Vol. 48    Advances in Nonlinear Partial Differential Equations and Stochastics
*eds. S. Kawashima and T. Yanagisawa*

Vol. 49    Propagation and Reflection of Shock Waves
*by F. V. Shugaev and L. S. Shtemenko*

Vol. 50    Homogenization
*eds. V. Berdichevsky, V. Jikov and G. Papanicolaou*

Vol. 51    Lecture Notes on the Mathematical Theory of Generalized Boltzmann Models
*by N. Bellomo and M. Lo Schiavo*

**SERIES ON ADVANCES IN MATHEMATICS FOR APPLIED SCIENCES**


Vol. 52    Plates, Laminates and Shells: Asymptotic Analysis and
           Homogenization
           *by T. Lewiński and J. J. Telega*

Vol. 53    Advanced Mathematical and Computational Tools in Metrology IV
           *eds. P. Ciarlini et al.*

Vol. 54    Differential Models and Neutral Systems for Controlling the
           Wealth of Nations
           *by E. N. Chukwu*

Vol. 55    Mesomechanical Constitutive Modeling
           *by V. Kafka*

Vol. 56    High-Dimensional Nonlinear Diffusion Stochastic Processes
           — Modelling for Engineering Applications
           *by Y. Mamontov and M. Willander*

Vol. 57    Advanced Mathematical and Computational Tools in Metrology V
           *eds. P. Ciarlini et al.*

Vol. 58    Mechanical and Thermodynamical Modeling of Fluid Interfaces
           *by R. Gatignol and R. Prud'homme*

Vol. 59    Numerical Methods for Viscosity Solutions and Applications
           *eds. M. Falcone and Ch. Makridakis*

Vol. 60    Stability and Time-Optimal Control of Hereditary Systems
           — With Application to the Economic Dynamics of the US
           (2nd Edition)
           *by E. N. Chukwu*

Vol. 61    Evolution Equations and Approximations
           *by K. Ito and F. Kappel*

Vol. 62    Mathematical Models and Methods for Smart Materials
           *eds. M. Fabrizio, B. Lazzari and A. Morro*

Vol. 63    Lecture Notes on the Discretization of the Boltzmann Equation
           *eds. N. Bellomo and R. Gatignol*

Vol. 64    Generalized Kinetic Models in Applied Sciences — Lecture Notes
           on Mathematical Problems
           *by L. Arlotti et al.*

Vol. 65    Mathematical Methods for the Natural and Engineering Sciences
           *by R. E. Mickens*

Vol. 66    Advanced Mathematical and Computational Tools in Metrology VI
           *eds. P. Ciarlini et al.*

Vol. 67    Computational Methods for PDE in Mechanics
           *by B. D'Acunto*

Vol. 68    Differential Equations, Bifurcations, and Chaos in Economics
           *by W. B. Zhang*

Vol. 69    Applied and Industrial Mathematics in Italy
           *eds. M. Primicerio, R. Spigler and V. Valente*

Vol. 70    Multigroup Equations for the Description of the Particle Transport
           in Semiconductors
           *by M. Galler*